The Audubon Society
Field Guide to
North American Trees

A Chanticleer Press Edition

The Audubon Society Field Guide to North American Trees

Western Region

Elbert L. Little,
former Chief Dendrologist,
U.S. Forest Service

Visual Key by
Susan Rayfield and Olivia Buehl

Alfred A. Knopf, New York

This is a Borzoi Book
Published by Alfred A. Knopf, Inc.

All rights reserved. Copyright 1980
under the International Union for the
protection of literary and artistic works
(Berne). Published in the United States
by Alfred A. Knopf, Inc., New York,
and simultaneously in Canada by
Random House of Canada Limited,
Toronto. Distributed by Random
House, Inc., New York.

Prepared and produced by
Chanticleer Press, Inc., New York.

Color reproductions by Cliche + Litho
Repro AG, Zurich, Switzerland.
Type set in Garamond by
Dix Type Inc., Syracuse, New York.
Printed and bound in Japan.

Published June 13, 1980
Reprinted twice
Fourth Printing, February 1987

Library of Congress Catalog Card
Number 79-3475
ISBN 0-394-50761-4

Trademark "Audubon Society" used by
publisher under license from the
National Audubon Society, Inc.

CONTENTS

THE AUDUBON SOCIETY

The National Audubon Society is among the oldest and largest private conservation organizations in the world. With over 550,000 members and more than 500 local chapters across the country, the Society works in behalf of our natural heritage through environmental education and conservation action. It protects wildlife in more than seventy sanctuaries from coast to coast. It also operates outdoor education centers and ecology workshops and publishes the prizewinning AUDUBON magazine, AMERICAN BIRDS magazine, newsletters, films, and other educational materials. For further information regarding membership in the Society, write to the National Audubon Society, 950 Third Avenue, New York, New York 10022.

AUTHOR AND
ACKNOWLEDGMENTS

Elbert L. Little has been associated
with the United States Forest Service
for many years, serving from 1942 to
1976 as the Dendrologist, or tree
identification specialist. Now retired,
he is a Research Associate at the U.S.
National Museum of Natural History,
Smithsonian Institution. An
international authority on tree
classification, identification, and
distribution, he has written numerous
technical and popular publications on
the trees of the United States and
tropical America, including *Checklist of
United States Trees* (1979), the 6-volume
Atlas of United States Trees (1971–81),
*Common Trees of Puerto Rico and the
Virgin Islands* (1964), and *Alaska Trees
and Shrubs* (1972). His field studies of
New World trees have extended from
Alaska to Patagonia. As a university
visiting professor, he has taught tree
identification to forestry students in
both Spanish and English. The U.S.
Department of Agriculture and the
American Forestry Association gave
him their Distinguished Service
Awards.

Many people have cooperated in the
preparation of this field guide. David
Cavagnaro, Sonja Bullaty, and Angelo
Lomeo, who spent months

photographing many of the species in the book, deserve special mention. Sincere appreciation is also due T. H. Everett, Senior Curator of Education, New York Botanical Garden, for his review of the manuscript and photographs. Thomas L. Gates, Director of the Plant Sciences Data Center, Mt. Vernon, Virginia, assisted us in locating many trees. The following individuals have also given of their time: Dr. Charles T. Mason Jr., University of Arizona Arboretum; Dr. Marshall C. Johnston and Scooter Cheatham, University of Texas, Austin; Dr. John M. Tucker, University of California, Davis; Dr. John Wanamaker, Principia College, Elsah, Illinois; Dr. Richard Spellenberg, New Mexico State University; Rodney Engard, Desert Botanical Garden, Phoenix, Arizona; Thom Peabody, Living Desert Reserve, Palm Desert, California; Robert Peril, Arizona Sonora Desert Museum; Dr. Dale Bever, Oregon State University, Corvallis; Dr. Brian Mulligan, University of Washington (Seattle) Arboretum; Dr. Mildred E. Mathias, University of California, Los Angeles; Dr. Steven Timbrook, Santa Barbara Botanical Garden; Dr. Louis C. Erickson, University of California, Riverside. My wife, Ruby Rice Little, typed the manuscript.

The dedicated staff of Chanticleer Press has contributed in many ways. Olivia Buehl located nearly every native tree species, assembled the photographs, and edited the manuscript. Susan Rayfield created the visual keys. Mary Beth Brewer coordinated the numerous maps and marginal artwork. My sincere thanks go to Paul Steiner, Gudrun Buettner, and Milton Rugoff for their support and encouragement, and to Carol Nehring, Helga Lose, Ray Patient, and Dean Gibson for their efforts in the production of the book.

INTRODUCTION

Trees are not only among the most beautiful works of nature, they are also useful in innumerable ways. Primitive man depended upon trees for many of the essentials of life, constructing homes, rafts, canoes, and many weapons, tools, and utensils from wood, as well as roof thatch, woven fabrics, and baskets from bark and leaves. Fruits and nuts constituted an important part of the diet, while medicines, spices, and dyes were also derived from trees.

Many cultures have built up a body of lore regarding trees. The ancient Celtic Druids worshiped holly because the evergreen leaves symbolized the return of the light and warmth of summer after the long, dark winter. Likewise, yews symbolized everlasting life to early Christian missionaries; to this day, yews are commonly associated with churchyards. Believing they could ward off rheumatism, American pioneers often carried buckeye seeds in their pockets. The ritual of a Christmas tree is with us today, and dowsers still believe a forked branch of Witch-hazel can locate underground water; however, most of the superstitions surrounding trees have disappeared. Man's fascination with trees continues in modern guise, in his desire to know how to identify and to learn more about trees.

This book is designed to enable you to identify trees quickly on a visual basis, by the grouping of similar-appearing species. The comprehensive text provides full, nontechnical descriptions of all species, supplemented by information on where they grow, their uses, and their lore.

Definition of a Tree: A tree is a woody plant with an erect perennial trunk at least 3″ (7.5 cm) in diameter at breast height (4½′, or 1.3 m), a definitely formed crown of foliage, and a height of at least 13′ (4 m). In contrast, a shrub is a small woody plant usually having several perennial stems branching from the base. Some species that are normally trees remain shrubs under severe environmental conditions, such as particularly cold or dry climates; these are included in this guide. Other species that are usually shrubs and rarely attain tree size are excluded. The trunk of a tree includes the fibrous inner wood and the outer covering, or bark. Leaves manufacture food by photosynthesis and may be either evergreen (remaining on the tree all year) or deciduous (falling off in winter or during dry periods).

Leaves: Each leaf is made up of a broad, flat blade and a leafstalk, which is usually short and narrow (or sometimes absent) and attached to the twig. Most leaves have one blade and are known as simple leaves. Compound leaves have the blade divided into three or more leaflets which may be arranged in two rows along an axis (pinnately compound), on side branches off an axis (bipinnately compound), or radiating like a fan (palmately compound). Whether simple or compound, leaves are attached to the stem in three different ways. Those attached singly at different levels

Alternate

(nodes) on the stem are called alternate; those attached at the same level in pairs are called opposite; and three or more attached at one level and arranged in a ring are called whorled. Alternate arrangement is by far the most common.

Opposite

Whorled

Leaves can be grasslike (linear); shaped like a lance (lanceolate); reverse lance-shaped, broadest near the tip and pointed at the base (oblanceolate); shaped like an egg (ovate); reverse egg-shaped, broadest toward the tip (obovate); oval, broadest in the middle (elliptical); with nearly parallel edges (oblong); rounded; or in the form of a spoon (spatulate).

Besides their overall shape, leaves are further distinguished by having edges without teeth (entire) or toothed; the teeth may point forward as in a saw (saw-toothed). The edges may also be wavy, turned under, or deeply divided into parts (lobed). In some species the leaves of young plants or of vigorous twigs are of a different shape; these are known as juvenile leaves. Leaves of conifers are either long and narrow (needlelike) or short, pointed, and often overlapping (scalelike).

Flowers: Flowers usually consist of four parts. The outermost calyx is composed of leaflike parts called sepals. The corolla consists of the colored petals. (The corolla may be regular, with uniform petals like the spokes of a wheel; irregular, with petals of unequal size; or tubular, with a tube and lobes formed by united petals.) The stamens, or male organs of the flower, have a filament, or stalk, and an anther, or enlarged part (usually yellow), composed of four or fewer pollen sacs containing tiny pollen grains. In the center of the flower is the pistil(s), or female organ, which has three parts: the stigma at the tip, which receives the pollen; the style, or stalk (sometimes absent); and the

Leaf Types

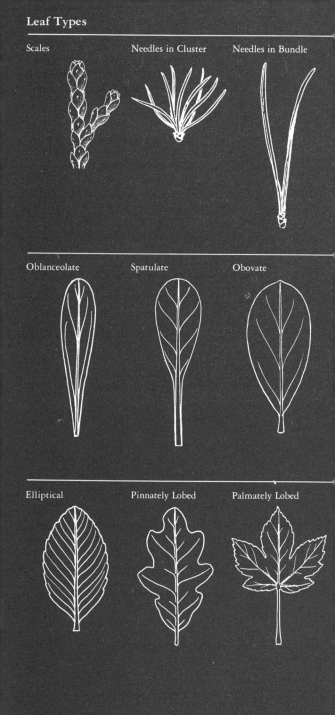

Scales

Needles in Cluster

Needles in Bundle

Oblanceolate

Spatulate

Obovate

Elliptical

Pinnately Lobed

Palmately Lobed

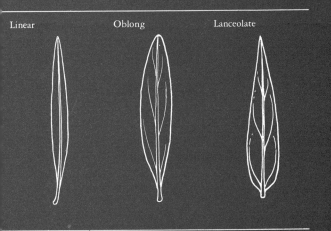

Linear   Oblong   Lanceolate

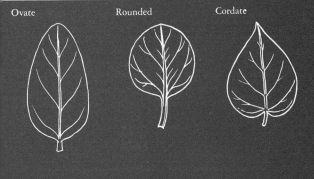

Ovate   Rounded   Cordate

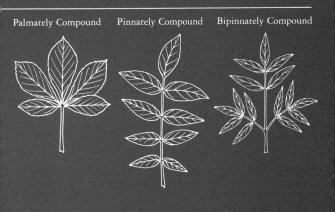

Palmately Compound   Pinnately Compound   Bipinnately Compound enlarged ovary at the base. The ovary has one to several cells, each containing one to many ovules, which are the rounded, whitish egg-bearing units. The mature ovary is the fruit; the mature ovules become seeds after fertilization, each capable of germinating into a tiny plant.

Most tree flowers are bisexual, possessing both pollen-bearing male stamens and female pistil(s). Others are of one sex only and, depending on species, may have both unisexual and bisexual flowers on the same plant. Some flowers are borne singly; more commonly, they are clustered either at the end of a twig (terminal) or at the base of a leaf or side of a twig (lateral). One type of flower cluster, found in some dogwoods, consists of tiny flowers and 4–7 petal-like bracts. Catkins, common in willows and poplars, are unbranched clusters, usually with a drooping axis, scales, and stalkless flowers of one sex.

Fruit: The fruit may be a berry, drupe, pome, multiple fruit, aggregate fruit, acorn or other nut, key (samara), achene, pod, capsule, or follicle. Commonly, the fruit develops from the ovary of a single pistil and is known as a simple fruit. Simple fruits may be dry (hard), as in walnuts, or fleshy with seeds in mealy or juicy pulp, as in most hawthorns. A fruit that develops from several pistils within a flower is known as an aggregate fruit, such as that of magnolias. A fruit that develops from several united flowers is known as a multiple fruit; an example is a mulberry. Fleshy fruits, which do not open, include the berry, which has several seeds, as in cacti; the drupe, which has a hard central stone, as in cherries; and the pome, which has a fleshy outer part and a hard or papery core containing seeds, as in apples. A

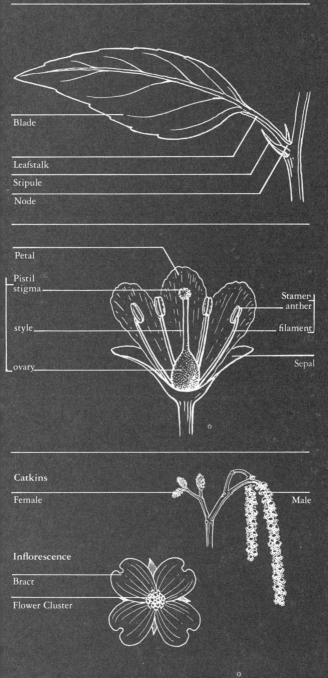

Blade

Leafstalk

Stipule

Node

Petal

Pistil
stigma

style

ovary

Stamen
anther

filament

Sepal

Catkins

Female

Male

Inflorescence

Bract

Flower Cluster

Types of Cones and Fruit

Cone

Nut

Acorn

Pod

Achene

Key

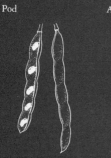

Drupe

Berry

Pome

Capsule

Aggregate of Capsules

Multiple of Nutlets

nut is a one-seeded fruit with a thick, hard shell (a nutlet is similar but smaller); an acorn is a distinctive pointed nut that has a scaly cup at the base. A key (samara)—characteristic of maples, elms, and ashes—is a one-seeded fruit with a thin, flat, dry wing; an achene is a small, dry seedlike fruit with a thin wall. The Legume Family is characterized by dry one-celled pods that split open, usually along two grooved lines; in contrast, a follicle opens along one line. Capsules are dry fruits with two or more cells, opening along as many lines as there are cells, as in some yuccas. While conifers technically lack flowers and fruit, they have separate male and female parts known as cones, which are usually on the same tree. The smaller, nonwoody, male cones, which resemble flowers, produce the pollen; the larger and more familiar female cones are composed of hard cone-scales, each commonly above a smaller scale or bract, with exposed seeds at the base of the cone-scale.

Major Western Habitats: The trees of western North America are distributed in various habitats throughout several climatic and altitudinal zones, from sea level to snow-capped mountain peaks, from subtropical to frigid climates, and from barren deserts to humid zones of high rainfall or snowfall. The vegetation of a summit a mile (1.6 km) above the base may be like that found near sea level roughly 1,600 miles (2,574 km) to the north. Conifers become dwarf, windswept shrubs at the timberline, as in Arctic regions. From the Mexican border through the Rocky Mountain region, the vegetation begins with almost treeless grasslands and deserts. Above the lower limit of trees are open woodlands of small, bushy evergreens such as junipers, pinyons, and oaks. Farther north are pine (especially Ponderosa Pine) and Douglas-fir forests.

In the far north the coniferous forests of White Spruce and Black Spruce, together with Paper Birch, extend to the northern limit of trees bordering the Arctic tundra. In general, conifers are more common than hardwoods in western forests.

Along the Pacific Coast are other distinct forest types, such as the evergreen oak woodlands in southern California. Giant Redwoods occupy the subtropical fog belt along the coast of northern California and adjacent Oregon. Northward to southern Alaska are forests of Sitka Spruce and Western Hemlock. Mixed coniferous forests of Douglas-fir, firs, pines, and other conifers occupy the inland ranges. Lodgepole Pine forms extensive forests in the northern Rockies and other areas. Quaking Aspen, the most widespread western hardwood, forms groves and thickets on burned-out areas. Cottonwoods, willows, and sycamores occupy wet soils of riverbanks and valleys, even in otherwise treeless regions. Alders make thickets along wet valleys and slopes to the north. The unusual trees of the southwestern deserts are adapted to scarcity of water. Their small leaves and other exposed surfaces reduce moisture loss by transpiration. The Joshua-tree and other yuccas have clusters of long, thick, bayonetlike evergreen leaves ending in a stout, sharp point. Cacti such as Saguaro and Jumping Cholla have clusters of formidable spines, tiny leaves or none, and thick, succulent, green stems that manufacture food by photosynthesis and also store water. Paloverdes, which shed their tiny leaves early, remain leafless most of the time. Their trunks, branches, and twigs stay green and smooth and manufacture food throughout the year. Mesquites have thin crowns of foliage; their long taproots obtain moisture from deep within the soil.

Geographic Scope:	This volume covers the native trees of western North America (north of Mexico), as well as common naturalized trees and a number of introduced species. Several rare subtropical species of the Mexican border region, which grow mainly in the extreme southerly parts of Texas and Arizona, have been omitted. Eastern tree species are treated in a companion guide by the same author. Our range is bounded by Alaska and the Yukon Territory in northwestern Canada, the eastern base of the Rocky Mountains in Alberta and eastern Montana, south to the Big Bend region of southwestern Texas (see map). Tree species of the western mountains are generally absent beyond the foothills in the grassland plains to the east.
Use of Color Photographs:	We have chosen to use color photographs rather than conventional drawings because they show bark, leaves, flowers, and fruit growing on the trees as you would find them in their natural setting, thus making identification much easier. In general, we have selected photographs that show the most typical examples of a species. Where a species is highly variable, we have tried to indicate its diversity, sometimes showing more than one example, such as Texas Mulberry, which may have both toothed and lobed leaves.
Organization of the Color Plates:	Unlike birds, mammals, reptiles, insects, and wildflowers, whose overall color and shape can be taken in at a glance, trees are too large and have too many components to be easily identified from just one photograph. Therefore, we have created three separate visual keys—leaves, flowers, and cones and fruit—each corresponding to a major identifying feature of the tree. The three keys are explained in the Thumb Tab Guide preceding the color plates.

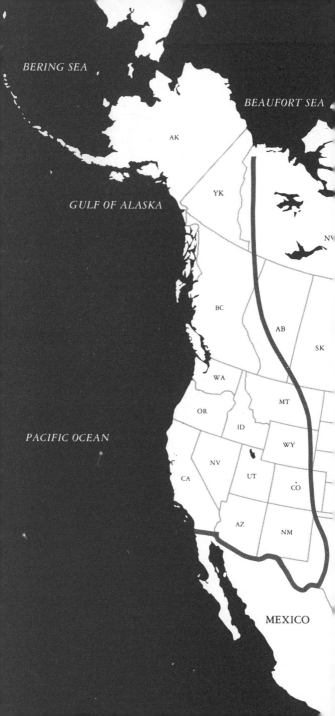

Leaf Visual Key: Since leaves are the most noticeable
feature of an unfamiliar tree, and since
they remain on the tree a relatively long
time, our primary visual key to
identification is through photographs of
leaves, paired with characteristic bark
photographs. These leaf photographs
are grouped by similar shape, regardless
of their family or genus relationships.
Many related trees, such as those in the
Maple Family, have similarly shaped
leaves, and therefore will appear near
each other in the photo section. Other
groups of trees such as oaks or sumacs,
whose species have widely differing
leaves, will be found in two or more
sections. Leaf photographs are arranged
in the following categories:

Needle-leaf Conifers
Scale-leaf Conifers
Untoothed Simple Leaves
Toothed Simple Leaves
Lobed Simple Leaves
Compound Leaves
Yuccas, Palms, and Cacti

Within each group, we have further
loosely organized the photographs
according to surface texture and vein
patterns—keeping in mind that leaves
can vary from one individual tree of a
species to another, and even on the
same tree.
The bark photograph for each tree
usually shows the bark of a mature
trunk of average size. Color, texture,
and surface pattern are often important
in distinguishing different species in
the same family, such as birches, while
they are a secondary characteristic in
others. It should be noted, however,
that many related species have similar
bark. Also the bark of a tree often
changes with age, as the outer layers
crack and are shed and inner layers with
different hues are exposed. Therefore,
bark photographs usually serve only as a
secondary tool for identification.

Flower Visual Key: Our flower key includes many photographs of the most common and interesting flowers, ranging from the tiny willows, scarcely ⅛″ long, to the yuccas, with large bell-shaped blossoms in clusters 2–3′ or more in length. Since most people notice color first, we have arranged flowers by color in the following groups:

Reddish
Yellow
Greenish
Cream-colored
White
Pink
Purple

Other important features to note in the flower photographs are shape, number of petals, and whether a flower is clustered or solitary.

Cones and Fruit Visual Key: Cones and fruit are also extremely important to identification, particularly since some remain on the tree or on the ground long after they have matured. We have grouped them by shape in the following easily recognizable categories:

Cones
Berrylike Fruit
Fleshy Fruit
Nuts
Acorns
Pods
Keys
Tufted Fruit
Balls and Capsules

Within these general groups the photographs are further arranged by color. Conelike fruits are placed with the true conifer cones because of their similar appearance. Some fruits that appear berrylike, such as Common Chokecherry, are grouped with the berries, even though they are actually drupes.

How to Read the Captions: The caption beneath each photograph gives the plate number, common name, measurements, and text page on which the species is described. Where helpful, other pertinent information is included, such as the number of pine needles in a bundle; if the tree is evergreen; whether the leaves are opposite instead of in the usual alternate arrangement; and if the leaves are pinnately, bipinnately, or palmately compound. In the flower section, where important, the sex of the tree flower is given: either male ($\male$) or female ($\female$). The measurements include cluster width (cw.) and cluster length (cl.), as well as width (w.) and length (l.) of a single flower.

Common Names: Each tree has at least two names, a common and a scientific one. The English common names in this guide include some that have been adapted from other languages, particularly Spanish names of trees in regions bordering Mexico. The main common names used here follow the latest U.S. Forest Service *Checklist of United States Trees (Native and Naturalized)* by the author of this guide (U.S. Department of Agriculture, Agriculture Handbook 541; 1979). Other English names in local use are also included, designated by quotation marks. Since a tree may have several common names, botanists have also assigned it a scientific name, used uniformly throughout the world.

Scientific Names: Scientific names are usually from Latin or, if derived from Greek or other languages, are in Latin form. Each scientific name consists of two words: the genus name (always capitalized), followed by the species name (lowercase). For example, all pines are in the genus *Pinus;* the species commonly known as Ponderosa Pine is designated *Pinus ponderosa,* indicating the genus *Pinus* and the species *ponderosa.* A related species in the same

genus, Jeffrey Pine, is known as *Pinus
jeffreyi.* The scientific name of a tree
may be followed by the name, usually
abbreviated, of the author who named
but not necessarily discovered it. For
example, in *Pinus ponderosa* Dougl., the
author is David Douglas (1798–1834),
the Scottish botanical collector who
discovered this pine and named it for
its heavy, or ponderous, wood. The
author's (or authors') name helps in
tracing the history of the scientific
name and the discovery of the species.
Varieties of trees are expressed with the
variety name preceded by "var.": for
example, *Pinus ponderosa* var. *arizonica,*
Arizona Pine.

Classification: Trees are classified in two major plant
groups: gymnosperms, which include
the conifers, and angiosperms, which
include all the flowering plants. The
conifers, or softwoods, are usually
evergreen, have narrow needlelike or
scalelike leaves, and bear exposed seeds,
usually in cones. The flowering plants,
which bear seeds enclosed in fruits, are
further divided into two groups.
Among the monocotyledons, only
yuccas and palms reach tree size; they
have large evergreen, bayonetlike or
fanlike leaves that are usually clustered
at the top of the trunk or on a few large
branches. Dicotyledons, the larger of
the two groups, contain all the
hardwoods, or broadleaf trees, which
are mostly deciduous and have net-
veined leaves.

Organization of The text contains a description of each
the Text: native tree species, as well as common
introduced species, found within the
geographical range of this guide.
Arrangement of families is botanical;
within each family, the listing is
alphabetical by scientific name.

Text Each species description begins with a
Descriptions: general statement about the shape and

appearance of the tree, as well as any special features that are useful for immediate identification.

Height and Diameter: Also given (in both the English and metric systems) are measurements for the height and diameter (at breast height) of mature specimens under favorable growing conditions. Naturally, height and trunk diameter vary greatly with age, location, and climate.

The description of the tree is subdivided into sections on leaves, bark, twigs, flowers, and fruit. Where flowers do not vary greatly from species to species, as among the oaks, they are discussed in the family description. Wherever possible, familiar terms are used in the description of trees; where technical terms are necessary, they are defined in the Glossary (pages 609–616). Important details are emphasized by italics.

Leaves: The leaf description includes information on size, shape, edges, texture, and color. Since most leaves are arranged alternately on the stem, leaf arrangement is usually mentioned only when it is opposite or whorled. Similarly, since most trees are deciduous, only the less common evergreen category is specified.

Bark: The bark description gives characteristics of color, texture, thickness, shagginess, and other relevant details. The barks of some species change significantly with age, and in such cases both young and mature trees are noted.

Twigs: Twig descriptions include color, stoutness, type of branching, hairiness, fragrance, and stickiness, when applicable.

Flowers: The flower description covers size, shape, and color as well as time of flowering and fragrance. If the tree has both male and female flowers, each is described. While many flowers are solitary, others are grouped in clusters; in these cases the description includes the size of the cluster and where it is placed on the twig.

Fruit: Fruit descriptions indicate size, shape, color, surface texture, dryness or pulpiness, shape and number of seeds, and time of maturity.

Habitat: The habitat, or place where a tree grows naturally, is given for each species and is often important in identification. Commonly associated trees and other vegetation are also mentioned.

Range: This section summarizes the boundaries of the natural range, or geographic distribution, of each tree species; these are given generally in a clockwise direction from northwest to east, then south, west, and north, together with the approximate altitudinal range. For introduced species the native continent or region is given, followed by information on the distribution in North America. Range maps for each native species supplement the range description. Condensed from the author's 6-volume *Atlas of United States Trees,* these maps reflect the most accurate and recent information on distribution available.

Comments: The description of each species concludes with notes on the uses of the tree and its products, cultivation, history, and lore. Frequently, too, the origin of common and scientific names is explained.

Drawings: Typical winter silhouettes of many species appear in the margins as a further aid to identification. The drawings show the characteristic shape and manner of branching of trees growing under ideal conditions. Since evergreen trees have foliage year-round, they are shown with their leaves or needles. Fruits not illustrated in color photographs are shown in small drawings beside the text. In addition to

the fruit drawings, we have included in
the margins drawings of the leaves of a
few rare or inaccessible species for
which photographs were not available.

Poisonous Trees: As noted in the comments section, a
few tree species are poisonous. Most
parts of yews, such as Pacific Yew
(*Taxus brevifolia*), including the seeds
and foliage, are toxic and can be fatal if
eaten. A single seed of Mescalbean
(*Sophora secundiflora*) could kill a person.
A safe rule is not to eat any unfamiliar
fruits and seeds.

Rare Species: Relatively few species of native trees are
rare or local in distribution. According
to lists proposed under the Endangered
Species Act of 1973, very few tree
species are threatened with extinction.
No tree of western North America has
been officially designated as
endangered. Cypresses are the rarest
group of native conifers, composed of
seven species and additional varieties of
very restricted range in scattered
groves. In nature, Monterey Cypress
(*Cupressus macrocarpa*) is confined to two
windswept groves along the Pacific
Ocean near Monterey, California, but is
widely grown as an ornamental around
the world. Some rare trees are also
protected by state laws, and there is
additional protection in public parks
and forests, nature preserves, and other
sheltered zones. However, the
destruction and disturbance of habitats
make the establishment of additional
preserves desirable. In the wild, one
may usually obtain small botanical
specimens without damage; of course,
collecting should never be done in a
park or arboretum.

HOW TO USE THIS GUIDE

Example 1
A Needleleaf
Tree:

You are in a mountain forest of many tall, narrow-crowned trees with light brown, scaly bark and short needles in bundles of 2.

1. Turning to the Thumb Tab Guide preceding the color plates, you find that the section called *Needle-leaf Conifers* (plates 1–54) contains several groups, 3 of which have needles in bundles or clusters: *Pines, Larches,* and *Cedars.*

2. When you turn to the color plates indicated in these groups, you quickly see that larches and cedars have clusters of many needles and that only pines have needles in bundles of 2. Reading the captions, you discover that your choice is narrowed to Pinyon and Ponderosa, Austrian, Bishop, Scotch, or Lodgepole pines. The needle lengths given in the captions indicate that 3 have longer needles and 1 has shorter needles. Your tree is either a Lodgepole or a Scotch pine.

3. Turning to the text described these species, you learn that only Lodgepole Pine has a narrow crown and light brown bark and is a widespread tree of the Western mountains, often growing in pure stands. The identification of your tree as a Lodgepole Pine is confirmed.

Example 2
A Tree with
Bell-shaped
Pink Flowers:

Along a stream in a desert grassland you see a small tree with large, bell-shaped flowers 1¼″ long, in unbranched clusters at the ends of twigs. The pink tubular corollas have 5 unequal lobes.

1. Referring to the Thumb Tab Guide, you find that the flower section (plates 316–393) is arranged according to color. Pink flowers appear in plates 382–393.

2. Looking through the photographs in the pink section, you find that only Desert-willow has bell-shaped tubular flowers and resembles your specimen.

3. The text confirms your identification of Desert-willow, which often forms thickets in moist soils of plains.

Example 3
A Tree with
Beanlike Fruit:

In a mountain canyon you find a small spreading tree with a short trunk and spiny twigs. The leaves have fallen, but short, narrow, beanlike fruits remain. These are slightly flat, long-pointed, light brown, and covered with bristly hairs. Many have split open to reveal small brown seeds.

1. Turn to the *Cones and Fruit* section (plates 400–540) of the Thumb Tab Guide, where you will see that *Pods* (plates 514–525) most closely resemble your specimen.

2. Looking at the color plates you find that four pods resemble your specimen: Littleleaf Leucaena, New Mexico Locust, Honeylocust, and Black Locust. The picture captions reveal that only New Mexico Locust and Black Locust have pods as small as your example.

3. Reading the text description, you find that New Mexico Locust, a native thicket-forming species, grows at much higher elevations than Black Locust, which is naturalized in the West. Your tree is identified as a New Mexico Locust.

Keys to the Color Plates

The color plates are arranged according to three keys: leaves, flowers, and fruit and cones. Within each key the photographs are organized as follows:

Leaf Key Needle-leaf Conifers
Scale-leaf Conifers
Untoothed Simple
Toothed Simple
Lobed Simple
Compound
Yuccas, Palms, and Cacti

Flower Key Reddish
Yellow
Greenish
Cream-colored
White
Pink
Purple

Fruit and Cones
Cones Key Berrylike Fruit
Fleshy Fruit
Nuts
Acorns
Pods
Keys
Tufted Fruit
Balls and Capsules

Thumb Tabs Each group of photographs within a visual key is represented by a silhouette that appears as a thumb tab at the left edge of each double-page of plates. A chart of the thumb tab organization appears on the pages preceding the color plates. The flower groups are indicated by one symbol, shown in various colors in the thumb tabs.

Captions The caption under each photograph contains the plate number, common name of the tree, measurement, and the page on which it is described. When pertinent, additional information is given, such as whether a tree is evergreen, if the leaves are opposite each other on the stem (rather than the usual alternate arrangement), or if the leaf is pinnately, bipinnately, or palmately compound. The flower captions include length (l.) or width (w.) of an individual flower, cluster length (cl.) or width (cw.), and, if a flower is not bisexual, the male (♂) or female (♀) symbol. The fruit captions indicate the width, unless the fruit is oblong, when length is given.

Part I
Color Keys

Typical Leaf Shapes		Plate Numbers
	pines	3–26
	spruces	27–33
	firs	34–40
	larches	43–45
	douglas-firs	41, 42
	hemlocks	49, 52
	redwood and cryptomeria	53, 54
	true cedars	46–48
	"cedars"	55–60
	cypresses	63–69
	junipers	70–78

Thumb Tab	Group	Plate Numbers
	Untoothed Simple Leaves	79–126
	Toothed Simple Leaves	127–213

Typical Leaf Shapes		Plate Numbers
	willows	81, 89, 97–99, 101–104
	catalpas	125, 126
	eucalyptus	86, 100
	oaks	90, 106–108
	magnolias	96, 113
	dogwoods	110, 111
	sumacs	115–117
	redbuds	122, 123
	willows	127–134, 136, 141–144, 150
	cherries	138, 145, 146, 148, 160–162, 176, 194

Thumb Tab	Group	Plate Numbers
	Toothed Simple Leaves	127–213

Typical Leaf Shapes		Plate Numbers
	crab apples and apples	149, 157
	elms	151. 152, 154, 156
	oaks	166, 168–170, 172–175, 177–179
	hawthorns	181, 184
	cercocarpuses	185, 186
	alders	187–192
	hophornbeams	199, 200
	mulberries	202–204
	birches	205, 206, 208
	cottonwoods, aspens, and poplars	135, 207, 209–213

	hawthorns	214–219
	sycamores	220, 223, 224
	maples	221, 225–231
	yellow-poplar	234
	ginkgo	235
	poplar	236
	California fremontia	237
	oaks	238–249
	cliffrose	250
	mulberries	251, 252

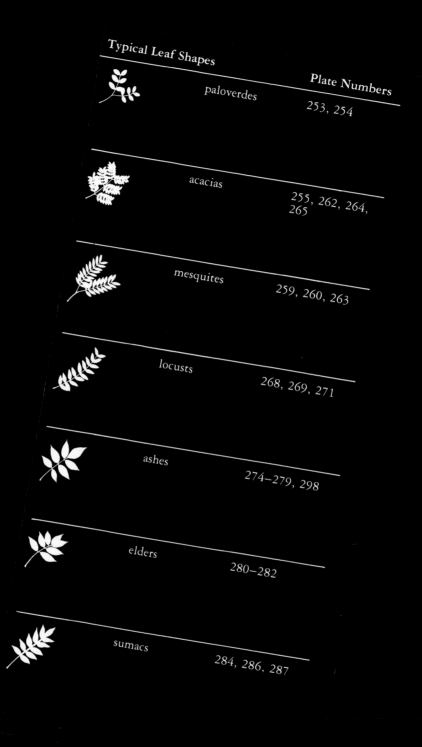

Typical Leaf Shapes

Plate Numbers

Leaf Shape	Name	Plate Numbers
	paloverdes	253, 254
	acacias	255, 262, 264, 265
	mesquites	259, 260, 263
	locusts	268, 269, 271
	ashes	274–279, 298
	elders	280–282
	sumacs	284, 286, 287

Thumb Tab	Group	Plate Numbers
	Compound Leaves	253–303
	Yuccas, Palms, and Cacti	304–315

Typical Leaf Shapes		Plate Numbers
	ailanthus	288
	mountain-ashes	289, 290
	walnuts	291–294
	buckeyes	302, 303
	yuccas	304–310
	palms	313–315
	cacti	311, 312

Thumb Tab	Flower Key	Plate Numbers
	Reddish Flowers	316–321
	Yellow Flowers	322–333
	Greenish Flowers	334–345
	Cream-colored Flowers	346–348, 352, 353
	White Flowers	349–351, 354–381
	Pink Flowers	382–393
	Purple Flowers	394–399

Thumb Tab	Cones/Fruit Key	Plate Numbers
	Cones	400–454
	Berrylike Fruit	455–489
	Fleshy Fruit	490–498
	Nuts	499–501
	Acorns	502–513
	Pods	514–525
	Keys	526–534
	Tufted Fruit	535, 536
	Balls and Capsules	537–540

The color plates on the following pages are numbered to correspond with the numbers preceding the text descriptions.

Needle-leaf and Scale-leaf Conifers

Most conifers, or softwoods, have narrow, needlelike, evergreen leaves. Pines, the largest group, are recognized by relatively long needles in bundles of 2–5; the number of needles per bundle is indicated in parentheses in the captions. The many clustered needles of larches are shed in autumn. Spruces have stiff, usually 4-angled, sharply pointed needles, while those of hemlocks are flat and blunt-tipped; both have rough twigs with peglike bases left after the needles have fallen. Firs have softer needles and a distinct fragrance of balsam. Douglas-firs have short-stalked needles and pointed conical buds. In Redwood the leaves are usually short, flat, needlelike, and in 2 rows. Yews have flat needles in 2 rows with short leafstalks forming lines on the twig.

Some conifers have small, short, scalelike, evergreen leaves arranged in pairs or in 3's. The leafy twigs of junipers and cypresses are slender and 4-angled or rounded, while those of Incense-cedar and Western Redcedar are flattened, and those of Port-Orford-cedar are intermediate. Giant Sequoia has very short, sharply pointed, scalelike leaves.

River-oak Casuarina and Athel Tamarisk, included here, are hardwoods with drooping, wirelike, evergreen twigs suggesting pine needles and with leaves reduced to tiny scales at joints.

Tamarisk, though not a conifer, has scalelike leaves resembling those of junipers.

1 River-oak Casuarina, 1/64″, *p. 336*

2 Athel Tamarisk, 1/16″, *p. 558*

3 Digger Pine, (3) 8–12″, *p. 289*

4   Jeffrey Pine, (3) 5–10″, *p. 279*

5   Torrey Pine, (5) 8–13″, *p. 292*

6   Coulter Pine, (3) 8–12″, *p. 275*

7 Apache Pine, (3) 8–12″, *p. 277*

8 Chihuahua Pine, (3) 2½–4″, *p. 281*

9 Sugar Pine, (5) 2¾–4″, *p. 280*

10 Western White Pine, (5) 2–4″, *p. 283*

11 Ponderosa Pine, (2–3) 4–8″, *p. 286*

12 Austrian Pine, (2) 3½–6″, *p. 285*

13   Southwestern White Pine, (5) 2½–3½", *p. 290*

14   Knobcone Pine, (3) 3–7", *p. 271*

15   Monterey Pine, (3) 4–6", *p. 288*

16 Bishop Pine, (2) 4–6″, *p. 284*

17 Scotch Pine, (2) 1½–2¾″, *p. 291*

19　Whitebark Pine, (5) 1½–2¾″, *p. 268*

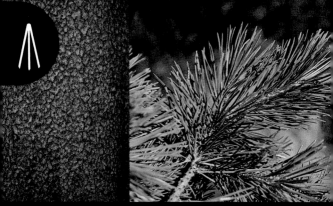

20　Lodgepole Pine, (2) 1¼–2¾″, *p. 273*

22 Parry Pinyon, (4) 1–2¼″, *p. 287*

23 Pinyon, (2) ¾–1½″, *p. 276*

24 Mexican Pinyon, (3) 1–2½″, *p. 272*

25   Foxtail Pine, (5) ¾–1½″, *p. 272*

26   Bristlecone Pine, (5) ¾–1½″, *p. 269*

27   Sitka Spruce, ⅝–1″, *p. 267*

28　White Spruce, ½–¾″, *p. 264*

29　Blue Spruce, ¾–1⅛″, *p. 267*

30　Black Spruce, ¼–⅝″, *p. 265*

31 Engelmann Spruce, ⅝–1″, *p. 263*

32 Norway Spruce, ½–1″, *p. 262*

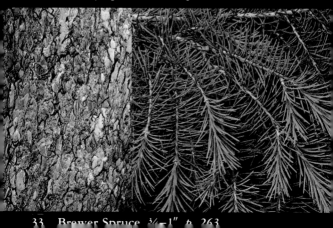

33 Brewer Spruce, ¾–1″, *p. 263*

34 California Red Fir, ¾–1⅜″, *p. 254*

35 Noble Fir, 1–1⅜″, *p. 254*

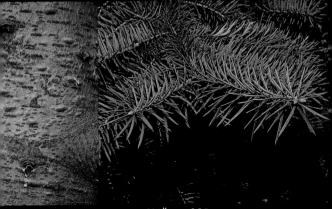

36 Bristlecone Fir, 1½–2¼″, *p. 250*

37 Subalpine Fir, 1–1¾", *p. 253*

38 Pacific Silver Fir, ¾–1½", *p. 249*

39 Grand Fir, 1¼–2", *p. 252*

40 White Fir, 1½–2½″, *p. 251*

41 Douglas-fir, ¾–1¼″, *p. 294*

42 Bigcone Douglas-fir, ¾–1¼″, *p. 293*

43 Tamarack, *deciduous, ¾–1″, p. 259*

44 Western Larch, *deciduous, 1–1½″, p. 261*

45 Subalpine Larch, *deciduous, 1–1¼″, p. 260*

46 Atlas Cedar, ¾–1″, *p. 256*

47 Deodar Cedar, 1–2″, *p. 257*

49 Mountain Hemlock, ¼–1″, *p. 297*

50 California Torreya, 1–2¾″, *p. 246*

2 Western Hemlock, ¼–¾″, *p. 296*

3 Redwood, ⅜–¾″, *p. 299*

55 Port-Orford-cedar, *opp.*, 1/16″, *p. 303*

56 Sawara False-cypress, *opp.*, 1/8″, *p. 305*

58 Incense-cedar, *opp.*, ⅛–½″, *p. 319*

59 Oriental Arborvitae, *opp.*, ¹⁄₁₆–⅛″, *p. 320*

60 Western Redcedar, *opp.*, ¹⁄₁₆–⅛″, *p. 321*

61　Giant Sequoia, ⅛–¼″, *p. 301*

62　Tamarisk, *deciduous,* ¹⁄₁₆″, *p. 559*

63　Monterey Cypress, *opp.,* ¹⁄₁₆″, *p. 310*

64 Sargent Cypress, *opp.*, ¹⁄₁₆″, *p. 311*

65 Gowen Cypress, *opp.*, ¹⁄₁₆″, *p. 308*

67 Arizona Cypress, *opp.,* ¹⁄₁₆″, *p. 306*

68 Baker Cypress, *opp.,* ¹⁄₁₆″, *p. 307*

70 Rocky Mountain Juniper, *opp.*, ¹⁄₁₆″, *p. 318*

71 Drooping Juniper, *opp.*, ¹⁄₁₆–¹⁄₈″, *p. 314*

73 Western Juniper, 1/16", *p. 315*

74 Alligator Juniper, *opp.*, 1/16–1/8", *p. 312*

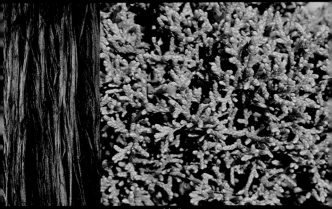

76 Redberry Juniper, *opp.*, ¹⁄₁₆″, *p. 313*

77 Pinchot Juniper, ¹⁄₁₆″, *p. 317*

78 Utah Juniper, *opp.*, ¹⁄₁₆″, *p. 316*

Untoothed Simple Leaves

Most broadleaf trees, or hardwoods, have simple leaves consisting of a single blade. All the leaves in this section have smooth edges, without teeth or lobes. Dogwoods, redbuds, California-laurel, and Desert-willow are included in this group, as are the untoothed species of oaks and willows.

79   Canotia, 1/16", *p. 527*

80   Smokethorn, 3/8–1", *p. 492*

81   Yewleaf Willow, 1/2–1 1/4", *p. 366*

2   Curlleaf Cercocarpus, *evergreen*, ½–1¼″, *p. 446*

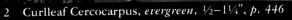

83   Olive, *evergreen*, *opp.*, 1½–3″, *p. 593*

85   Desert-willow, 3–6″, *p. 599*

86   Bluegum Eucalyptus, *evergreen*, 4–12″, *p. 569*

88 Lyontree, *evergreen, opp.*, 4–7″, *p. 459*

89 Hinds Willow, 1½–3¼″, *p. 356*

90 Silverleaf Oak, *evergreen,* 2–4″, *p. 403*

91   Arizona Madrone, *evergreen,* 1½–3″, *p.* 576

92   Giant Chinkapin, *evergreen,* 2–5″, *p.* 388

93   California-laurel, *evergreen,* 2–5″, *p.* 434

94 Gum Bumelia, 1–3″, *p. 581*

95 Camphor-tree, *evergreen,* 2½–4″, *p. 433*

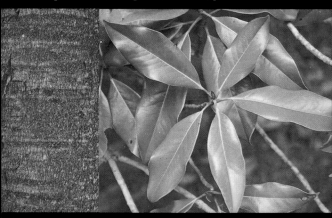

96 Southern Magnolia, *evergreen,* 5–8″, *p. 430*

97 Tracy Willow, 1¼–2″, *p. 367*

98 Arroyo Willow, 2½–4″, *p. 359*

99 Sitka Willow, 2–4″, *p. 365*

00 **Red-ironbark Eucalyptus,** *evg.,* 2½–5″, *p. 570*

101 **Feltleaf Willow,** 2–4″, *p. 346*

103　Scouler Willow, 2–5″, *p. 363*

104　Hooker Willow, 1½–4½″, *p. 357*

106   Mexican Blue Oak, *evergreen,* 1–2″, *p. 408*

107   Gray Oak, ¾–2″, *p. 402*

108   Toumey Oak, *evergreen,* ½–1″, *p. 412*

109   Buttonbush, *opp.*, 2½–6″, *p. 601*

110   Red-osier Dogwood, *opp.*, 1½–3½″, *p. 573*

111   Pacific Dogwood, *opp.*, 2½–4½″, *p. 572*

12 Osage-orange, 2½–5″, *p. 422*

113 Saucer Magnolia, 5–8″, *p. 431*

114 Pacific Madrone, *evergreen, 2–4½″, p. 577*

115　Laurel Sumac, *evergreen,* 2–4", *p. 521*

116　Sugar Sumac, *evergreen,* 1½–3¼", *p. 522*

117　Lemonade Sumac, *evergreen,* 1–2½", *p. 519*

18 Crapemyrtle, *opp.*, 1–2″, *p.* 567

119 Catalina Cherry, *evergreen*, 2–4″, *p.* 468

20 Wavyleaf Silktassel *evg., opp.*, 2–3¼″, *p.* 575

121 Singleleaf Ash, *opp.*, 1½–2″, *p. 584*

122 California Redbud, 1½–3½″, *p. 491*

123 Eastern Redbud, 2½–4½″, *p. 490*

24 Royal Paulownia, *opp.*, 6–16″, *p.* 595

25 Southern Catalpa, *opp.*, 5–10″, *p.* 597

26 Northern Catalpa, *opp.*, 6–12″, *p.* 598

Toothed Simple Leaves

The largest group of broadleaf trees has simple leaves with teeth along the edges. Often the teeth are uniform and pointed forward—saw-toothed—as in most willows. Some trees, such as alders, birches, and elms, have doubly saw-toothed leaves with alternating large and small teeth. In some oaks the teeth are bristle-tipped, while in others they are blunt and wavy, as in cottonwoods and poplars. Other trees with toothed simple leaves are some hawthorns, the cherries and plums, ceanothuses, and Toyon.

127 Sandbar Willow, 1½–4″, *p. 354*

128 Weeping Willow, 2½–5″, *p. 350*

129 River Willow, 2–6″, *p. 355*

130   White Willow, 2–4½", *p. 348*

131   Black Willow, 3–5", *p. 361*

132   Pacific Willow, 2–5", *p. 358*

133 Bonpland Willow, 3–6″, *p. 352*

134 Peachleaf Willow, 2–4½″, *p. 348*

135 Narrowleaf Cottonwood, 2–5″, *p. 340*

136 Mackenzie Willow, 2½–4″, *p. 360*

137 Peach, 3½–6″, *p. 471*

138 Pin Cherry, 2½–4½″, *p. 470*

139   Pacific Bayberry, *evergreen,* 2–4½″, p. 368

140   Torrey Vauquelinia, *evergreen,* 1½–4″, p. 479

141   Northwest Willow, 1–2″, p. 364

142　Arroyo Willow, 2½–4″, p. 359

143　Littletree Willow, 1–3″, p. 349

145 Sour Cherry, 2–3½", *p. 464*

146 Black Cherry, 2–5", *p. 472*

48 Bitter Cherry, 1–2½", *p. 466*

49 Oregon Crab Apple, 1½–3½", *p. 460*

50 Pussy Willow, 1½–4¼", *p. 353*

151   Siberian Elm, ¾–2″, *p. 419*

152   American Elm, 3–6″, *p. 416*

153   Blueblossom, *evergreen*, ¾–2″, *p. 547*

154 English Elm, 2–3¼″, *p. 418*

155 Japanese Zelkova, 1–3½″, *p. 420*

157 Apple, 2–3½", *p. 461*

158 Pear, 1½–3", *p. 475*

60 Catalina Cherry, *evergreen,* 2–4″, *p. 468*

161 Mahaleb Cherry, 1¼–2¾″, *p. 469*

163　Sour Orange, *evergreen*, 2½–5½", p. 509

164　Common Camellia, *evergreen*, 2–4", p. 556

66   Engelmann Oak, *evergreen*, 1–2¾", *p. 397*

67   Toyon, *evergreen*, 2–4", *p. 458*

68   Interior Live Oak, *evergreen*, 1–2", *p. 414*

169   Island Live Oak, *evergreen,* 1¼–3½", *p. 411*

170   Chinkapin Oak, *4–6", p. 407*

171   Tanoak, *evergreen,* 2½–5", *p. 390*

72 Emory Oak, *evergreen,* 1–2½", p. 396

73 Arizona White Oak, *evergreen,* 1½–3", p. 392

175 Coast Live Oak, *evergreen,* ¾–2½", *p. 391*

176 Hollyleaf Cherry, *evergreen,* 1–2", *p. 467*

78 Dunn Oak, *evergreen,* 1/2–1 1/4", *p. 396*

79 Turbinella Oak, *evergreen,* 5/8–1 1/2", *p. 413*

181 Tracy Hawthorn, 1–1½″, *p. 457*

182 Klamath Plum, 1–2½″, *p. 473*

84   Fleshy Hawthorn, 2–2½″, *p. 456*

85   Hairy Cercocarpus, *evergreen*, ⅜–1″, *p. 445*

86   Birchleaf Cercocarpus, *evergreen*, 1–1¼″, *p. 444*

187   Red Alder, 3–5″, *p. 377*

188   Mountain Alder, 1½–4″, *p. 381*

189   Speckled Alder, 2–4″, *p. 379*

190 Arizona Alder, 1½–3¼", *p. 375*

191 White Alder, 2–3½", *p. 376*

193 Cascara Buckthorn, 2–6″, *p. 549*

194 Sweet Cherry, 3–6″, *p. 463*

196 European Beech, 2–4″, *p. 389*

197 European Linden, 2–4″, *p. 551*

199   Chisos Hophornbeam, 1½–2½", *p. 385*

200   Knowlton Hophornbeam, 1–2½", *p. 386*

202 Texas Mulberry, 1–2½", *p. 424*

203 Red Mulberry, 4–7", *p. 425*

205 Water Birch, 1–2″, *p. 382*

206 Paper Birch, 2–4″, *p. 383*

208 European White Birch, 1¼–2¾″, *p. 384*

209 Fremont Cottonwood, 2–3″, *p. 342*

211 Quaking Aspen, 1¼–3″, *p. 344*

212 Lombardy Poplar, 1½–3″, *p. 343*

Lobed Simple Leaves

Some broadleaf trees have leaves that are shallowly or deeply cut into narrow or broad lobes. Included here are maples and sycamores, many oaks and hawthorns, as well as the oddly shaped leaves of mulberries, California Fremontia, Ginkgo, Sweetgum, and Yellow-poplar.

214   Washington Hawthorn, 1½–2½", *p. 454*

215   Fireberry Hawthorn, 1½–2", *p. 449*

216   Oneseed Hawthorn, 1–2", *p. 453*

17 Columbia Hawthorn, 1–3½″, *p. 450*

18 Cerro Hawthorn, 1–2½″, *p. 452*

19 Black Hawthorn, 1–3″, *p. 451*

220 London Planetree, 5–10″, *p. 438*

221 Planetree Maple, *opp.,* 3½–6″, *p. 536*

222 Sweetgum, 3–6″, *p. 436*

23 Arizona Sycamore, 6–9″, *p. 440*

24 California Sycamore, 6–9″, *p. 439*

226   Canyon Maple, *opp.*, 2–3¼", *p. 531*

227   Red Maple, *opp.*, 2½–4", *p. 537*

228   Norway Maple, *opp.*, 4–7", *p. 535*

229 Silver Maple, *opp.*, 4–6″, *p. 538*

230 Rocky Mountain Maple, *opp.*, 1½–4½″, *p. 530*

232   Chinese Parasoltree, 6–12″, *p. 553*

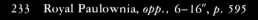

233   Royal Paulownia, *opp.*, 6–16″, *p. 595*

234   Yellow-poplar, 3–6″, *p. 429*

235   Ginkgo, *1–2″, p. 243*

236   White Poplar, *2½–4″, p. 339*

237   California Fremontia, *evergreen,* ½–1½″, *p. 554*

238 Mohr Oak, 1–3″, *p. 406*

239 Sandpaper Oak, *evergreen,* ¾–2″, *p. 409*

240 Chisos Oak, 3–4″, *p. 400*

241 McDonald Oak, 1½–2¾", *p. 405*

242 Blue Oak, 1¼–4", *p. 394*

244 California Scrub Oak, *evergreen,* ⅝–1″, *p. 395*

245 Valley Oak, 2–4″, *p. 404*

246 Gambel Oak, 2–6″, *p. 398*

247 Graves Oak, 2–4″, *p. 401*

248 Oregon White Oak, 3–6″, *p. 399*

249 California Black Oak, 3–8″, *p. 403*

250 Cliffrose, *evergreen,* ¼–⅝", *p. 448*

251 Texas Mulberry, 1–2½", *p. 424*

Compound Leaves

Compound leaves are composed of 3 or more small leaflets. When the leaflets are in 2 rows along a central stalk, the leaf is pinnately compound, as in walnuts, ashes, and most sumacs. When the central stalk has side branches, the leaf is bipinnately compound; examples include acacias, mesquites, and paloverdes. When the leaflets are all attached to the end of the leafstalk and spread like the fingers of a hand, the leaf is palmately compound, as in buckeyes and Common Hoptree.

253 Yellow Paloverde, *bipinnate,* ¾–1″, *p. 489*

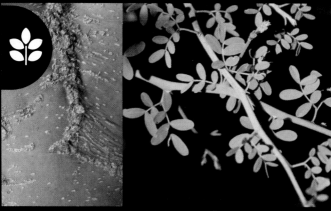

254 Blue Paloverde, *bipinnate,* 1″, *p. 488*

255 Gregg Catclaw, *bipinnate,* 1–3″, *p. 483*

256   Texas Lignumvitae, *opp., pin.,* 1–3″, p. 507

257   Elephant-tree, *pinnate,* 1–1¼″, p. 514

259   Honey Mesquite, *bipinnate,* 3–8″, p. 499

260   Screwbean Mesquite, *bipinnate,* 2–3″, p. 501

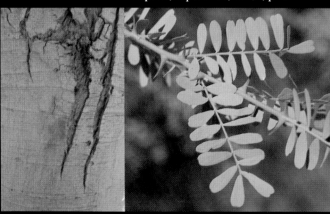

261   Tesota, *evergreen, pinnate,* 1–2¼″, p. 497

262   Roemer Catclaw, *bipinnate,* 1½–4″, *p. 485*

263   Velvet Mesquite, *bipinnate,* 5–6″, *p. 502*

264   Huisache, *bipinnate,* 2–4″, *p. 482*

265   Green Wattle, *evergreen, bipinnate, 3–6″, p. 481*

266   Silktree, *bipinnate, 6–15″, p. 486*

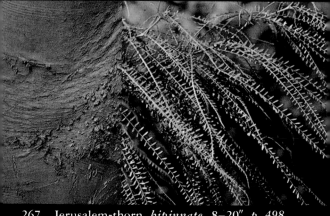

267   Jerusalem-thorn, *bipinnate, 8–20″, p. 498*

268 Black Locust, *pinnate,* 6–12", p. 504

269 Honeylocust, *pinnate-bipinnate,* 4–8", p. 494

270 Littleleaf Leucaena, *eve, bipin,* 3–4", p. 496

271 New Mexico Locust, *pinnate,* 4–10″, *p. 503*

272 Mescalbean, *evergreen, pinnate,* 3–6″, *p. 505*

273 Carob, *evergreen, pinnate,* 4–8″, *p. 487*

274　Velvet Ash, *opp., pinnate,* 3–6″, p. 592

275　Fragrant Ash, *opp., pinnate,* 3–7″, p. 585

276　Oregon Ash, *opp., pinnate,* 5–12″, p. 589

277   Two-petal Ash, *opp.*, *pinnate*, 1½–4½″, *p. 586*

278   Goodding Ash, *opp.*, *pinnate*, 1–3¼″, *p. 587*

80 Mexican Elder, *opp., pinnate, 5–7″, p. 605*

81 Blue Elder, *opp., pinnate, 5–7″, p. 604*

283 Western Soapberry, *pinnate*, 5–8″, *p. 542*

284 Peppertree, *evergreen, pinnate*, 6–12″, *p. 523*

285 Mexican-buckeye, *pinnate*, 5–12″, *p. 543*

286 Prairie Sumac, *pinnate, 9", p. 520*

287 Smooth Sumac, *pinnate, 12", p. 518*

288 Ailanthus, *pinnate, 12–24", p. 512*

289   European Mountain-ash, *pinnate,* 4–8″, *p. 476*

290   Greene Mountain-ash, *pinnate,* 4–9″, *p. 477*

291   Northern California Walnut, *pin.,* 7–12″, *p. 371*

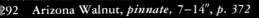

295 Chinaberry, *bipinnate,* 8–18″, *p. 516*

296 Lyontree, *evergreen, opp., pinnate,* 4–7″, *p. 459*

297 Silk-oak, *evergreen, pinnate,* 6–12″, *p. 427*

298 Lowell Ash, *opp.*, *pinnate*, 3½–6″, *p. 590*

299 Boxelder, *opp.*, *pinnate*, 6″, *p. 533*

300 Rocky Mountain Maple, *opp.*, *palm.*, 3″, *p. 530*

301 Common Hoptree, *palmate, 2–4", p. 510*

302 California Buckeye, *opp., palmate, 6–12", p. 539*

303 Horsechestnut, *opp., palmate, 7–17", p. 540*

Yuccas, Palms, and Cacti

Yuccas and palms differ from other trees in having large, evergreen, parallel-veined leaves clustered at the top of a stout trunk or at the ends of large branches. Yuccas have long, narrow, bayonetlike leaves, usually ending in a sharp point. Palms have fanlike leaves, as in washingtonias, or pinnately compound leaves, as in Canary Island Date. Cacti, including Saguaro and Jumping Cholla, have stout, green, fleshy trunks with few branches, clusters of spines, and either very small leaves or none at all.

304   Schott Yucca, *evergreen*, 16–32″, *p. 333*

305   Giant Dracaena, *evergreen*, 12–36″, *p. 327*

306   Torrey Yucca, *evergreen*, 24–42″, *p. 334*

307 Faxon Yucca, *evergreen*, 20–32″, *p. 331*

308 Mohave Yucca, *evergreen*, 18–24″, *p. 332*

309 Soaptree Yucca, *evergreen*, 12–30″, *p. 330*

310   Joshua-tree, *evergreen*, 8–14″, *p. 328*

311   Jumping Cholla, ½–1″, *p. 563*

312   Saguaro, *p. 561*

13 California Washingtonia, *evg.*, 6–10', *p. 324*

14 Mexican Washingtonia, *evergreen*, 5½–9', *p. 326*

Flowers

Tree flowers vary greatly in color, shape, size, and arrangement. Some, such as elm flowers, are a fraction of an inch wide; others, like those of the yuccas, grow in showy clusters several feet long. Some flowers are borne singly; others form branched clusters or drooping catkins. The flowers shown here are grouped by color in the following order: red, yellow, green, cream-colored, white, pink, and purple. Most flowers are bisexual. Those that are only male or only female are indicated in the captions.

316   Silver Maple ♀, *cw. ½″, p. 538*

317   Red Maple ♂, *cw. ½–¾″, p. 537*

318   Red Maple ♀, *l. ⅛″, p. 537*

19　Pussy Willow ♂, *cl.* 1–2½″, *p. 353*

20　Speckled Alder ♂, *cl.* 1½–3″, *p. 379*

322   **Russian-olive,** *l.* ⅜″, *p.* 565

323   **Cliffrose,** *w.* ¾–1″, *p.* 448

325   Yellow Paloverde, *w.* ½″, *p.* 489

326   Blue Paloverde, *w.* ¾″, *p.* 488

327   Jerusalem-thorn, *w.* ¾″, *p.* 498

328   Silk-oak, *cl. 3–5″, p. 427*

329   Gregg Catclaw, *cl. 1–2″, p. 483*

330   Screwbean Mesquite, *cl. 2″, p. 501*

31 Huisache, *cw.* ½″, *p. 482*

32 Littleleaf Leucaena, *cw.* ¾–1″, *p. 496*

334 Osage-orange ♂, *cw*. ¾–1″, *p. 422*

335 Norway Maple ♂, *cw*. 3″, *p. 535*

337   **European Beech** ♂, *cw.* ¾″, *p. 389*

338   **Hinds Willow** ♀, *cl.* ¾–1½″, *p. 356*

340   Little Walnut ♂, *cl. 2–4″, p. 373*

341   English Oak ♂, *cl. 2–5″, p. 410*

343　Common Hoptree, *cw. 2½", p. 510*

344　Ailanthus ♂, *cl. 6–10", p. 512*

345　Smooth Sumac, *cl. 8", p. 518*

346   Pacific Red Elder, *cl.* 4″, *p. 603*

347   Mexican Elder, *cw.* 2–8″, *p. 605*

348   Blue Elder, *cw.* 4–8″, *p. 604*

349　Laurel Sumac, *cl. 2–6″, p. 521*

350　Toyon, *cw. 4–6″, p. 458*

352 Bluegum Eucalyptus, *w.* 2″, *p. 569*

353 Buttonbush, *cw.* 1–1½″, *p. 601*

55   Hollyleaf Cherry, *cl.* 2½″, *p. 467*

56   Black Cherry, *cl.* 4–6″, *p. 472*

57   Common Chokecherry, *cl.* 4″, *p. 474*

358 Sweet Cherry, *w.* 1–1¼″, *p. 463*

359 Klamath Plum, *w.* ⅝″, *p. 473*

360 Washington Hawthorn, *w.* ½″, *p. 454*

361 Oregon Crab Apple, *w.* ¾–1", *p. 460*

362 Pin Cherry, *w.* ½", *p. 470*

364   Western Serviceberry, *w.* ¾–1¼", *p. 443*

365   Garden Plum, *w.* ¾–1", *p. 465*

367 American Plum, *w. ¾–1″, p. 462*

368 Apple, *w. 1¼″, p. 461*

369 Pear, *w. 1¼″, p. 475*

370   Common Manzanita, *l.* ⁵⁄₁₆", *p.* 579

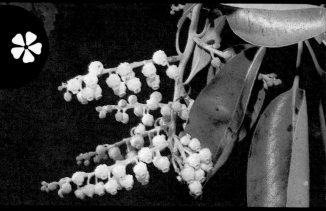

371   Pacific Madrone, *l.* ¼", *p.* 577

372   Texas Madrone, *l.* ¼", *p.* 578

73 Black Locust, *l.* ¾", *p. 504*

74 Southern Magnolia, *w.* 6–8", *p. 430*

376 Desert-willow, *l.* 1¼″, *p.* 599

377 Northern Catalpa, *l.* 2–2¼″, *p.* 598

379   Sugar Sumac, *cl. 2″, p. 522*

380   Horsechestnut, *cl. 10″, p. 540*

382   California Buckeye, *cl.* 4–8″, *p. 539*

383   Tamarisk, *cl.* ¾–2″, *p. 559*

85 New Mexico Locust, *l.* ¾", *p.* 503

86 Silktree, *cw.* 1½–2", *p.* 486

388   Peach, *w.* 1–1¼″, *p. 471*

389   Eastern Redbud, *l.* ½″, *p. 490*

391 Jumping Cholla, *w.* 1″, *p. 563*

392 Crapemyrtle, *cl.* 2½–6″, *p. 567*

394 Chinaberry, *w.* ¾″, *p. 516*

395 Royal Paulownia, *l.* 2″, *p. 595*

397 Mescalbean, *cl. 2–4½", p. 505*

398 Smokethorn, *l. ½", p. 492*

Cones and Fruit

This group of photographs is divided into several categories. First are the cones of pines, spruces, and many other conifers, as well as the conelike fruit of alders. Berrylike fruit includes true berries, such as hollies, as well as juniper "berries," cherries, and other small fleshy fruit. The larger plums, Apple, Pear, and Peach are in the fleshy fruit category. Nuts include the fruits of Horsechestnut, Giant Chinkapin, and Mexican-buckeye. Oaks are characterized by their acorns, while pods typify mesquites, acacias, locusts, and other members of the legume family. The papery winged keys are represented by maples, elms, ashes, ailanthus, and Common Hoptree. Balls and capsules include several oddities such as Sweetgum, Buttonbush, and cottonwoods.

400   Black Spruce, ⅝–1¼″, *p. 265*

401   Engelmann Spruce, 1½–2½″, *p. 263*

402   Sitka Spruce, 2–3½″, *p. 267*

403 Blue Spruce, 2¼–4″, *p. 267*

404 Brewer Spruce, 2½–4″, *p. 263*

405 White Spruce, 1½–2½″, *p. 264*

406   Sugar Pine, 11–18″, *p. 280*

407   Southwestern White Pine, 6–9″, *p. 290*

408   Western White Pine, 5–9″, *p. 283*

409 Limber Pine, 3–6", *p. 278*

410 Whitebark Pine, 1½–3¼", *p. 268*

411 Knobcone Pine, 3¼–6", *p. 271*

412   Bishop Pine, 2–3½″, *p. 284*

413   Coulter Pine, 8–12″, *p. 275*

414   Jeffrey Pine, 5–10″, *p. 279*

15 Foxtail Pine, 3½–5″, *p. 272*

16 Digger Pine, 6–10″, *p. 289*

418 Monterey Pine, 3–6″, *p. 288*

419 Ponderosa Pine, 2–6″, *p. 286*

421 Pinyon, 1½–2″, *p. 276*

422 Chihuahua Pine, 1½–2½″, *p. 281*

424 Western Larch, 1–1½″, *p. 261*

425 Bigcone Douglas-fir, 4–6″, *p. 293*

427 **Bristlecone Fir,** 2½–4″, *p. 250*

428 **Noble Fir,** 4½–7″, *p. 254*

430   Singleleaf Pinyon, 2–3″, *p. 282*

431   Parry Pinyon, 1½–2½″, *p. 287*

433 Mountain Hemlock, 1–3″, *p. 297*

434 Western Hemlock, ¾–1″, *p. 296*

435 Tamarack, ½–¾″, *p. 259*

436 Giant Sequoia, 1¾–2¾″, *p. 301*

437 Redwood, ½–1⅛″, *p. 299*

438 Cryptomeria, ½–¾″, *p. 298*

439　White Alder, ⅜–¾″, *p. 376*

440　Mountain Alder, ⅜–⅝″, *p. 381*

442 **Baker Cypress**, ⅜–¾″, *p. 307*

443 **MacNab Cypress**, ¾–1″, *p. 309*

45   Sargent Cypress, ¾–1″, *p. 311*

46   Tecate Cypress, 1–1¼″, *p. 308*

47   Arizona Cypress, ¾–1¼″, *p. 306*

448   Port-Orford-cedar, ⅜″, *p. 303*

449   Bluegum Eucalyptus, ¾–1″, *p. 569*

450   Red-ironbark Eucalyptus, ⅜″, *p. 570*

51   Alaska-cedar, ½″, *p. 304*

52   Western Redcedar, ½″, *p. 321*

53   Incense-cedar, ¾–1″, *p. 319*

454 Gowen Cypress, ¾″, *p. 308*

455 Rocky Mountain Juniper, ¼″, *p. 318*

456 California Juniper, ½–¾″, *p. 312*

57 Alligator Juniper, ½″, *p. 312*

8 Utah Juniper, ¼–⅝″, *p. 316*

9 Western Juniper, ¼–⅜″, *p. 315*

460   Blue Elder, ¼″, *p. 604*

461   Oneseed Juniper, ¼″, *p. 315*

462   Elephant-tree, ¼″, *p. 514*

463　White Mulberry, ⅜–¾", *p. 423*

464　Black Cherry, ⅜", *p. 472*

465　Sweet Cherry, ¾–1", *p. 463*

466   Common Chokecherry, ¼–⅜″, *p. 474*

467   Catalina Cherry, ½–1″, *p. 468*

468   Hollyleaf Cherry, ½–⅝″, *p. 467*

469   Pin Cherry, ¼″, *p. 470*

470   Tracy Hawthorn, ⅜″, *p. 457*

471   Oneseed Hawthorn, ⁵⁄₁₆–⅜″, *p. 453*

472   Fireberry Hawthorn, ⅜–½″, *p. 449*

473   Columbia Hawthorn, ⅜″, *p. 450*

474   Netleaf Hackberry, ¼–⅜″, *p. 415*

475   Washington Hawthorn, 1/4", *p. 454*

476   Toyon, 1/4–3/8", *p. 458*

478 Common Manzanita, ⁵⁄₁₆–½″, *p. 579*

479 Lemonade Sumac, ½″, *p. 519*

481 Pacific Yew, ¼″, *p. 245*

482 Redberry Juniper, ⁵⁄₁₆″, *p. 313*

484 Texas Lignumvitae, ½", *p. 507*

485 Southern Magnolia, 3–4", *p. 430*

487   **Pacific Red Elder,** ⁵⁄₁₆″, *p. 603*

488   **Pacific Madrone,** ⅜–½″, *p. 577*

489   **English Holly,** ¼–⅜″, *p. 525*

490 American Plum, ¾–1", *p. 462*

491 Garden Plum, 1–2", *p. 465*

492 Peach, 2–3", *p. 471*

93 Apple, 2–3½″, *p. 461*

94 Pear, 2½–4″, *p. 475*

496　**California Torreya,** 1–1½″, *p. 246*

497　**Olive,** 1″, *p. 593*

99   Mexican-buckeye, 1½–2″, *p. 543*

00   Horsechestnut, 2–2½″, *p. 540*

502 Netleaf Oak, ½–¾", *p. 410*

503 Dunn Oak, ¾–1¼", *p. 396*

05 Tanoak, ¾–1¼", *p. 390*

06 Turbinella Oak, ⅝–1", *p. 413*

508 Arizona White Oak, ¾–1″, *p. 392*

509 Oregon White Oak, 1–1¼″, *p. 399*

11   Emory Oak, ½–¾″, p. 396

512   Valley Oak, 1¼–2¼″, p. 404

514 Littleleaf Leucaena, 6–10″, *p. 496*

515 New Mexico Locust, 2½–4½″, *p. 503*

17 Honey Mesquite, 3½–8″, *p. 499*

18 Honeylocust, 6–16″, *p. 494*

19 Black Locust, 2–4″, *p. 504*

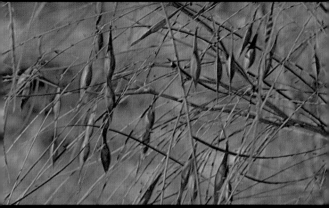

520 Jerusalem-thorn, 2–4″, *p. 498*

521 Yellow Paloverde, 2–3″, *p. 489*

522 Mescalbean, 1–5″, *p. 505*

526 Red Maple, ¾–1″, *p. 537*

527 Silver Maple, 1½–2½″, *p. 538*

529 Ailanthus, 1½″, _p. 512_

530 Fragrant Ash, ¾–1″, _p. 585_

532 Common Hoptree, ⅞″, *p. 510*

533 Knowlton Hophornbeam, 1–1½″, *p. 386*

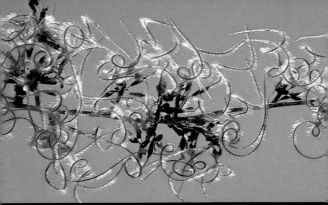

535 Birchleaf Cercocarpus, ⅜″, *p. 444*

536 Curlleaf Cercocarpus, ¼″, *p. 446*

538   Sweetgum, 1–1¼″, *p. 436*

539   Buttonbush, ¾–1″, *p. 601*

Part II
Family and Species
Descriptions

The numbers preceding the species
descriptions in the following pages
correspond to the plate numbers in the
color section.

GINKGO FAMILY
(Ginkgoaceae)

The following species is the only one of its family worldwide.

235 Ginkgo
"Maidenhair-tree"
Ginkgo biloba L.

Description: Planted, deciduous tree with straight
trunk and open, pyramid-shaped
crown, becoming wide-spreading and
irregular; without flowers or fruit.
Height: 50–70′ (15–21 m).
Diameter: 2′ (0.6 m).
Leaves: *3–5 in cluster* on spurs, or
alternate; 1–2″ (2.5–5 cm) long, 1½–
3″ (4–7.5 cm) wide. Oddly *fan-shaped,*
slightly thickened, slightly wavy on
broad edge, often 2-lobed, with fine,
forking parallel veins but no midvein.
Dull light green, turning yellow and
shedding in autumn. Long, slender
leafstalk.
Bark: gray, becoming rough and deeply
furrowed.
Twigs: light green to light brown;
hairless, long and stout, with *many
spurs* or short side twigs bearing
crowded leaf scars.
Seeds and Male Cones: male and female
on separate trees in early spring. Seeds
1″ (2.5 cm) long; *elliptical,* naked,
yellowish, with thin juicy pulp of *bad
odor* and large thick-walled edible
kernel; *1–2 at end of long stalk;*
maturing and shedding in autumn.
Male or pollen cones ¾″ (19 mm) long.
Habitat: Lawns and along streets in moist soils,
in humid temperate regions.
Range: Apparently native in SE. China.
Planted in E. United States and on
Pacific Coast.

Ginkgo is best known as a living fossil
related to conifers and the sole survivor

of its ancient and formerly widespread family. This sacred tree has long been cultivated and possibly preserved from extinction by Buddhist priests on temple grounds in China, Japan, and Korea. Female trees are objectionable because of the litter of seeds, which reek like rancid butter; they should not be touched because the smell lingers after contact. The seeds or Ginkgo nuts are eaten in the Orient. The name "Maidenhair-tree" alludes to the resemblance of the leaves to that fern. This hardy tree is resistant to smoke, dust, wind, ice, insect pests, and disease.

YEW FAMILY
(Taxaceae)

Trees or sometimes shrubs, slightly
aromatic and resinous, without flowers
or fruit; mostly in northern temperate
regions. About 20 species worldwide,
including 4 native tree and 1 shrub
species in North America in the genera
yew (*Taxus*) and torreya (*Torreya*).
Leaves: evergreen, alternate, spreading
in 2 rows; needlelike, flattened, stiff,
with short leafstalk extending down
twig.
Twigs: slender, much-branched, with 2
lines below each leaf, rough with scars
from fallen leaves, ending in scaly
buds.
Seeds and Male Cones: mainly on
separate plants, usually solitary from
scaly bud at leaf base. Seeds naked, not
in cones, nutlike, elliptical, hard with
soft outer coat or cuplike base. Male
cones rounded, with several pollen-sacs
on a stalk.

51, 481 **Pacific Yew**
"Western Yew"
Taxus brevifolia Nutt.

Description: Poisonous, nonresinous, evergreen tree
with angled trunk often twisted or
irregular and with broad crown of
slender, horizontal branches; sometimes
shrubby.
Height: 50' (15 m).
Diameter: 2' (0.6 m).
Needles: *evergreen;* spreading in *2 rows;*
½–¾" (12–19 mm) long, ¹⁄₁₆" (1.5
mm) or more wide. *Flattened,* short-
pointed at both ends, soft and *flexible,*
short-stalked. Deep *yellow-green* above,
light green with 2 *broad, whitish bands*
beneath.
Bark: purplish-brown, very thin,
smooth, with red-brown papery scales.
Twigs: green, becoming light brown;

slender and slightly drooping, with 2
lines below each leaf.
Seeds and Male Cones: on separate
trees. *Elliptical seeds* ¼″ (6 mm) long;
stalkless, blunt-pointed, 2- to 4-
angled, brown; nearly enclosed by
scarlet cup ⅜″ (10 mm) in diameter;
soft, juicy, and sweet; scattered and
single on leafy twigs. Male or pollen
cones ⅛″ (3 mm) in diameter; pale
yellow, short-stalked, single at leaf
bases.

Habitat: Moist soils of stream banks and
canyons; in understory of coniferous
forests.

Range: Extreme SE. Alaska south along coast
to central California; also SE. British
Columbia south in Rocky Mountains to
central Idaho; from sea level in north to
7000′ (2134 m) in south.

The strong wood has been used for
archery bows, poles, canoe paddles, and
small cabinetwork; however, the
limited supply and small dimensions
restrict use. Most parts of yew plants,
including seeds and foliage, are
poisonous and, if eaten, can be fatal.
However, the red, juicy cup around the
seed is reported to be edible, provided
the poisonous seed is not chewed or
swallowed. Birds eat these cups and
scatter the seeds.

50, 496   **California Torreya**
"California-nutmeg" "Stinking-cedar"
Torreya californica Torr.

Description: *Strongly aromatic* tree with conical or
rounded crown and rows of slender,
spreading branches.
Height: 16–70′ (5–21 m).
Diameter: 8″–2′ (0.2–0.6 m),
sometimes larger.
Needles: *evergreen;* spreading in *2 rows;*
1–2¾″ (2.5–7 cm) long, less than ⅛″
(3 mm) wide. Mostly paired; *flattish*

and slightly curved; *long, sharp point* at tip, short-pointed and almost stalkless at base; *stiff. Shiny dark green* above, green with *2 narrow, whitish lines* beneath.

Bark: gray-brown, thin, irregularly fissured into narrow scaly ridges.

Twigs: mostly paired, slender, with 2 lines below base of each leaf; yellow-green, turning reddish-brown.

Seeds and Male Cones: on separate trees. Seeds 1–1½″ (2.5–4 cm) long; *elliptical; fleshy outer layer green* with purplish markings and shedding; inner layer yellow-brown, thick-walled, stalkless; scattered and single on leafy twigs; maturing in 2 seasons. Male or pollen cones ⅜″ (10 mm) long; elliptical, pale yellow, single at leaf bases.

Habitat: Mixed evergreen forests along mountain streams, especially in shady canyon bottoms; also on exposed slopes.

Range: Mountains of central and N. California including Coast Ranges and western slope of Sierra Nevada; at 3000–6500′ (914–1981 m), also down almost to sea level near coast.

The name "California-nutmeg" refers to the resemblance of the aromatic seeds, with a deeply folded seed coat, to those of the unrelated commercial spice, nutmeg (*Myristica fragrans* Houtt.). "Stinking-cedar" alludes to the disagreeable resinous odor of crushed foliage and other parts. Indians used to make bows from the strong wood.

PINE FAMILY
(Pinaceae)

Large to very large trees, without
flowers or fruit, including pines (*Pinus*),
larches (*Larix*), spruces (*Picea*),
hemlocks (*Tsuga*), firs (*Abies*), and
Douglas-firs (*Pseudotsuga*). Resinous,
mostly evergreen with straight axis and
narrow crown, usually with soft,
lightweight wood. About 200 species
worldwide in north temperate and
tropical mountain regions. In North
America, 61 native and 1 naturalized
species; others southward.
Leaves: mostly alternate or whorled,
sometimes of 2 forms, very narrow and
needlelike.
Cones: pollen and seeds borne on same
plant in separate cones. Male cones
small and herbaceous; female cones
large and woody, composed of many
spirally arranged flattened cone-scales
each above a bract. Usually 2 naked
seeds at base of a cone-scale, mostly
with wing at end.

The large genus of pines (*Pinus*) is
characteristic of acid soils, often
sprouting after fire. They occur nearly
throughout temperate continental
North America (although not native in
Kansas) and also from Mexico to
Nicaragua, West Indies, and Eurasia.
They are among the most important
native softwood timbers, producing
lumber and pulpwood. Usually large
trees, pines are resinous and evergreen.
Leaves are of 2 kinds: mostly scalelike
on long twigs; and needles on dwarf
twigs or spurs in bundles of 2–5 (rarely
1), with sheath of bud-scales at base.
Bark varies from furrowed into ridges
to fissured into scaly plates. Twigs are
stout, ending in compound bud with
many bud-scales. Pollen and seeds are
on same plant: male cones are
numerous, clustered, small, and
herbaceous; female cones are solitary to

a few in a cluster, large and woody,
composed of many flattened cone-scales
each above a bract, maturing the second
year, with 2 seeds at base of cone-scale,
mostly with long wing. Two groups
(subgenera) include soft pines with
needles commonly 5 (rarely 4–1) in a
bundle, with the sheath shedding.
Hard pines have needles mostly 2 or 3
in a bundle and the sheath persistent.

38 Pacific Silver Fir
"Amabilis Fir"
"Cascades Fir"
Abies amabilis Dougl. ex Forbes

Description:

Large fir with *beautiful, spirelike, conical
crown* of short, down-curving branches
and *flat, fernlike foliage.*
Height: 80–150' (24–46 m).
Diameter: 2–4' (0.6–1.2 m).
Needles: *evergreen:* crowded and
spreading forward in 2 rows; curved
upward on upper twigs; ¾–1½" (2–4
cm) long. Flat; *shiny dark green* and
grooved above, *silvery-white* beneath.
Bark: light gray, smooth; becoming
scaly and reddish-gray or reddish-
brown.

Cones: 3–6" (7.5–15 cm) long;
cylindrical, upright on topmost twigs,
purple: cone-scales with fine hairs, bracts
short and hidden; paired, long-winged
seeds.

Habitat: Cool, wet regions, including coastal fog
belt and interior mountain valleys; in
coniferous forests.

Range: Pacific Coast from extreme SE. Alaska
south to W. Oregon; local in NW.
California; to 1000' (305 m) in north;
to 6000' (1829 m) in south.

The common name refers to the silvery
lower surface of the foliage with the
word "Pacific" added to avoid confusion
with another silver fir native to Europe.
David Douglas (1798–1834), the

Scottish botanical explorer and discoverer of this species, named it *amabilis,* meaning "lovely." Although beautiful when young, this tree does not attain a pleasing shape in maturity.

36, 427   **Bristlecone Fir**
"Santa Lucia Fir" "Silver Fir"
Abies bracteata D. Don ex Poiteau

Description:  The rarest of the firs, with *narrow, conical, spirelike crown* of short, slightly drooping branches.
Height: 40–100′ (12–30 m).
Diameter: 1–3′ (0.3–0.9 m).
Needles: *evergreen;* spreading almost at right angles in 2 rows; 1½–2¼″ (4–6 cm) long, ⅛″ (3 mm) wide. *Flat, sharp-pointed, stiff. Shiny dark green* above, with *2 broad, whitish bands* beneath.
Bark: light reddish-brown, smooth, becoming scaly and slightly fissured at base.
Twigs: stout, light reddish-brown, hairless.
Cones: 2½–4″ (6–10 cm) long; *egg-shaped, purple-brown, upright* on topmost twigs; cone-scales thin, rounded, hairless, finely toothed each with bract ending in *very long,* spreading, yellow-brown *bristle;* paired, long-winged seeds.

Habitat: Steep, rocky slopes and canyons; in mixed evergreen forests.

Range: Santa Lucia Mountains of S. California; at 2000–5000′ (610–1524 m); locally at 600′ (183 m).

The entire natural range of this rare species is limited to a coastal strip about 60 miles (97 km) long, within Los Padres National Forest. Aromatic resin from the trunk was used as incense in the early Spanish mission nearby.

40 White Fir
"Silver Fir" "Concolor Fir"
Abies concolor (Gord. & Glend.) Hildebr.

Description: Very large fir, widespread in western
mountains, with *narrow, pointed crown*
of short, symmetrical, horizontal
branches; 2 geographic varieties.
Height: 70–160' (21–49 m).
Diameter: 1½–4' (0.5–1.2 m).
Needles: *evergreen;* spreading almost at
right angles in 2 rows; curved upward
on upper twigs; 1½–2½" (4–6 cm)
long. *Flat, flexible,* almost stalkless;
with tip short-pointed, rounded, or
notched. *Light blue-green* with *whitish
lines* on both surfaces.

Bark: light gray, smooth, becoming
very thick near base and deeply
furrowed into scaly, *corky ridges.*
Twigs: light brown or gray; stout,
hairless.

Cones: 3–5" (7.5–13 cm) long;
cylindrical; greenish, purple or yellow;
upright on topmost twigs; cone-scales
finely hairy, with short, hidden bracts;
paired, long-winged seeds.

Habitat: Moist, rocky mountain soils; in pure
stands and with other firs.

Range: Extreme SE. Idaho southeast to New
Mexico, west to California, and north
to SW. Oregon; local in NW. Mexico.
At 5500–11,000' (1676–3353 m) in
south; to 2000' (610 m) in north.

Rocky Mountain White Fir (var.
concolor), of the Rocky Mountain
region, grows in the warmest and driest
climate of all native firs. California
White Fir (var. *lowiana* (Gord.) Lemm.),
the Pacific Coast variety, is grown for
ornament, shade, and Christmas trees.
The scientific name, meaning "of
uniform color," refers to both needle
surfaces. The winged seeds of this and
other firs are eaten by songbirds and
various mammals, especially squirrels
and chipmunks. Deer and grouse feed
on the foliage; porcupines gnaw the bark.

39 Grand Fir

"Lowland White Fir" "Lowland Fir"
Abies grandis (Dougl. ex D. Don) Lindl.

Description: One of the tallest true firs, with *narrow, pointed crown* of stout, curved, and slightly drooping branches.
Height: 100–200' (30–61 m).
Diameter: 1½–3½' (0.5–1 m).
Needles: *evergreen;* spreading almost at right angles in 2 rows; crowded and curved upward on upper twigs; 1¼–2" (3–5 cm) long. *Flat, flexible; shiny dark green* above, *silvery white* beneath.
Bark: brown; *smooth, with resin blisters,* becoming deeply furrowed into narrow scaly ridges.
Twigs: brown; slender, with fine hairs when young.

Cones: 2–4" (5–10 cm) long; *cylindrical, upright* on topmost twigs, *green or brown;* cone-scales hairy, bracts short and hidden. Paired, long-winged seeds.

Habitat: Valleys and mountain slopes in cool, humid climate; in coniferous forests.

Range: S. British Columbia south along coast to California; also south in Rocky Mountain region to central Idaho; to 1500' (457 m) along coast; to 6000' (1829 m) inland.

Common and scientific names refer to the large size; the champion in Olympic National Park, Washington, is 231' (70.4 m) tall with a circumference of 20'8" (6.3 m). Like those of related species, the smooth bark of small trunks has swellings or blisters; when pinched or opened, fragrant, transparent resin or balsam squirts out.

37 Subalpine Fir
"Alpine Fir" "Rocky Mountain Fir"
Abies lasiocarpa (Hook.) Nutt.

Description:

The most widespread western true fir, with dense, *long-pointed, spirelike crown* and rows of horizontal branches reaching nearly to base; shrubby at timberline.

Height: 50–100' (15–30 m).

Diameter: 1–2½' (0.3–0.8 m).

Needles: *evergreen:* spreading almost at right angles in 2 rows; crowded and curved upward on upper twigs; 1–1¾" (2.5–4.5 cm) long. *Flat; dark green,* with *whitish lines* on both surfaces.

Bark: gray, smooth, with resin blisters, becoming fissured and scaly.

Twigs: gray, stout, with rust-colored hairs.

Cones: 2¼–4" (6–10 cm) long; *cylindrical, upright* on topmost twigs, dark *purple;* cone-scales finely hairy with short, hidden bracts; paired, long-winged seeds.

Habitat: Subalpine zone of high mountains to timberline; forming spruce-fir forest with Engelmann Spruce, and with other conifers.

Range: Central Yukon and SE. Alaska southeast to S. New Mexico; at 8000–12,000' (2438–3658 m) in south; to sea level in north.

The spires of Subalpine Fir add beauty to the Rocky Mountain peaks. When weighted down to the ground with snow, the lowest branches sometimes take root, forming new shoots. The bark of this and related firs is browsed by deer, elk, bighorn sheep, and moose; the leaves are eaten by grouse, and the seeds are consumed by songbirds and mammals. The scientific name, meaning "hairy-fruited," refers to the cones. Corkbark Fir (var. *arizonica* (Merriam) Lemm.), a variety from Arizona to Colorado, has thin, whitish, corky bark.

34 California Red Fir

"Red Fir" "Silvertip"
Abies magnifica A. Murr.

Description: Large, handsome fir with an open
conical *crown rounded at tip* and short,
nearly horizontal branches.
Height: 60–120' (18–37 m).
Diameter: 1–4' (0.3–1.2 m).
Needles: *evergreen;* spreading in 2 rows;
crowded and curved upward on upper
twigs; ¾–1⅛" (2–3.5 cm) long.
4-sided, blue-green with whitish lines.
Bark: *thick, reddish-brown, deeply
furrowed* into narrow ridges.
Twigs: stout, brown, with fine hairs
when young.
Cones: 6–8" (15–20 cm) long;
cylindrical, purplish-brown, *upright* on
topmost twigs; cone-scales with fine
hairs, with yellowish *bracts mostly short
and hidden* (exposed in a variation,
Shasta Red Fir), pointed and finely
toothed; paired, long-winged
seeds.

Habitat: High mountains with dry summers and
deep snow in winter; often in pure
stands, also in mixed conifer forests.

Range: Cascade Mountains of SW. Oregon
south to Coast Ranges of California and
through Sierra Nevada to central
California and extreme W. Nevada. At
6000–9000' (1829–2743 m) in south;
to 4500' (1372 m) in north.

Named for its characteristic bark, this
magnificent conifer forms almost pure
forests at high altitudes along the
western slopes of Sierra Nevada. It is
common in Yosemite National Park.
Early mountaineers prepared their beds
by cutting and overlapping 2 rows of
the plushy, aromatic boughs.

35, 428 Noble Fir
"Red Fir" "White Fir"
Abies procera Rehd.

Description: The largest native true fir, with conical *crown rounded at tip* and with short, nearly horizontal branches.
Height: 100–150' (30–46 m) often much taller.
Diameter: 2½–4' (0.8–1.2 m).
Needles: *evergreen;* spreading in 2 rows; 1–1⅜" (2.5–3.5 cm) long. *Flat,* grooved above, often notched, *blue-green with whitish lines;* shorter, crowded and curved upward, *4-sided, pointed* on upper twigs.
Bark: gray-brown and smooth, becoming brown to red-brown; slightly thickened, and furrowed into irregular, long, scaly plates.
Twigs: stout, brown, with rust-colored hairs when young.
Cones: 4½–7" (11–18 cm) long; *cylindrical, upright* on topmost twigs, *green becoming purplish brown;* cone-scales with fine hairs and mostly *covered by large papery bracts,* finely toothed, long-pointed, and bent downward; paired, long-winged seeds.

Habitat: Moist soils in high mountains with short, cool growing season and deep snow in winter. Associated with other conifers; not in pure stands.

Range: Cascade Mountains and Coast Ranges from Washington south to NW. California; at 3000–7000' (914–2134 m); occasionally at 200–8800' (61–2682 m).

A handsome tree with large, showy cones mostly covered by papery bracts, Noble Fir was named by the Scottish botanical explorer David Douglas (1798–1834). It is the tallest true fir; the champion in the Gifford Pinchot National Forest in southwestern Washington is 278' (85 m) high, has a trunk circumference of 28' (8.6 m), and has a crown spread of 47' (14 m).

46, 429   Atlas Cedar
"Atlas Mountain Cedar"
Cedrus atlantica (Endl.) Manetti ex Carr.

Description:  Large ornamental tree with straight trunk and pointed, pyramidal crown becoming broad and flattened and pale blue-green or silvery foliage.
Height: 80' (24 m).
Diameter: 3' (0.9 m).
Needles: *evergreen; 10−20 clustered* and crowded on spurs or alternate on leading twigs; ¾−1" (2−2.5 cm) long; 3-angled; pale *blue-green* or silvery.
Bark: gray, smooth, becoming irregularly furrowed into flat ridges.
Twigs: spreading, long and stout, gray, finely *hairy*, with *many spurs* or short side twigs.
Cones: 2−3" (5−7.5 cm) long, 1½−2" (4−5 cm) in diameter; *barrel-shaped* with *flat top, upright*, almost *stalkless*, light brown; composed of many hard cone-scales; maturing second year, axis remaining attached; paired, broad-winged seeds.

Habitat: Moist soils in parks and gardens and around homes; in temperate regions.

Range: Native to NW. Africa. Planted in eastern states and Pacific region.

The genus *Cedrus*, from the ancient Greek name, consists of 4 species of true cedars, all native in the Old World. Atlas Cedar is distinguished by the short, pale blue-green needles, silvery or whitish in varieties, and by the relatively small cones. The species name refers to the location of the Atlas Mountains of northwestern Africa, near the Atlantic Ocean. The common name refers to its native home, where the heavy, aromatic wood is used for construction and cabinetmaking.

47 Deodar Cedar
"Deodar" "California Christmas-tree"
Cedrus deodara (D. Don) G. Don

Description:

Large planted tree with straight trunk, regular, pyramidal crown with *curved or drooping tip,* and graceful, drooping branches down to base.
Height: 80' (24 m).
Diameter: 3' (0.9 m).
Needles: *evergreen; 10–20 clustered* and crowded on spurs or alternate on leading twigs; 1–2" (2.5–5 cm) long; 3-angled; *dark blue-green.*
Bark: gray, becoming brown and deeply furrowed.
Twigs: gray, *densely hairy,* with *many spurs* or short side twigs.
Cones: 3–4" (7.5–10 cm) long, 2–3" (5–7.5 cm) in diameter; *elliptical, rounded at tip, upright,* almost *stalkless,* reddish-brown; composed of many hard cone-scales; maturing second year, axis remaining attached; paired, broad-winged seeds.

Habitat: Moist soils in parks and gardens and around homes; in warm, temperate regions.

Range: Native of the Himalayas. Planted mainly in southeastern, Gulf, and Pacific states, and along the Mexican border; especially popular in California.

This large, handsome ornamental is distinguishable from other true cedars by the slightly drooping branches and long needles. The large developing cones are usually present on lower branches, even through winter. Several cultivated varieties differ in having yellow foliage or compact, weeping, and creeping habits. Deodar Cedar is often decorated as a living Christmas tree. In India, the durable aromatic wood is commercially important for construction and has been used for incense. *Deodar,* the Hindu name, means "timber of the gods."

48 Cedar-of-Lebanon
Cedrus libani A. Rich.

Description:

Large planted, cone-bearing, evergreen tree with straight, stout trunk and narrow, pointed crown, becoming irregular and broad or flattened with *spreading, horizontal branches.*
Height: 80' (24 m).
Diameter: 3' (0.9 m).
Needles: *evergreen;* 1–1¼" (2.5–3 cm) long. *10–15 clustered* and crowded on spurs or alternate on leading twigs; 3-angled; mostly dark green.
Bark: dark gray, becoming thick and furrowed into scaly plates.
Twigs: abundant, spreading, long and stout, mostly hairless or slightly hairy, with *many spurs* or short side twigs.
Cones: 3–4½" (7.5–11 cm) long, 1¾–2½" (4.5–6 cm) in diameter; *barrel-shaped* with flat top, reddish-brown, *upright,* almost *stalkless,* resinous;

composed of many hard cone-scales; maturing in second year, axis remaining attached; paired broad-winged seeds.

Habitat: Moist soils in parks and gardens and around homes; in temperate regions.

Range: Native of Asia Minor from S. Turkey to Lebanon and Syria at high altitudes. Planted mainly in SE. United States and on Pacific Coast; a hardy variety also in Northeast.

Cedar-of-Lebanon is a handsome, picturesque ornamental. Developing large cones often are conspicuous on lower branches. Its association with the Bible and the Holy Land make this species of special interest. The fragrant durable wood is used for construction timbers, lumber, furniture, and paneling in its native lands. The northeastern range was extended by collecting seed from native trees at their highest altitude.

43, 435 Tamarack
"Hackmatack" "Eastern Larch"
Larix laricina (Du Roi) K. Koch

Description: Deciduous tree with straight, tapering trunk and thin, open, conical crown of horizontal branches; a shrub at timberline.
Height: 40–80' (12–24 m).
Diameter: 1–2' (0.3–0.6 m).
Needles: *deciduous;* ¾–1" (2–2.5 cm) long, ⅟₃₂" (1 mm) wide. Soft, very slender, 3-angled; crowded in cluster *on spur twigs,* also *scattered* and alternate on leader twigs. Light blue-green, turning yellow in autumn before shedding.
Bark: reddish-brown, scaly, thin.
Twigs: orange-brown, stout, hairless, with *many spurs* or short side twigs.
Cones: ½–¾" (12–19 mm) long; *elliptical,* rose red turning brown, *upright, stalkless;* falling in second year; several overlapping rounded cone-scales; paired, brown, long-winged seeds.

Habitat: Wet, peaty soils of bogs and swamps; also in drier upland loamy soils; often in pure stands.

Range: Across N. North America near northern limit of trees from Alaska east to Labrador, south to N. New Jersey, and west to Minnesota; local in N. West Virginia and W. Maryland. From near sea level to 1700–4000' (518–1219 m) southward.

One of the northernmost trees, the hardy Tamarack is useful as an ornamental in very cold climates. Indians used the slender roots to sew together strips of birch bark for their canoes. Roots bent at right angles served the colonists as "knees" in small ships, joining the ribs to deck timbers. The durable lumber is used as framing for houses, railroad cross-ties, poles, and pulpwood. The larch sawfly defoliates stands in infrequent years, causing damage or death.

45 Subalpine Larch
"Alpine Larch" "Tamarack"
Larix lyallii Parl.

Description: Deciduous tree with straight trunk,
short branches, and irregular, spreading
crown.
Height: 30–50' (9–15 m).
Diameter: 1–2' (0.3–0.6 m).
Needles: *deciduous;* 1–1¼" (2.5–3 cm)
long, ¹⁄₃₂" (1 mm) wide. Many *crowded*
in cluster *on spur twigs;* also alternate
and scattered on *leader twigs; 4-angled,*
stiff, short-pointed. *Pale blue-green,*
turning yellow in autumn before
shedding.
Bark: dark red-brown, thin, becoming
fissured into irregular scaly plates.
Twigs: 2 kinds: long leaders densely
covered with whitish hairs, and many
short spurs.
Cones: 1½–2" (4–5 cm) long; *elliptical,
upright,* nearly stalkless, composed of
many densely hairy, dark reddish-
purple cone-scales shorter than dark
purple, 3-toothed bracts; paired, brown,
long-winged seeds.

Habitat: At timberline on rocky soils in
Engelmann Spruce–Subalpine Fir
forest; locally in pure stands.

Range: SE. British Columbia and SW. Alberta
south to W. Montana and west to NE.
Washington; at 4000–8000' (1219–
2438 m).

Subalpine Larch is seldom seen because
of its isolated timberline location in
high mountains. David Lyall (1817–
95), a Scottish surgeon and naturalist,
discovered this species in 1858. For
most of the year the branches are bare,
except for the blackened dead cones.

44, 424 Western Larch
"Hackmatack" "Western Tamarack"
Larix occidentalis Nutt.

Description: Very large deciduous tree with narrow,
conical crown of horizontal branches.
Height: 80–150' (24–46 m).
Diameter: 1½–3' (0.5–0.9 m),
sometimes larger.
Needles: *deciduous;* 1–1½" (2.5–4 cm)
long, ⅟₃₂" (1 mm) wide. *Crowded* in
cluster *on spur twigs;* also alternate and
scattered on leader twigs; *3-angled,*
stiff, sharp-pointed. *Light green,* turning
yellow in autumn before falling.
Bark: reddish-brown, scaly, becoming
deeply furrowed into flat ridges with
many overlapping plates.
Twigs: 2 kinds: long leaders (orange-
brown and hairy when young), and
many short spurs.
Cones: 1–1½" (2.5–4 cm) long;
elliptical, brown, *upright* on short stalks;
many rounded, hairy cone-scales shorter
than *long-pointed bracts;* paired, pale
brown, long-winged seeds.

Habitat: Mountain slopes and valleys on porous,
gravelly, sandy, and loamy soils; with
other conifers.

Range: SE. British Columbia south to NW.
Montana and N. Oregon; at 2000–
5500' (610–1676 m) in north; to
7000' (2134 m) in south.

Western Larch often follows or survives
fires, later being replaced by other
conifers. The natural sugar, or galactan,
in the gum and wood resembles a
slightly bitter honey and can be made
into medicine and baking powder.
Grouse eat the buds and leaves. The
wood is used for construction,
paneling, flooring, utility poles,
plywood, and pulpwood.

32 Norway Spruce
Picea abies (L.) Karst.

Description: Large introduced, cone-bearing tree
with straight trunk and pyramid-
shaped crown of spreading branches.
Height: 80′ (24 m).
Diameter: 2′ (0.6 m).
Needles: *evergreen;* ½–1″ (1.2–2.5 cm)
long. Stiff, *4-angled, sharp-pointed;*
spreading on all sides of twig from very
short leafstalks; shiny dark green with
whitish lines.
Bark: reddish-brown, scaly.
Twigs: reddish-brown, slender,
drooping, mostly hairless, rough, with
peglike bases.
Cones: 4–6″ (10–15 cm) long;
cylindrical, light brown, *hanging down;*
cone-scales numerous, thin, slightly
pointed, *irregularly toothed,* opening and
shedding year after maturing; paired
long-winged seeds.

Habitat: Moist soils in humid, cool, temperate
regions.

Range: Native of N. and central Europe, at
high altitudes. Widely planted in SE.
Canada, NE. United States, Rocky
Mountains, and Pacific Coast region.
Escaped in Northeast and perhaps
naturalized locally.

Norway Spruce has been widely
cultivated for ornament, shade,
shelterbelts, Christmas trees, and forest
plantations. The showy cones are the
largest of the spruces. Numerous
horticultural varieties include trees with
a narrow columnar shape, drooping or
weeping branches, dwarf habit, and
yellowish or variegated needles. It is
the common spruce of northern Europe.

33, 404 Brewer Spruce
"Weeping Spruce"
Picea brewerana Wats.

Description: Large tree crowned with long, slender,
horizontal branches ending in *ropelike,
drooping branches;* trunk enlarged and
buttressed at the base and tapering
above.
Height: 70–100' (21–30 m).
Diameter: 1½–3' (0.5–0.9 m).
Needles: *evergreen;* spreading on all sides
of twig; ¾–1" (2–2.5 cm) long.
*Flattish; blunt-pointed, 4–6 whitish
lines* above, *shiny dark green* beneath.
Bark: reddish-brown, thin, scaly.
Twigs: reddish-brown, long and
slender, finely hairy, rough with
peglike leaf bases.
Cones: 2½–4" (6–10 cm) long;
cylindrical, short-stalked, dull *orange-
brown;* cone-scales thin, *rounded,* not
toothed; paired, brown, long-winged
seeds.

Habitat: High mountain ridges near timberline;
with Red Fir and other conifers; seldom
in pure stands.

Range: Chiefly in Siskiyou Mountains of SW.
Oregon and NW. California; at 3300–
7500' (1006–2286 m).

The weeping habit serves to reduce
breakage of branches by heavy snowfall.
Rare even as a cultivated ornamental,
this local species is found in 5 National
Forests and a special preserve, the
Brewer Spruce Natural Area. It is
named for its discoverer, William
Henry Brewer (1828–1910), a professor
of agriculture at Yale University and
co-author of *Botany of California.*

31, 401 Engelmann Spruce
Picea engelmannii Parry ex Engelm.

Description: Large tree with dark or blue-green
foliage and a dense, narrow, conical

crown of short branches spreading in close rows.

Height: 80–100' (24–30 m).

Diameter: 1½–2½' (0.5–0.8 m).

Needles: *evergreen;* ⅝–1" (1.5–2.5 cm) long. Spreading on all sides of twig from very short leafstalks; *4-angled,* sharp-pointed, slender, flexible; with disagreeable skunklike odor when crushed; *dark or blue-green,* with whitish lines.

Bark: grayish- or purplish-brown; thin, with loosely attached scales.

Twigs: brown, slender, hairy, rough, with peglike leaf bases.

Cones: 1½–2½" (4–6 cm) long; *cylindrical,* shiny light brown; hanging at end of leafy twig; cone-scales long, thin, and flexible, narrowed and *irregularly toothed;* paired, blackish, long-winged seeds.

Habitat: Dominant with Subalpine Fir in subalpine zone up to timberline; also with other conifers.

Range: Central British Columbia and SW. Alberta southeast to New Mexico; chiefly in Rocky Mountains; at 8000–12,000' (2438–3659 m) in south; down to 2000' (619 m) in north.

Its resonant qualities make the wood of Engelmann Spruce valuable for piano sounding boards and violins. This species was named after George Engelmann (1809–84), the German-born physician and botanist of St. Louis and authority on conifers.

28, 405 White Spruce
"Canadian Spruce" "Skunk Spruce"
Picea glauca (Moench) Voss

Description: Tree with rows of horizontal branches forming a conical crown; smaller and shrubby at tree line.
Height: 40–100' (12–30 m).
Diameter: 1–2' (0.3–0.6 m).

Needles: *evergreen:* ½–¾" (12–19 mm) long. Stiff, *4-angled, sharp-pointed;* spreading mainly on upper side of twig, from very short leafstalks; blue-green, with whitish lines; exuding *skunklike odor* when crushed.

Bark: gray or brown, thin, smooth or scaly; cut surface of inner bark whitish.

Twigs: orange-brown, slender, hairless, rough, with peglike bases.

Cones: 1½–2½" (4–6 cm) long; *cylindrical, shiny light brown,* hanging at end of twigs, falling at maturity; cone-scales thin and flexible, *margins nearly straight* and without teeth; paired brown, long-winged seeds.

Habitat: Many soil types in coniferous forests; sometimes in pure stands.

Range: Across N. North America near northern limit of trees from Alaska and British Columbia east to Labrador, south to Maine, and west to Minnesota; local in NW. Montana, South Dakota, and Wyoming; from near sea level to timberline at 2000–5000' (610–1524 m).

This is the foremost pulpwood and generally the most important commercial tree species of Canada. As well as providing lumber for construction, the wood is valued for piano sounding boards, violins, and other musical instruments. White Spruce and Black Spruce are the most widely distributed conifers in North America after Common Juniper, which rarely reaches tree size. Various kinds of wildlife, including deer, rabbits, and grouse, browse spruce foliage in winter.

30, 400 Black Spruce
"Bog Spruce" "Swamp Spruce"
Picea mariana (Mill.) B.S.P.

Description: Tree with open, irregular, conical crown of short, horizontal or slightly

drooping branches; a prostrate shrub at timberline.

Height: 20–60′ (6–18 m).

Diameter: 4–12″ (0.1–0.3 m).

Needles: *evergreen;* ¼–⅝″ (6–15 mm) long. Stiff, *4-angled,* sharp-pointed; spreading on all sides of twig from very short leafstalks; ashy blue-green with whitish lines.

Bark: gray or blackish, thin, scaly; brown beneath; cut surface of inner bark yellowish.

Twigs: brown, slender, hairy, rough, with peglike bases.

Cones: ⅝–1¼″ (1.5–3 cm) long; *egg-shaped* or rounded, dull gray; *curved downward* on short stalk and *remaining attached,* often *clustered* near top of crown; cone-scales stiff and brittle, rounded and finely toothed; paired, brown, long-winged seeds.

Habitat: Wet soils and bogs including peats, clays, and loams; in coniferous forests; often in pure stands.

Range: Across N. North America near northern limit of trees from Alaska and British Columbia east to Labrador, south to N. New Jersey, and west to Minnesota; at 2000–5000′ (610–1524 m).

Black Spruce is one of the most widely distributed conifers in North America. Uses are similar to those of White Spruce; however, the small size limits lumber production. The lowest branches take root by layering when deep snows bend them to the ground, forming a ring of small trees around a large one. Spruce gum and spruce beer were made from this species and Red Spruce.

29, 403 Blue Spruce
"Colorado Spruce"
"Silver Spruce"
Picea pungens Engelm.

Description: Large tree with blue-green foliage and a conical crown of stout, horizontal branches in rows.
Height: 70–100' (21–30 m).
Diameter: 1½–3' (0.5–0.9 m).
Needles: *evergreen;* spreading on all sides of twig from very short leafstalks; ¾–1⅛" (2–2.8 cm) long. *4-angled, sharp-pointed,* stiff; with resinous odor when crushed; dull *blue-green* or bluish, with whitish lines.
Bark: gray or brown; furrowed and scaly.
Twigs: yellow-brown, stout, hairless, rough, with peglike leaf bases.
Cones: 2¼–4" (6–10 cm) long; *cylindrical,* mostly stalkless, shiny light brown; cone-scales long, thin, and flexible, narrowed and *irregularly toothed;* paired, long-winged seeds.

Habitat: Narrow bottomlands along mountain streams; often in pure stands.

Range: Rocky Mountain region from S. and W. Wyoming and E. Idaho south to N. and E. Arizona and S. New Mexico; at 6000–11,000' (1829–3353 m).

Cultivated varieties of Blue Spruce include several with dramatic bluish-white and silvery-white foliage. It is a popular Christmas tree and is also used in shelterbelts.

27, 402 Sitka Spruce
"Coast Spruce" "Tideland Spruce"
Picea sitchensis (Bong.) Carr.

Description: The world's largest spruce, with tall, straight trunk from buttressed base, and broad, open, conical crown of horizontal branches.
Height: 160' (49 m).

Diameter: 3–5' (0.9–1.5 m), sometimes much larger.

Needles: *evergreen;* spreading on all sides of twig; ⅝–1" (1.5–2.5 cm) long. *Flattened* and slightly *keeled, sharp-pointed; dark green.*

Bark: gray, smooth, thin; becoming dark purplish-brown with scaly plates.

Twigs: brown, stout, hairless, rough, with peglike bases.

Cones: 2–3½" (5–9 cm) long; *cylindrical,* short-stalked, light orange-brown; hanging at ends of twigs; opening and falling at maturity; cone-scales long, stiff, *thin,* rounded, and *irregularly toothed.* Paired, brown, long-winged seeds.

Habitat: Coastal forests in fog belt, a narrow strip of high rainfall and cool climate; in pure stands and with Western Hemlock.

Range: Pacific Coast from S. Alaska and British Columbia to NW. California; to timberline at 3000' (914 m) in Alaska; below 1200' (366 m) in California.

The main timber tree in Alaska, Sitka Spruce produces high-grade lumber for many uses and wood pulp for newsprint. Special products are piano sounding boards, boats, and food containers. It was formerly used in aircraft construction.

19, 410　Whitebark Pine
"Scrub Pine" "White Pine"
Pinus albicaulis Engelm.

Description: Tree with short, twisted or crooked trunk and irregular, spreading crown; a shrub at timberline; *foliage has sweetish taste and odor.*

Height: 20–50' (6–15 m).

Diameter: 1–2' (0.3–0.6 m).

Needles: *evergreen; 5 in bundle,* with sheath shedding first year; crowded at ends of twigs; 1½–2¾" (4–7 cm) long.

Stout, stiff, short-pointed; *dull green,*
with faint white lines on all surfaces.
Bark: *whitish-gray, smooth,* thin,
becoming scaly.
Twigs: stout, tough and flexible,
brown; with fine hairs when young.
Cones: 1½–3¼" (4–8 cm) long; *egg-
shaped* or rounded, almost stalkless,
purple to brown; shedding at maturity but
not opening; cone-scales very thick, with
sharp edge ending in raised, *stout point.*
Large, elliptical, dark brown, *thick-
walled, wingless,* edible *seeds.*

Habitat: Dry, rocky soils on exposed slopes and
ridges in subalpine zone to timberline;
sometimes forms pure stands and
thickets.

Range: Central British Columbia, east to SW.
Alberta, south to W. Wyoming and
west to central California. At 4500–
7000' (1372–2134 m) in north; at
8000–12,000' (2438–3658 m) in
south.

American Indians gathered the cones
and ate the seeds of this species. A bird
called Clark's nutcracker tears open the
cones to eat the seeds; in northern
Eurasia, another nutcracker uses a
similar method to obtain the seeds of a
closely related species. Whitebark Pine
is considered the most primitive native
pine because its cones do not open until
they decay.

26, 423 **Bristlecone Pine**
"Foxtail Pine" "Hickory Pine"
Pinus aristata Engelm.

Description: Tree with very short needles crowded
into mass suggesting a foxtail and a
broad, irregular crown of spreading
branches; a low shrub at timberline.
Height: 20–40' (6–12 m).
Diameter: 1–2½' (0.3–0.8 m).
Needles: *evergreen; 5 in bundle,* with
sheath shedding after first year; ¾–1½"

(2–4 cm) long. *Crowded in long, dense mass* curved against twig; stout, stiff, blunt-pointed; persisting 10–20 years. *Dark green,* with white lines on inner surfaces; often with whitish resin dots on outer surface.

Bark: whitish-gray, smooth; becoming reddish-brown and furrowed into irregular, scaly ridges.

Cones: 2½–3½″ (6–9 cm) long; *cylindrical,* dark *purplish-brown,* almost stalkless; opening at maturity; cone-scales 4-sided, with *slender, curved bristle;* seeds brown mottled with black and with detachable wing.

Habitat: Exposed, dry, rocky slopes and ridges of high mountains in subalpine zone to timberline; often in pure stands.

Range: Colorado and N. New Mexico west to E. California; at 7500–11,500′ (2286–3505 m).

The oldest known dated living trees are Bristlecone Pines more than 4600 years old, protected at Inyo National Forest near Bishop, in eastern California. Other very old Bristlecone Pines are found at Wheeler Peak Scenic Area, in the Humboldt National Forest of eastern Nevada. Although these trees are classed among the oldest known living things, some shrubs and trees that spread in colonies or clumps from the same root system may be older. The age of a tree is dated by counting the annual rings of wood produced by the cambium layer inside the bark. Wood cells formed in the spring are generally large, while those formed in the summer are smaller; the contrast in cell size from one year to the next is visible as a line. In an old Bristlecone Pine, small cores of wood are removed at different places around the trunk to locate the oldest ring in the center. Two varieties have been named. Colorado Bristlecone Pine (var. *aristata*) is found in Colorado, northern New Mexico, and northern Arizona.

Intermountain Bristlecone Pine (var. *longaeva* (D.K. Bailey) Little), found from Utah to Nevada and eastern California, has cones with rounded bases and fine bristles.

14, 411 Knobcone Pine
Pinus attenuata Lemm.

Description: Pine with narrow, pointed crown of slender, nearly horizontal branches turned up at ends, becoming irregular with age, and with *abundant cones remaining closed many years.*
Height: 30–80' (9–24 m).
Diameter: 1–2½' (0.3–0.8 m).
Needles: *evergreen; 3 in bundle;* 3–7" (7.5–18 cm) long; slender, stiff, yellow-green.
Bark: gray and smooth, becoming dark gray and fissured into large, scaly ridges.
Cones: 3¼–6" (8–15 cm) long; egg-shaped, *clustered in many rings* or whorls, stalkless and turned back, *shiny yellow-brown;* cone-scales raised and keeled, ending in short, *stout spine.* Blackish seeds about ¼" (6 mm) long; narrow wing 1–1¼" (2.5–3 cm) long.

Habitat: Forms almost pure stands on poor, coarse, rocky, mountain soils.

Range: SW. Oregon south to S. California; local in N. Baja California, Mexico; at 1000–2000' (305–610 m) in north; 1500–4000' (457–1219 m), sometimes higher, in south.

The whorls of many knobby, closed cones help identify this species. Since the cones may become imbedded within the wood of the expanding trunk, this species has been called "the tree that swallows its cones." When fires kill the trees, cones as much as 30 years old are opened by the heat and shed their seeds. The abundant seedlings then begin a new forest.

25, 415 Foxtail Pine
Pinus balfouriana Grev. & Balf.

Description: Pine with *very short needles* crowded against twigs suggesting a foxtail, and irregular crown of short, spreading branches; a shrub at timberline.
Height: 20–50' (6–15 m).
Diameter: 1–2' (0.3–0.6 m).
Needles: *evergreen; 5 in bundle,* with sheath shedding after first year; ¾–1½" (2–4 cm) long. *Crowded in long, dense mass* curved against twig; stout, stiff, sharp-pointed; dark green, with white lines on inner surfaces; persisting 10–20 years.
Bark: whitish-gray and smooth, becoming reddish-brown and deeply furrowed into irregular ridges.
Cones: 3½–5" (9–13 cm) long; *cylindrical* but tapering, almost stalkless, dark *reddish-brown;* opening at maturity; thick, 4-sided cone-scales, with *tiny prickle.* Purple, mottled seeds more than ¼" (6 mm) long; wing about ⅞" (22 mm), remaining attached.

Habitat: Exposed, dry, rocky slopes and ridges of high mountains in subalpine zone to timberline; with Whitebark Pine.

Range: Local in mountains of N. California and Sierra Nevada of E. central California; at 6000–11,500' (1829–3505 m).

Foxtail Pine is closely related to Bristlecone Pine, which is sometimes also called Foxtail Pine because of the similar foliage. Bristlecone Pine, inhabiting high mountains farther east, has shorter cones, with each cone-scale ending in a slender bristle.

24, 432 Mexican Pinyon
"Nut Pine" "Piñón"
Pinus cembroides Zucc.

Description: Small, resinous tree with short trunk and spreading crown of low, horizontal

branches and thick-walled, edible seeds; often shrubby.

Height: 16–20′ (5–6 m).

Diameter: 1′ (0.3 m).

Needles: *evergreen; 3 in bundle* (rarely 2), with *sheath shedding* after first year; 1–2½″ (2.5–6 cm) long. *Slender, flexible;* green, with white lines on all surfaces or only 2 inner surfaces.

Bark: light gray and smooth; becoming dark gray or reddish-brown and furrowed into scaly plates.

Cones: ¾–2″ (2–5 cm) long; *egg-shaped* or rounded, almost stalkless, dull orange or *reddish*-brown, *resinous;* opening and shedding; cone-scales thick, slightly 4-angled, sometimes with tiny prickle. *Seeds* large, oblong or elliptical, dark brown, *wingless, thick-walled,* oily, edible.

Habitat: Dry, rocky slopes of mesas, plateaus, and mountains; with junipers and evergreen oaks.

Range: Central and Trans-Pecos Texas west to SE. Arizona; also Mexico; at 5000–7500′ (1524–2286 m); locally to 2500′ (762 m).

The hard seeds are the main commerical pinyon nuts (*piñones*) of Mexico. However, in the United States this species has limited distribution and usually bears light cone crops; other species with thin-walled seeds are more common. Rodents, especially "packrats," eat the seeds.

20, 420 Lodgepole Pine
"Tamarack Pine" "Shore Pine"
Pinus contorta Dougl. ex Loud.

Description: Widely distributed pine that may grow tall with narrow, dense, conical crown, or remain small with broad, rounded crown; 3 geographic varieties.

Height: 20–80′ (6–24 m).

Diameter: 1–3′ (0.3–0.9 m).

Needles: *evergreen; 2 in bundle;* 1¼–2¾" (3–7 cm) long. *Stout,* slightly *flattened* and often *twisted;* yellow-green to dark green.

Bark: light brown, thin, and scaly; or in Shore Pine (the coastal variety), dark brown, thick, furrowed into scaly plates.

Cones: ¾–2" (2–5 cm) long; *egg-shaped,* stalkless, *oblique* or 1-sided at base, shiny *yellow-brown; remaining closed* on tree many years (or in varieties opening); cone-scales raised, rounded, *keeled,* with tiny, *slender prickle.*

Habitat: High mountains on mostly well-drained soils, often in pure stands; Shore Pine in peat bogs, muskegs, and dry, sandy sites.

Range: SE. Alaska and central Yukon south on Pacific Coast to N. California, south through Sierra Nevada to S. California, and south in Rocky Mountains to S. Colorado; also local in Black Hills of South Dakota and N. Baja California; coastal variety from sea level to 2000' (610 m); inland varieties at 1500–3000' (457–914 m) in north and at 7000–11,500' (2134–3505 m) in south.

Lodgepole Pine is one of the most widely distributed New World pines and the only conifer native in both Alaska and Mexico. Its name refers to the use by American Indians of the slender trunks as poles for their conical tents or teepees. Shore Pine (var. *contorta*), the Pacific Coast variety, is a small tree with spreading crown, thick, furrowed bark, short leaves, and oblique cones pointing backward, opening at maturity but remaining attached. Sierra Lodgepole Pine (var. *murrayana* (Grev. & Balf.) Engelm.), of the Cascade Mountains of southwestern Washington and western Oregon, the Sierra Nevada of central California, and south to northern Baja California, is a tall, narrow tree with thin, scaly bark, relatively broad leaves, and symmetrical, lightweight cones opening at maturity and shedding within a few years. Lodgepole Pine or Rocky Mountain Lodgepole Pine (var. *latifolia* Engelm.), of the Rocky Mountain region, is a tall, narrow tree with thin, scaly bark, long needles, and cones often oblique and pointing outward. This variety is adapted to forest fires, with cones that remain tightly closed on the trees many years until a fire destroys the forest. When the heat causes the cones to open, the seeds fall to the bare ground to begin a new forest.

6, 413 Coulter Pine
"Bigcone Pine" "Pitch Pine"
Pinus coulteri D. Don

Description: Straight-trunked tree with rows of nearly horizontal branches formed annually, an open, thin, irregular crown, and very large, heavy cones.
Height: 40–70′ (12–21 m).
Diameter: 1–2½′ (0.3–0.8 m).
Needles: *evergreen; 3 in bundle;* crowded at ends of stout, brown twigs; 8–12″ (20–30 cm) long. *Very stout,* stiff, sharp-pointed; light *gray-green,* with many white lines.
Bark: dark gray, thick, deeply furrowed into scaly ridges, becoming slightly shaggy; blackish-gray, very rough, divided into rectangular plates on branches.
Cones: 8–12″ (20–30 cm) long; *egg-shaped,* bent down on very stout stalk, *very heavy,* slightly *shiny yellow-brown,* resinous; opening gradually and remaining on tree; cone-scales very long, thick, sharply keeled, with very long, stout spine flattened and curved forward. *Seeds very large,* elliptical, dark brown, *thick-walled,* edible, *with detachable wing.*

Habitat: Dry, rocky slopes and ridges in foothills and mountains; with other conifers.

Range: Central and S. California; also N. Baja California; at 3000–6000′ (914–1829 m); rarely at 1000–7000′ (305–2134 m).

This pine has the heaviest cones of all pines in the world, often weighing 4–5 pounds (1.8–2.3 kilos). The lightweight, soft wood serves for rough lumber and fuel. Indians once gathered and ate the large seeds; now squirrels and other wildlife consume the annual crop. It was discovered in 1831 by Thomas Coulter (1793–1843), the Irish botanist and physician, who collected plants in Mexico and California.

23, 421 Pinyon
"Two-leaf Pinyon" "Colorado Pinyon"
Pinus edulis Engelm.

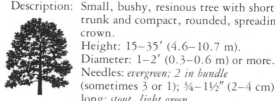

Description: Small, bushy, resinous tree with short trunk and compact, rounded, spreading crown.
Height: 15–35′ (4.6–10.7 m).
Diameter: 1–2′ (0.3–0.6 m) or more.
Needles: *evergreen; 2 in bundle* (sometimes 3 or 1); ¾–1½″ (2–4 cm) long; *stout, light green.*
Bark: gray to reddish-brown, rough, furrowed into scaly ridges.
Cones: 1½–2″ (4–5 cm) long; *egg-shaped, yellow-brown,* resinous or sticky; opening and shedding; with thick, blunt cone-scales; *seeds* large, *wingless,* slightly thick-walled, *oily,* edible.

Habitat: Open, orchardlike woodlands, alone or with junipers; on dry, rocky foothills, mesas, plateaus, and lower mountain slopes.

Range: Southern Rocky Mountain region from Utah and Colorado south to New Mexico and Arizona; local in SW. Wyoming, extreme NW. Oklahoma, Trans-Pecos Texas, SE. California, and

Mexico. Mostly at 5000–7000′ (1524–2134 m).

The edible seeds, known as pinyon nuts, Indian nuts, pine nuts, and *piñones* (Spanish), are a wild, commercial nut crop. Eaten raw, roasted, and in candies, they were once a staple food of southwestern Indians. Pinyon ranks first among the native nut trees of the United States that are not also cultivated. Every autumn, local residents, especially Navajo Indians and Spanish-Americans, harvest quantities for the local and gourmet markets. However, most of these oily seeds are promptly devoured by pinyon jays, wild turkeys, woodrats or "packrats," bears, deer, and other wildlife. Small pinyons are popular Christmas trees. This species is the most common tree on the south rim of Grand Canyon National Park.

7 **Apache Pine**
"Arizona Longleaf Pine"
Pinus engelmannii Carr.

Description: Medium-sized tree with straight axis, open, rounded crown of few, large, spreading branches, 1 row added a year, and very long needles.
Height: 50–70′ (15–21 m).
Diameter: 2′ (0.6 m).
Needles: *evergreen; 3 in bundle* (sometimes 4–5); crowded at end of very stout twigs; 8–12″ (20–30 cm) long. *Stout,* spreading widely or drooping, *dull green.*
Bark: dark brown or blackish-gray; rough, thick, deeply furrowed into scaly ridges.
Cones: 4–5½″ (10–14 cm) long; *egg-shaped* or conical, almost stalkless, shiny *light brown;* opening and shedding at maturity, leaving a few cone-scales on twig; cone-scales 4-sided, thick, raised,

with prominent *keel* and stout, *short spine* straight or curved backward.

Habitat: Rocky ridges and slopes of mountains; with Arizona Pine and Chihuahua Pine.

Range: SE. Arizona and extreme SW. New Mexico; also N. Mexico; at 5000–8200' (1524–2499 m).

This Mexican pine extends northward to the Chiricahua Mountains and a few other ranges near the border. As the name implies, it grows in Apache Indian country. The seedlings pass through a grasslike stage with a short stem and very long needles to 15" (38 cm), as does Longleaf Pine (*Pinus palustris* Mill.) in the Southeast.

18, 409 Limber Pine
"White Pine"
"Rocky Mountain White Pine"
Pinus flexilis James

Description: Medium-sized tree with short trunk and broad, rounded crown of annual rows of stout branches nearly down to ground; or a windswept, deformed shrub at timberline.
Height: 40–50' (12–15 m).
Diameter: 2–3' (0.6–0.9 m).
Needles: *evergreen; 5 in bundle,* with sheath shedding first year; 2–3½" (5–9 cm) long. Slender, long-pointed, not toothed; light or dark green, with *white lines* on all surfaces.
Bark: light gray and smooth; becoming dark brown and furrowed into scaly ridges or rectangular plates.
Twigs: slender, very tough and flexible.
Cones: 3–6" (7.5–15 cm) long; *egg-shaped, yellow-brown,* short-stalked; opening at maturity; cone-scales thick, rounded, ending in *blunt point;* seeds large and edible, with very short wing.
Habitat: Dry, rocky slopes and ridges of high mountains up to timberline; often in pure stands.

Range: Rocky Mountain region chiefly, from
SE. British Columbia and SW. Alberta
south to N. New Mexico and west to S.
California; also local in NE. Oregon,
SW. North Dakota, Black Hills of
South Dakota, and W. Nebraska; at
5000–12,000' (1524–3658 m).

The names refer to the very tough and
flexible twigs, which can sometimes be
twisted into a knot. Plants on exposed
ridges and at timberline are shaped by
the wind into stunted shrubs with
crooked or twisted branches that are
bent over and are longer on one side.
Birds and mammals, especially
squirrels, consume the large seeds.

4, 414 **Jeffrey Pine**
"Western Yellow Pine" "Bull Pine"
Pinus jeffreyi Grev. & Balf.

Description: Large tree with straight axis and open,
conical crown of spreading branches and
with large cones. Both *bark and twigs*
give off *odor of lemon* or vanilla when
crushed.
Height: 80–130' (24–39 m).
Diameter: 2–4' (0.6–1.2 m),
sometimes much larger.
Needles: *evergreen; 3 in bundle;* 5–10"
(13–25 cm) long. *Stout,* stiff; light
gray-green or blue-green, with broad
white lines on all surfaces.
Bark: purplish-brown, thick, furrowed
into narrow scaly plates.
Twigs: stout, hairless, *gray-green* with
whitish bloom, smooth.
Cones: 5–10" (13–25 cm) long; *conical*
or egg-shaped, light *reddish-brown,*
almost stalkless; opening and shedding
at maturity, leaving a few cone-scales
on twig; cone-scales numerous, raised,
and *keeled,* ending in long, *bent-back
prickle.*
Habitat: Dry slopes of mountains, especially
from lava flows and granite; best

developed on deep, well-drained soils; often forming pure stands and with other conifers.

Range: SW. Oregon south through the Sierra Nevada (especially eastern slopes) to W. Nevada and S. California; also N. Baja California; mostly at 6000–9000' (1829–3048 m); less frequently down to 3500' (1067 m) and up to 10,000' (3048 m).

The odor of crushed twigs defies exact description. The scent has been likened not only to lemons and vanilla, but also to violets, pineapples, and apples. This species was named for its discoverer, John Jeffrey, the 19th century Scottish botanical explorer who collected seeds and plants in Oregon and California for introduction into Scotland.

9, 406   Sugar Pine
Pinus lambertiana Dougl.

Description: Large, very tall tree with a straight trunk unbranched for a long span and open, conical crown of long, nearly horizontal branches, bearing *giant cones* near the ends; becoming flat-topped.
Height: 100–160' (30–49 m).
Diameter: 3–6' (0.9–1.8 m), sometimes much larger.
Needles: *evergreen; 5 in bundle,* with sheath shedding first year; 2¾–4" (7–10 cm) long. Twisted, *slender,* stiff, sharp-pointed; blue-green, with white lines on all surfaces.
Bark: brown or gray, furrowed into irregular scaly ridges; gray and smooth on branches.
Cones: 11–18" (28–46 cm) long; *cylindrical,* shiny *light brown;* hanging down on long stalk near ends of upper branches; cone-scales thick, *rounded,* ending in *blunt point,* spreading widely; seeds large, long-winged, edible.
Habitat: Many kinds of mountain soils; not

forming pure stands but occurring in mixed coniferous forests.

Range: Mountains from W. Oregon south through Sierra Nevada to S. California; also in N. Baja California; at 1100–5400' (335–1646 m) in north, 2000–7800' (610–2377 m) in Sierra Nevada, and 4000–10,500' (1219–3200 m) in south.

A major lumber species, Sugar Pine is one of the most beautiful and largest pines and has been called the "king of pines." The trunk diameter occasionally reaches 6–8' (1.8–2.4 m); the current champion is 10' (3 m) in diameter, and the tallest tree recorded was 241' (73.5 m) high. No other conifer has such long cones, reaching a maximum of 21" (53 cm). Sugar Pine provided early settlers of California with wood for their houses, especially shingles or shakes, and with fences. Forty-niners made ample use of the wood for flumes, sluice boxes, bridges, and mine timbers. American Indians gathered and ate the large, sweet seeds. The common name refers to the sweetish resin that exudes from cut or burned heartwood which was also eaten by Indians.

8, 422 Chihuahua Pine
"Yellow Pine"
Pinus leiophylla Schiede & Deppe

Description: Pine with trunk bearing short, leafy twigs, *very thick bark,* and thin, open, spreading crown of upturned branches.
Height: 30–80' (9–24 m).
Diameter: 1–2' (0.3–0.6 m).
Needles: *evergreen; 3 in bundle,* with sheath shedding first year; 2½–4" (6–10 cm) long. *Stout,* stiff; *blue-green,* with white lines on all surfaces.
Bark: dark brown or blackish, as much as 2–3" (5–7.5 cm) thick, and deeply

furrowed into broad, scaly ridges.
Cones: 1½–2½" (4–6 cm) long;
narrowly *egg-shaped,* shiny *light brown,*
long-stalked; usually opening but
remaining attached; *cone-scales flattened,*
mostly with tiny *prickle.*

Habitat: Rocky ridges and slopes of mountains;
with Arizona Pine and Apache Pine.

Range: SW. New Mexico, E. central and SE.
Arizona, and Mexico; at 5000–7800'
(1524–2377 m).

Unlike most pines, this species often
produces new shoots or sprouts from
cut stumps. The cones mature in 3
growing seasons, instead of the usual 2;
cones in 3 stages of development, as
well as many old, open cones, are
usually present. Trees native to the
United States and others in
northwestern Mexico are placed in a
separate variety (var. *chihuahuana*
(Engelm.) Shaw), characterized by 3 stout
needles in a bundle. The typical variety
(var. *leiophylla*), of wider distribution in
Mexico, has 5 slender needles in a
bundle.

21, 430 Singleleaf Pinyon
"Nut Pine" "Singleleaf Pinyon Pine"
Pinus monophylla Torr. & Frém.

Description: Slow-growing, small pine with
spreading, rounded, gray-green crown
and low, horizontal branches; often
shrubby.
Height: 16–30' (5–9 m).
Diameter: 1–1½' (0.3–0.5 m).
Needles: *evergreen; 1 in bundle* (rarely 2),
sheath shedding after first year; 1–2¼"
(2.5–6 cm) long. *Stout, stiff, sharp-
pointed;* straight or slightly curved; dull
gray-green, with many whitish lines;
resinous.
Bark: dark brown or gray, smoothish,
becoming furrowed into scaly plates
and ridges.

Cones: 2–3″ (5–7.5 cm) long; *egg-shaped* or rounded, dull *yellow-brown,* almost stalkless, resinous; opening and shedding; with thick, 4-angled cone-scales, often with tiny prickle; *seeds* large, *wingless, thin-walled, mealy,* edible.

Habitat: Dry, gravelly slopes of mesas, foothills, and mountains; in open, orchardlike, pure stands and with junipers.

Range: SE. Idaho and N. Utah south to NW. Arizona and west to S. California; also N. Baja California; at 3500–7000′ (1067–2134 m).

This species is easily recognized by the needles borne singly, instead of in bundles of 2–5, as in other native pines. The large, edible, mealy seeds are sold locally as pinyon or pine nuts and used to be a staple food of Indians in the Great Basin region. Many kinds of birds and mammals, especially woodrats or "packrats," also consume the seeds.

10, 408 Western White Pine
"Mountain White Pine"
"Idaho White Pine"
Pinus monticola Dougl. ex D. Don

Description:

Large to very large tree with straight trunk and narrow, open, conical crown of horizontal branches.
Height: 100′ (30 m).
Diameter: 3′ (0.9 m), sometimes much larger.
Needles: *evergreen; 5 in bundle,* with sheath shedding first year; 2–4″ (5–10 cm) long. Slightly stout; *blue-green,* with whitish lines on inner surfaces.
Bark: gray and thin, smooth, becoming furrowed into rectangular, scaly plates.
Cones: 5–9″ (13–23 cm) long; narrowly *cylindrical, yellow-brown,* mostly long-stalked; opening and shedding at maturity; cone-scales thin,

rounded, ending in small point,
spreading widely; long-winged seeds.

Habitat: Moist mountain soils; in mixed forests
and occasionally in almost pure stands.

Range: Northern Rocky Mountains from
British Columbia southeast to NW.
Montana; also along Pacific Coast south
through the Sierra Nevada to central
California; to 3500′ (1067 m) in north;
at 6000–9800′ (1829–2987 m) in
south.

An important timber tree, Western
White Pine is also a leading match
wood, because of its uniformly high
grade without knots, twisted grain or
discoloration. It is one of the world's
largest pines; the champion near
Medford, Oregon, is 239′ (72.8 m)
tall. White pine blister rust, caused by
an introduced fungus (*Cronartium
ribicola*), is a serious disease of this and
other 5-needle white pines; a resistant
strain is being developed.

16, 412 Bishop Pine
"Santa Cruz Island Pine"
"Prickle-cone Pine"
Pinus muricata D. Don

Description:  Tree with conical, rounded, or irregular
crown of stout, spreading branches and
with numerous *spiny cones remaining
closed many years.*
Height: 40–80′ (12–24 m).
Diameter: 2–3′ (0.6–0.9 m).
Needles: *evergreen; 2 in bundle;* 4–6″
(10–15 cm) long. *Stout,* slightly
flattened, stiff, blunt-pointed, *dull
green.*
Bark: dark gray, *very thick,* furrowed
into scaly plates; smoothish on
branches.
Cones: 2–3½″ (5–9 cm) long; *conical* or
egg-shaped, shiny yellow-brown,
stalkless; *oblique* or 1-sided at base;
many *clustered in rings* or whorls; cone-

scales raised and keeled, those on outer part much enlarged, ending in *stout, flattened,* straight or curved *spine.*

Habitat: Low hills and plains along coast in fog belt; in scattered groves and with other pines.

Range: Coast of central and N. California and Santa Cruz and Santa Rosa islands; also local in Baja California; a variety on Cedros Island, Mexico; near sea level.

The numerous cones remain closed, even when enclosed by the bark and wood of the expanding trunk. Fossil cones from the Pleistocene, or glacial, epoch indicate that this pine was associated with extinct vertebrates, including the woolly mammoth. Its common name apparently refers to the discovery of this local pine in 1835 near the mission of San Luis Obispo (Saint Louis, Bishop of Toulouse) in California.

12 Austrian Pine
"European Black Pine"
Pinus nigra Arnold

Description: Introduced, cone-bearing, resinous tree with straight trunk and dense, rounded, spreading crown of dark green foliage.
Height: 60′ (18 m).
Diameter: 2′ (0.6 m).
Needles: *evergreen; 2 in bundle;* 3½–6″ (9–15 cm) long; crowded, *stiff, shiny dark green.*
Bark: dark gray, thick, rough, furrowed into irregular scaly plates.
Twigs: light brown or gray; stout, hairless.

Cones: 2–3″ (5–7.5 cm) long, 1–1¼″ (2.5–3 cm) wide; *egg-shaped, shiny yellow-brown,* almost stalkless; opening and shedding after maturity; cone-scales numerous, raised and keeled, ending in *short prickle.*

Habitat: Hardy except in coldest, hottest, and driest regions.

Range: Native of central and S. Europe and Asia Minor; also local in NW. Africa. Planted across the United States; escaped locally but apparently not naturalized.

Austrian Pine is one of the more common introduced ornamental trees, with several geographical and horticultural varieties. Used also in shelterbelts and screens, it is fast-growing and tolerant of city smoke, dust, and dry soil.

11, 419   Ponderosa Pine
"Western Yellow Pine"
"Blackjack Pine"
Pinus ponderosa Dougl. ex Laws.

Description: Large to very large tree with broad, open, conical crown of spreading branches; 3 distinct geographic varieties.
Height: 60–130' (18–39 m).
Diameter: 2½–4' (0.8–1.2 m), sometimes larger.
Needles: *evergreen;* usually *2 or 3 in bundle* (2–5 in varieties); generally 4–8" (10–20 cm) long. *Stout,* stiff, *dark green.*
Bark: blackish, rough, and furrowed into ridges; on trunks of small trees (blackjacks), becoming yellow-brown and irregularly furrowed into large, flat, scaly plates.
Cones: 2–6" (5–15 cm) long; *conical* or egg-shaped, almost stalkless, light *reddish*-brown; opening and shedding at maturity, leaving a few cone-scales on twig; cone-scales raised and *keeled,* ending in short, *sharp prickle;* small, long-winged seeds.

Habitat: Mostly in mountains in pure stands, forming extensive forests; also in mixed coniferous forests.

Range: Widely distributed; S. British Columbia east to SW. North Dakota, south to Trans-Pecos Texas, and west to S. California; also N. Mexico; from sea level in north to 9000' (2743 m) in south; the best developed stands at 4000–8000' (1219–2438 m).

This is the most widely distributed and common pine in North America. The typical variety, Ponderosa Pine or Pacific Ponderosa Pine (var. *ponderosa*), has long needles, 3 in a bundle, and large cones, and occurs in the Pacific Coast region. Rocky Mountain Ponderosa Pine or Interior Ponderosa Pine (var. *scopulorum* Engelm.) with short needles, 2 in a bundle, and small cones, is found in the Rocky Mountain region. Arizona Pine or Arizona Ponderosa Pine (var. *arizonica* (Engelm.) Shaw), occurring mainly in southeastern Arizona, has 5 slender needles in bundle. David Douglas, the Scottish botanical explorer, found this pine in 1826 and named it for its ponderous, or heavy, wood. This valuable timber tree is the most commercially important western pine. Its lumber is especially suited for window frames and panel doors. Quail, nutcrackers, squirrels, and many other kinds of wildlife consume the seeds; and chipmunks store them in their caches, thus aiding dispersal.

22, 431 **Parry Pinyon**
"Four-needle Pinyon" "Nut Pine"
Pinus quadrifolia Parl. ex Sudw.

Description: Small resinous tree with spreading, rounded crown and low, horizontal branches; often shrubby.
Height: 16–30' (5–9 m).
Diameter: 1–1½' (0.3–0.5 m).
Needles: *evergreen; 4 in bundle* (sometimes 3 or 5), sheath shedding after first year; 1–2¼" (2.5–6 cm)

long. *Stout, stiff,* sharp-pointed; *bright green* with whitish inner surfaces.

Bark: light gray and smooth, becoming reddish-brown and furrowed into scaly ridges.

Cones: 1½–2½″ (4–6 cm) long; *egg-shaped* or nearly round, almost stalkless, dull *yellow-brown,* resinous; opening and shedding; cone-scales thick, 4-angled, often with tiny prickle; large, wingless, edible seeds.

Habitat: Dry, gravelly slopes of foothills and mountains; in woodlands or with junipers.

Range: S. California and N. Baja California; at 4000–6000′ (1219–1829 m).

The edible seeds are not gathered commercially because of the tree's limited distribution. Rodents (especially woodrats), other mammals, and birds consume the small annual crop.

15, 418 Monterey Pine
"Insignis Pine"
Pinus radiata D. Don

Description: Tree with straight trunk, narrow, irregular, open crown, and many closed cones grouped in rings.

Height: 50–100′ (15–30 m).

Diameter: 1–3′ (0.3–0.9 m).

Needles: *evergreen; 3 in bundle;* 4–6″ (10–15 cm) long; slender, *shiny green.*

Bark: dark reddish-brown, thick, deeply furrowed into scaly ridges or plates.

Cones: 3–6″ (7.5–15 cm) long; *conical* or egg-shaped, *very oblique* or 1-sided at base and pointed at tip, *shiny brown;* many *clustered in rings* or whorls on short stalks and turned back; *remaining closed* on tree many years; cone-scales thick, slightly raised and *rounded,* those on outer side much enlarged, ending in tiny prickle; small, long-winged seeds.

Habitat: Coarse soils, usually sandy loams, on slopes; in pure stands or with Monterey and Gowen cypresses and Coast Live Oak.

Range: Rare; at 3 localities on coast of central California in fog belt to about 6 miles (9.7 m) inland; also a variety on Guadalupe Island, Mexico; to nearly 1000′ (305 m).

Although rare in its native California, Monterey Pine is one of the world's most valuable pines and is the most common commercially planted one in the southern hemisphere (where pines are not native), especially in New Zealand, Australia, Chile, and South Africa. Like those of Knobcone and Bishop pines, the cones of Monterey Pine remain closed until opened by the heat of a forest fire; the abundant seeds are then discharged and begin a new forest. The cones may also burst open in hot weather with a snapping sound.

3, 416 Digger Pine
"Bull Pine" "Gray Pine"
Pinus sabiniana Dougl.

Description: Tree with crooked, forking trunk and branches; open, very thin, irregular, broad, or rounded crown; and very large, heavy cones.
Height: 40–70′ (12–21 m), sometimes much larger.
Diameter: 2–4′ (0.6–1.2 m).
Needles: *evergreen; 3 in bundle;* 8–12″ (20–30 cm) long. Slender and *drooping;* dull *gray-green,* with many white lines.
Bark: dark gray, thick, deeply and irregularly furrowed into scaly ridges, becoming slightly shaggy; light gray, smooth on branches.
Cones: mostly 6–10″ (15–25 cm) long; *egg-shaped* and *slightly 1-sided, brown,* bent down on long stalks; opening late and remaining on tree many years;

cone-scales very long, thick, sharply keeled and 4-sided, narrowed into very *large, stout,* straight or slightly *curved spine.* Very large, elliptical, thick-walled, edible *seeds with detachable wing.*

Habitat: Dry slopes and ridges in foothills and low mountains; with oaks and other conifers.

Range: N. to S. California through the Coast Ranges and the Sierra Nevada; mostly at 1000–3000' (305–914 m); rarely down to 100' (30 m) and up to 6000' (1829 m).

The soft, lightweight wood of this common and widespread pine is not durable; the crooked, forking trunks also make the wood impractical to use except as fuel. The common name refers to the Digger Indians (a pioneer term grouping all California Indian tribes together), who dug up roots for food and harvested quantities of the large seeds.

13, 407   **Southwestern White Pine**
"Mexican White Pine"
"Border White Pine"
Pinus strobiformis Engelm.

Description: Tree with straight trunk and narrow, conical crown of horizontal branches.

Height: 50–80' (15–24 m).
Diameter: 1½–3' (0.5–0.9 m).
Needles: *evergreen; 5 in bundle,* sheath shedding first year; 2½–3½" (6–9 cm) long. Slender, *finely toothed* at least near tip; *bright green,* with white lines on inner surfaces only.
Bark: gray and smooth; becoming dark gray or brown and deeply furrowed into narrow, irregular ridges.
Cones: 6–9" (15–23 cm) long; *cylindrical, yellow-brown,* short-stalked; opening at maturity; cone-scales slightly thickened, *very long,* the thin, narrow tip spreading and *curved back;*

seeds large, edible, very short-winged.

Habitat: Dry, rocky slopes and canyons in high mountains; a minor component of coniferous forests.

Range: Trans-Pecos Texas west to E. central Arizona; also N. Mexico; at 6500–10,000′ (1981–3048 m).

The large seeds are consumed by wildlife and were eaten by southwestern Indians. This species of the Mexican border region was formerly considered a southern variety of Limber Pine, which has a broader, more northern distribution, is smaller, and has smooth-edged needles with white lines on all surfaces and shorter cones with thick, rounded, blunt-pointed cone-scales.

17 Scotch Pine
"Scots Pine"
Pinus sylvestris L.

Description: Beautiful, large, introduced tree with crown of spreading branches that become rounded and irregular and rich blue-green foliage.

Height: 70′ (21 m).
Diameter: 2′ (0.6 m) and much larger with age.
Needles: *evergreen; 2 in bundle;* 1½–2¾″ (4–7 cm) long. Stiff, slightly flattened, *twisted* and spreading; blue-green.
Bark: *reddish-brown,* thin; becoming gray and shedding in papery or *scaly plates.*
Cones: 1¼–2½″ (3–6 cm) long; *egg-shaped,* pale *yellow-brown,* short-stalked; opening at maturity; cone-scales thin, flattened, often with minute prickle.

Habitat: Various soils from loams to sand; tolerating city smoke.

Range: Native across Europe and N. Asia, south to Turkey. Naturalized locally in SE. Canada and NE. United States; cultivated on the West Coast.

The native pine of the Scottish Highlands, this is the most widely distributed pine in the world and one of the most important European timber trees. In the United States, native pines are better adapted for forestry plantations, but Scotch Pine is commonly grown for shelterbelts, ornament, and Christmas trees.

5, 417   Torrey Pine
"Del Mar Pine" "Soledad Pine"
Pinus torreyana Parry ex Carr.

Description: Medium-sized tree with open, spreading crown of branches; shrubby on exposed sites.
Height: 30–50′ (9–15 m).
Diameter: 1–2′ (0.3–0.6 m).
Needles: *evergreen; 5 in bundle,* crowded in large clusters at ends of very stout twigs; 8–13″ (20–33 cm) long. *Stout,* stiff, *dark gray-green,* with white lines.
Bark: blackish, deeply furrowed into broad, flat, scaly ridges; gray and smoothish on branches.
Cones: 4–6″ (10–15 cm) long; broadly *egg-shaped, nut-brown,* bent-down on stout stalk; maturing in 3 seasons, opening late and *remaining on tree* a few years; cone-scales thick, keeled, and 4-sided, ending in stout, *straight spine. Seeds very large,* elliptical, *thick-walled,* edible, with *short, detachable wing.*

Habitat: Limited to 2 areas on dry, sandy bluffs and slopes; in pure stands.

Range: S. California (San Diego County) and Santa Rosa Island; to 500′ (152 m).

One of the rarest native conifers; only several thousand trees exist. A large part of the mainland grove north of San Diego is within Torrey Pines State Park. When cultivated elsewhere for shade and commercial forestry in moist, warm climates, the trees grow rapidly to large size.

Washoe Pine
Pinus washoensis Mason & Stockwell

Description: Medium-sized tree with tapering trunk and spreading crown of branches.
Height: 60' (18 m).
Diameter: 3' (0.9 m).
Needles: *evergreen; 3 in bundle;* 4–6" (10–15 cm) long; stout, stiff, *gray-green.*
Bark: blackish-brown or yellow-brown; furrowed into scaly ridges or plates.
 Cones: 2–4" (5–10 cm) long; *conical* or egg-shaped, *nut-brown,* dull or slightly shiny, stalkless; opening at maturity but persistent about a year, leaving a few cone-scales on twig; cone-scales raised and *keeled,* ending in *stout, sharp prickle,* usually curved back.

Habitat: Rocky slopes and ridges; in small, pure stands or in coniferous forests.

Range: W. Nevada (Washoe County) and NE. California; at 7000–8400' (2134–2560 m).

This rare and local tree was the last native pine species to be discovered. Found in 1938 on Mount Rose, Nevada, it was named in 1945 in honor of the Washoe Indians who once hunted in the area.

42, 425 **Bigcone Douglas-fir**
"Bigcone-spruce"
Pseudotsuga macrocarpa (Vasey) Mayr

Description:  Evergreen tree with open, broad, conical crown of long, spreading branches.
Height: 40–80' (12–24 m).
Diameter: 2–3' (0.6–0.9 m).
Needles: *evergreen;* spreading mostly in 2 rows; ¾–1¼" (2–3 cm) long. Flattened, *sharp-pointed,* almost stalkless; *blue-green* or blue-gray.
Bark: dark reddish-brown, thick,

deeply furrowed into broad, scaly
ridges.
Twigs: dark reddish-brown, slender,
often drooping, slightly hairy when
young, ending in *dark brown, pointed,
scaly, hairless bud.*
Cones: 4–6" (10–15 cm) long; *narrowly
egg-shaped, brown,* short-stalked; with
many thick, stiff, rounded cone-scales
each above a short, *3-pointed bract.*
Paired, long-winged seeds.

Habitat: Dry slopes and canyons in various soils;
with chaparral, Canyon Live Oak, and
in mixed coniferous forests.

Range: Mountains of S. California; at 900–
8000' (274–2438 m).

Bigcone Douglas-fir is native only in
the mountains of southern California,
mostly within national forests, and
beyond the range of Douglas-fir. This
species is distinguishable from its
relative by its much larger cones
(referred to in both common and
scientific names) and by its sharp-
pointed, bluish needles. The trees
recover from fire damage and injuries
by sprouting vigorously from trunks
and branches.

41, 426 **Douglas-fir**
"Douglas-spruce" "Oregon-pine"
Pseudotsuga menziesii (Mirb.) Franco

Description: Large to very large tree with narrow,
pointed crown of slightly drooping
branches; 2 distinct geographic
varieties: Coast and Rocky Mountain.
Height: 80–200' (24–61 m).
Diameter: 2–5' (0.6–1.5 m),
sometimes much larger.
Needles: *evergreen;* spreading mostly in
2 rows; ¾–1¼" (2–3 cm) long.
Flattened, mostly rounded at tip,
flexible; dark yellow-green or blue-
green; very short, twisted leafstalks.
Bark: reddish-brown, very thick,

deeply furrowed into broad ridges;
often corky.

Twigs: orange, turning brown; slender,
hairy, ending in *dark red,* conical,
pointed, scaly, hairless *bud.*

Cones: 2–3½" (5–9 cm) long; *narrowly
egg-shaped, light brown,* short-stalked;
with many thin, rounded cone-scales
each above a long, protruding, *3-pointed
bract;* paired, long-winged seeds.

Habitat: Coast Douglas-fir forms vast forests on
moist, well-drained soils; often in pure
stands. Rocky Mountain Douglas-fir is
chiefly on rocky soils of mountain
slopes; in pure stands and mixed
coniferous forests.

Range: Central British Columbia south along
Pacific Coast to central California; to
2700' (823 m) in north and to 6000'
(1829 m) in south; also in Rocky
Mountains to SE. Arizona and Trans-
Pecos Texas; down to 2000' (610 m) in
north and at 8000–9500' (2438–2896
m) in south; also local in mountains of
N. and central Mexico.

Coast Douglas-fir (var. *menziesii*), the
typical Douglas-fir of the Pacific Coast,
is a very large tree with long, dark
yellow-green needles and large cones
with spreading bracts. Rocky Mountain
Douglas-fir (var. *glauca* (Beissn.) Franco),
of the Rocky Mountain region, is a
medium-sized to large tree with
shorter, blue-green needles and smaller
cones with bracts bent upward. One of
the world's most important timber
species, Douglas-fir ranks first in the
United States in total volume of
timber, in lumber production, and in
production of veneer for plywood. It is
one of the tallest trees as well as a
popular Christmas tree. David Douglas
(1798–1834), the Scottish botanical
collector, who sent seeds back to
Europe in 1827, is commemorated in
the common name. The foliage is
consumed by grouse and by deer and
elk; birds and mammals eat the seeds.

52, 434 Western Hemlock
"Pacific Hemlock"
"West Coast Hemlock"
Tsuga heterophylla (Raf.) Sarg.

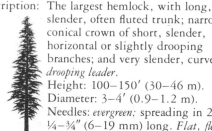

Description: The largest hemlock, with long,
slender, often fluted trunk; narrow,
conical crown of short, slender,
horizontal or slightly drooping
branches; and very slender, curved, and
drooping leader.
Height: 100–150' (30–46 m).
Diameter: 3–4' (0.9–1.2 m).
Needles: *evergreen;* spreading in 2 rows;
¼–¾" (6–19 mm) long. *Flat, flexible,
rounded at tip,* very short-stalked. *Shiny
dark green* above, with *2 broad, whitish
bands* and indistinct green edges, often
with tiny teeth beneath.
Bark: reddish-brown to gray-brown,
thick, deeply furrowed into broad, scaly
ridges; cut surface of inner bark red.
Twigs: very slender, yellow-brown,
finely hairy, *rough with peglike bases.*
Cones: ¾–1" (2–2.5 cm) long;
elliptical, brown, stalkless; with many
rounded, elliptical cone-scales; hanging
down at ends of twigs; paired, long-
winged seeds.

Habitat: Moist, acid soils, especially flats and
lower slopes; in dense pure stands and
with Sitka Spruce and other conifers.

Range: S. Alaska southeast along Pacific coast
to NW. California; also SE. British
Columbia south in Rocky Mountains to
N. Idaho and NW. Montana; to 2000'
(610 m) along coast; to 6000' (1,829
m) inland.

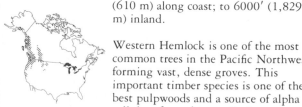

Western Hemlock is one of the most
common trees in the Pacific Northwest,
forming vast, dense groves. This
important timber species is one of the
best pulpwoods and a source of alpha
cellulose for making cellophane, rayon
yarns, and plastics. Indians of
southeastern Alaska used to make coarse
bread from the inner bark.

49, 433 Mountain Hemlock
"Black Hemlock" "Alpine Hemlock"
Tsuga mertensiana (Bong.) Carr.

Description: Tree with tapering trunk, conical crown of slender horizontal or drooping branches, and very slender, curved, *drooping leader;* a prostrate shrub at timberline.
Height: 30–100' (9–30 m).
Diameter: 1–3' (0.3–0.9 m).
Needles: *evergreen;* usually *crowded at ends of short side twigs; spreading on all sides and curved upward;* ¼–1" (0.6–2.5 cm) long. Short-stalked, flattened above and *half-round, stout, and blunt; blue-green,* with whitish lines on both surfaces.
Bark: gray to dark brown; thick, deeply furrowed into scaly ridges.
Twigs: light reddish-brown, mostly short, slender, finely hairy, *rough with peglike* bases.
Cones: 1–3" (2.5–7.5 cm) long; *cylindrical,* purplish turning brown, stalkless, hanging down, with many *rounded cone-scales;* paired, long-winged seeds.

Habitat: Moist, coarse or rocky soils, from sheltered valleys to exposed ridges; with firs and in mixed coniferous forests.

Range: S. Alaska southeast along Pacific Coast to British Columbia and in mountains to central California; also SE. British Columbia south in Rocky Mountains to N. Idaho and NE. Oregon; to 3500' (1067 m) in Alaska and at 5500–11,000' (1676–3353 m) in south.

Mountain Hemlock is a characteristic species of high mountains, varying greatly in size from a large tree at low altitudes to a dwarf, creeping shrub at timberline. Hemlock groves provide cover, nesting sites, and seeds for birds, as well as foliage for mountain goats and other hoofed browsers. This species honors its discoverer, Karl H. Mertens (1796–1830), the German naturalist.

REDWOOD FAMILY
(Taxodiaceae)

Medium-sized to very large, aromatic and resinous trees without flowers or fruit. About 15 species in temperate North America, east Asia, and Tasmania; 4 native species in North America, 2 eastern and 2 western.
Leaves: needlelike or scalelike, often of both kinds, mostly evergreen (deciduous in baldcypress, *Taxodium*), mainly alternate in spirals.
Twigs: often very slender and shedding with leaves.
Cones: pollen and naked seeds borne on same plant. Male cones small and herbaceous, with 2–9 pollen-sacs. Female cones hard or woody, rounded or slightly elliptical, at ends of twigs, with many flat cone-scales lacking bracts; 2–9 naked seeds angled or narrowly winged at base of cone-scale.

54, 438 **Cryptomeria**
"Japanese-cedar" "Sugi"
Cryptomeria japonica (L. f.) D. Don

Description:  Introduced, cone-bearing, evergreen tree with straight trunk tapering from enlarged base, conical or pyramid-shaped, narrow crown of horizontal branches, and aromatic resinous foliage.
Height: 50–80′ (15–24 m).
Diameter: 2′ (0.6 m).
Leaves: *evergreen;* spreading on all sides of twig; ¼–⅝″ (6–15 mm) long. *Awl-shaped or needlelike,* curved forward and inward, the base extending down twig; *slightly stiff* and blunt, somewhat flattened and *4-angled;* dull blue-green.
Bark: reddish-brown, fibrous, fissured, peeling off in long strips.
Twigs: dull green, very slender, hairless, mostly shedding with leaves.
Cones: ½–¾″ (12–19 mm) in diameter; *rounded,* dull, short-stalked;

opening at maturity in autumn but
remaining attached; many cone-scales
with bristly *point in center* and *3—5
points at end;* 2—5 slightly 3-angled,
narrowly winged seeds under cone-
scale.

Habitat: Moist soils in humid, warm temperate
regions.

Range: Native of Japan and China. Planted in
E. United States and in Pacific states.

Cryptomeria has several cultivated
varieties and is seen mainly as an
ornamental. In Hawaii the trees are also
grown in forest plantations and for
windbreaks. It is propagated by seeds
and is fast-growing. The national tree
of Japan, where it is called "Sugi," it is
one of the most important native
timbers and is widely planted in
forests, as a street tree, and around
temples.

53, 437 **Redwood**
"Coast Redwood"
"California Redwood"
Sequoia sempervirens (D. Don) Endl.

Description: The world's tallest tree, with reddish-

brown trunk much enlarged and
buttressed at base and often with
rounded swellings or burls and slightly
tapering; crown short, narrow, irregular
and open with horizontal or drooping
branches.
Height: 200—325' (61—99 m).
Diameter: 10—15' (3—4.6 m),
sometimes larger.
Leaves: *evergreen;* of 2 kinds. Mostly
needlelike and *unequal,* ⅜—¾" (10—19
mm) long; flat and slightly curved, stiff
and sharp-pointed, extending down
twig at base; *dark green* above, whitish-
green beneath; spreading in *2 rows.*
Leaves on leaders scalelike, as short as ¼"
(6 mm); *keeled,* concave, *spreading*
around twig.

Bark: *reddish-brown,* very tough and fibrous, *very thick, deeply furrowed* into broad, scaly ridges; inner bark cinnamon-brown.

Twigs: slender, dark green, forking in 1 plane, ending in scaly bud.

Cones: ½–1⅛" (1.2–3 cm) long; *elliptical,* reddish-brown, with *many flat,* short-pointed *cone-scales;* hanging down at end of leafy twig; maturing in 1 season; 2–5 seeds under cone-scale, light brown, 2-winged.

Habitat: Mostly alluvial soils on flats and benches or terraces; forms pure stands in luxuriant dense forests; also with Douglas-fir, Port-Orford-cedar, and mixed conifers.

Range: Extreme SW. Oregon south to central California in fog belt, a coastal strip 5–35 miles (8–56 km) wide; from sea level to 3000' (914 m).

The world's tallest tree is a Redwood 368' (112 m) high. The age of these trees at maturity is 400–500 years; the maximum age counted in annual rings is 2200. Circles of trees grow from sprouts around stumps and dead trunks. The genus name commemorates the Indian named Sequoyah (also spelled Sequoia) (1770?–1843), the inventor of the Cherokee alphabet. Existing stands of Redwood occupy only a fraction of the large area in California and Oregon where they originally grew before the arrival of European settlers. Virgin forests remain in several state parks, as well as in the Redwoods National Park and along the Redwoods Highway. But there is still some question concerning the status of the species outside of these parks. The Redwood industry maintains that selective logging, leaving seed trees, and planting in tree farms assure the future of this species. Conservationists feel that every effort should be made to maintain this magnificent tree at its present levels.

61, 436 Giant Sequoia
"Sierra Redwood" "Bigtree"
Sequoiadendron giganteum (Lindl.) Buchholz

Description:

One of the world's largest trees with fibrous, reddish-brown trunk much enlarged and buttressed at base, fluted into ridges, and conspicuously narrowed or tapered above; narrow, conical crown of short, stout, horizontal branches reaches nearly to base. Giant trees have tall, bare trunk and irregular, open crown.
Height: 150–250' (46–76 m).
Diameter: 20' (6 m), sometimes larger.
Leaves: evergreen; crowded and *overlapping;* ⅛–¼" (3–6 mm) long, to ½" (12 mm) on leaders. *Scalelike;* ovate or lance-shaped, sharp-pointed; *blue-green* with 2 whitish lines.
Bark: *reddish-brown,* fibrous, *very thick,* deeply furrowed into scaly ridges.
Twigs: much-branched, slender, drooping; blue-green turning brown.
Cones: 1¾–2¾" (4.5–7 cm) long; *elliptical,* reddish-brown; *many flat,* short-pointed *cone-scales;* maturing in 2 seasons; hanging down at end of leafy twig and remaining attached; 3–9 seeds under cone-scale, light brown, 2-winged, falling gradually.

Habitat: Granitic and other rocky soils in scattered groves in moist mountain sites, usually canyons or slopes; in coniferous forests.

Range: Western slope of Sierra Nevada, central California; at 4500–7500' (1372–2286 m); rarely at 3000–8900' (914–2713 m).

This rare species ranks among the world's oldest trees; felled trees show annual rings indicating up to 3200 years of age. Almost all Giant Sequoias are protected in Yosemite, Kings Canyon, and Sequoia national parks, in 4 national forests, and in state parks and forests. It is a popular, large ornamental tree in moist, cool temperate climates along the Pacific Coast and around the world. The lumber is no longer used, although many trees were cut and wasted in the early logging days. Seedlings and saplings are killed by forest fires, but the very thick bark of mature trees offers resistance. Douglas squirrels cut and store quantities of mature cones, and sparrows, finches, and chipmunks destroy many seedlings.

CYPRESS FAMILY
(Cupressaceae)

Resinous, evergreen trees and shrubs without flowers or fruits, mostly with straight axis and narrow crown, usually with soft, lightweight wood. The family includes cypresses (*Cupressus*), Junipers (*Juniperus*), and some "cedars." About 130 species worldwide; 26 native and 1 naturalized tree species and 1 native shrub species in North America.

Leaves: opposite or whorled, usually of 2 forms, mostly small and scalelike or awl-shaped and producing flattened or angled twigs.

Cones: pollen and seeds borne mostly on same plant. Male cones small and herbaceous; female cones woody (berrylike in juniper, *Juniperus*), usually composed of few cone-scales opposite or whorled and flattened or attached at mid-point; 1–2 naked seeds at a cone-scale, often with 2 lateral wings.

55, 448 **Port-Orford-cedar**
"Oregon-cedar" "Lawson Cypress"
Chamaecyparis lawsoniana (A. Murr.) Parl.

Description:  Large evergreen tree with enlarged base, narrow, pointed, spirelike crown, and horizontal or drooping branches.
Height: 70–200′ (21–61 m).
Diameter: 2½–4′ (0.8–1.2 m).
Leaves: *opposite* in 4 rows; ¹⁄₁₆″ (1.5 mm) long. *Scalelike;* dull green above, whitish beneath, with gland-dot.
Bark: reddish-brown, very thick, deeply furrowed into long fibrous ridges.
Twigs: *very slender, flattened,* regularly branched and spreading horizontally in *fernlike spray.*
Cones: ⅜″ (10 mm) in diameter; many in clusters, reddish-brown, often with a bloom; with *8 or 10 blunt cone-scales;*

maturing in 1 season; 2–4 seeds under a cone-scale.

Habitat: Sandy and clay loams, also rocky ridges with other conifers, sometimes in pure stands.

Range: SW. Oregon and NW. California in narrow coastal belt; also local in Mount Shasta area; to 5000' (1524 m).

Port-Orford-cedar is adapted to the humid climate of the Pacific Coast with its wet winters and frequent summer fog. Logs of the aromatic wood are exported to Japan for woodenware and toys and for construction of shrines and temples; a special use is for arrow shafts. Many horticultural varieties are grown as ornamentals and shade trees, especially in European countries with moist climates. Varieties include columnar, drooping, and dwarf forms and others with foliage of varying shades, ranging from silvery or steel-blue to bright green, and yellowish. The names honor Port Orford, Oregon, located in the center of the range, and Peter Lawson and his sons, Scottish nurserymen who introduced this species into cultivation in 1854.

57, 451 **Alaska-cedar**
"Alaska Yellow-cedar"
"Nootka Cypress"
Chamaecyparis nootkatensis (D. Don) Spach

Description: Tree with narrow crown and horizontal or slightly drooping branches.
Height: 50–100' (15–30 m).
Diameter: 1–4' (0.3–1.2 m).
Leaves: evergreen; *opposite* in 4 rows; ⅛' (3 mm) long. *Scalelike,* pointed and *spreading.* Bright *yellow-green,* generally without gland-dot.
Bark: gray-brown, thin, with long narrow fissures, *fibrous and shreddy.*
Twigs: slightly stout, *flattened or 4-angled,* regularly branched and

spreading horizontally, becoming
reddish-brown.
Cones: ½″ (12 mm) in diameter;
rounded, reddish-brown, with *4 or 6
rounded cone-scales* ending in long point;
maturing in 2 seasons; 2–4 seeds under
a cone-scale.

Habitat: Wet mountain soils; mainly in mixed
conifer forests, sometimes in pure
stands.

Range: Pacific Coast region from S. and SE.
Alaska southeast to mountains of W.
Oregon and extreme NW. California;
local farther inland; at 2000–7000′
(610–2134 m); to sea level farther
north.

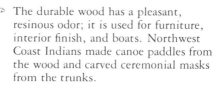

The durable wood has a pleasant,
resinous odor; it is used for furniture,
interior finish, and boats. Northwest
Coast Indians made canoe paddles from
the wood and carved ceremonial masks
from the trunks.

56 Sawara False-cypress
"Sawara Cypress" "Retinospora"
Chamaecyparis pisifera (Sieb. & Zucc.) Endl.

Description: Large introduced, cone-bearing
evergreen tree with straight trunk and
open, narrow, pyramid-shaped crown.
Height: 70′ (21 m).
Diameter: 2′ (0.6 m).
Leaves: *evergreen;* nearly ⅛″ (3 mm)
long. Mostly *scalelike, sharp-pointed,*
overlapping, *opposite in 4 rows;* top pair
flattened, side pair keeled; shiny green
above, whitish-green beneath. Leaves
on young plants ¼″ (6 mm) long;
needlelike. spreading in 4 rows; blue-
green, whitish beneath.
Bark: reddish-brown, fibrous and
shreddy, peeling in long thin strips.
Twigs: slender, spreading in *horizontal,
flattened, fernlike sprays.*
Cones: ¼″ (6 mm) in diameter; *rounded,*
whitish-green, turning dark brown,

short-stalked; usually composed of 10
cone-scales with tiny point in center;
1–2 broad-winged seeds under cone-
scale.

Habitat: Moist soils in humid temperate
regions.

Range: Native of Japan. Planted across the
United States.

Sawara False-cypress is commonly
grown as an ornamental. Cultivated
varieties include those characterized by
golden-yellow foliage, very small
leaves, weeping threadlike twigs, dwarf
spreading shrub habit, and very
feathery or mosslike, spreading,
needlelike leaves. In Japan this is a
timber tree as well as an ornamental.

67, 447   Arizona Cypress
Cupressus arizonica Greene

Description: Evergreen tree with conical crown and
stout, horizontal branches.

Height: 40–70′ (12–21 m).
Diameter: 1–2′ (0.3–0.6 m).
Leaves: opposite in 4 rows; ¹⁄₁₆″ (1.5
mm) long. Scalelike, keeled; *dull gray-
green,* mostly with gland-dot that
exudes whitish resin.
Bark: varying from gray or dark brown,
rough, and furrowed, to reddish-brown,
smooth, thin, and peeling.
Twigs: 4-angled, slightly stout,
branching almost at right angles.
Cones: ¾–1¼″ (2–3 cm) in diameter;
dark reddish-brown with a bloom,
becoming gray; with 6 or 8 rounded,
short-pointed, hard scales, sometimes
slightly warty; usually remaining closed
and attached. Many brown seeds ⅛–
³⁄₁₆″ (3–5 mm) long.

Habitat: Coniferous woodlands on coarse, rocky
soils; in pure stands or with pinyons
and junipers.

Range: Local and rare from Trans-Pecos Texas
west to SW. New Mexico, Arizona,

and S. California; also N. Mexico; at 3000–5000' (914–1524 m).

Arizona Cypress is often grown for Christmas trees. The durable wood is used locally for fenceposts. As many as 5 varieties, based on minor differences of foliage and bark, have been distinguished.

68, 442 Baker Cypress
"Siskiyou Cypress" "Modoc Cypress"
Cupressus bakeri Jeps.

Description: Medium-sized to large evergreen tree with narrow, conical, open crown.
Height: 30–100' (9–30 m).
Diameter: 2' (0.6 m).
Leaves: opposite in 4 rows; more than 1/16" (1.5 mm) long. Scalelike; *dull gray-green,* with gland-dot that exudes whitish resin.
Bark: reddish-brown, smoothish, *often peeling* in thin curling plates.
Twigs: 4-angled, slender, crowded, and irregularly arranged.
Cones: 3/8–3/4" (10–19 mm) in diameter; nearly round, gray or dull brown, with 6 *or 8 rounded, short-pointed,* hard, *warty scales.* Light tan seeds 1/8–3/16" (3–5 mm) long.
Habitat: Dry Ponderosa Pine forests, mostly on volcanic soil or serpentine.
Range: SW. Oregon and N. California, especially Siskiyou Mountains; at 3800–6000' (1158–1829 m).

The northernmost and hardiest New World cypress, this rare tree is the only one of its genus native to Oregon. Its names honor Milo Samuel Baker (1868–1961), the California botanist who discovered this species in 1898.

65, 454 Gowen Cypress
Cupressus goveniana Gord.

Description: Typically a small evergreen tree with
conical or spreading crown; 3 rare
geographic varieties.
Height: 15–30′ (4.6–9 m).
Diameter: ½–1½′ (0.15–0.5 m).
Leaves: opposite in 4 rows; less than
1⁄16″ (1.5 mm) long. Scalelike; mostly
bright green and without gland-dot.
Bark: brown to gray, smooth or
becoming rough and shreddy, fibrous.
Twigs: slender, 4-angled.
Cones: *small*, usually less than ¾″
(2 cm) in diameter; nearly round,
gray-brown to dull gray, with *6–10
rounded*, hard *cone-scales;* remaining
closed and attached. Many dull brown
to shiny black seeds ⅛″ (3 mm) long.

Habitat: Coastal Redwood forests.

Range: NW. and central California; usually
near sea level.

This species was named after James
Robert Gowen, a 19th century British
horticulturist. The typical variety is
confined to 2 groves within Point Lobos
Reserve and Del Monte Forest near
Monterey and Carmel. Mendocino
Cypress (var. *pigmaea* Lemm.) varies from
a low shrub of 3′ (0.9 m) on sandy soil
(in Mendocino White Plains) to a large
tree 100′ (30 m) high (in Mendocino
and Sonoma counties). Santa Cruz
Cypress (var. *abramsiana* (C.B. Wolf)
Little) with large cones to 1¼″ (3 cm)
long is local in Ponderosa Pine forest to
2500′ (762 m) and in the Santa Cruz
Mountains of the Coast Ranges (Santa
Cruz and San Mateo counties).

69, 446 Tecate Cypress
Cupressus guadalupensis Wats.

Description: Small evergreen tree with forked trunk
and irregular, spreading crown.

Height: 20–30' (6–9 m).
Diameter: 1–1½' (0.3–0.5 m).
Leaves: opposite in 4 rows, forming 4-angled, slender twigs; 1/16" (1.5 mm) long. Scalelike; *bright green*, mostly without gland-dot.
Bark: reddish-brown, smooth, often *peeling* in thin, curling plates.
Cones: 1–1¼" (2.5–3 cm) in diameter; dull brown or gray, with *6–10 rounded*, hard *cone-scales*, usually ending in point; remaining closed and attached; many dark brown seeds.

Habitat: Chaparral zone of coastal mountains, canyons, and rocky slopes.

Range: SW. California (Orange and San Diego counties) and nearby Baja California; at 2000–4000' (610–1219 m).

The common name refers to Mt. Tecate near the Mexican border, where this cypress was first found in California. The scientific name alludes to Guadalupe Island off the west coast of Baja California, where a related variety is native.

66, 443 MacNab Cypress
Cupressus macnabiana A. Murr.

Description: Small evergreen tree or shrub with stout trunk, several forks, and spreading crown broader than high.
Height: 30' (9 m).
Diameter: 1' (0.3 m).
Leaves: opposite in 4 rows; 1/16" (1.5 mm) long. Scalelike; dull *gray-green*, with *gland-dot* that exudes whitish resin; very *fragrant*.
Bark: gray, rough, furrowed, fibrous.
Twigs: slender, spreading in 1 plane in *flat sprays*.
Cones: ¾–1" (2–2.5 cm) in diameter; angular, brown or gray, with *6 or 8 irregular scales* with prominent *raised point*; remaining closed and attached; many irregular brown seeds.

Habitat: Dry, rocky soils in foothills and lower
mountain zones in chaparral and with
Digger Pine.

Range: Mountains of N. California, including
Coast Ranges and Sierra Nevada
foothills; to 2600' (792 m).

MacNab Cypress is perhaps the most
widely distributed cypress in California.
The wood makes durable fenceposts.
This species commemorates James
MacNab (1810–78), the Scottish
horticulturist and curator of the
Edinburgh Botanic Garden, who made
large horticultural collections in the
United States and Canada in 1834.

63, 444 Monterey Cypress
Cupressus macrocarpa Hartw.

Description: Medium-sized evergreen tree with *large
cones* and with symmetrical crown when
young or where protected; often
becoming irregular and flat-topped
when exposed to high winds.
Height: 60–80' (18–24 m).
Diameter: 2' (0.6 m); rarely 4' (1.2 m).
Leaves: opposite in 4 rows; more than
1/16" (1.5 mm) long. Blunt, scalelike;
bright green, usually without gland-dot.
Bark: gray, rough, fibrous.
Twigs: 4-angled, *stout,* irregularly
arranged.
Cones: 1–1⅜" (2.5–3.5 cm) long;
longer than broad, brown, with *8–12*
rounded, *stout-pointed,* hard *scales;*
remaining closed and attached. Many
irregular seeds ¼–5/16" (6–8 mm) long,
shiny dark brown.

Habitat: Exposed granitic headlands and
sheltered areas on seacoast, subject to
salt spray and winds.

Range: Confined to 2 groves near Monterey and
Carmel in W. central California.

The gnarled, picturesque Monterey
Cypresses growing on sea cliffs are a

favorite photographic subject. The 2 native groves are protected within Point Lobos Reserve and Del Monte Forest at Point Cypress. It is widely planted as an ornamental, hedge, and windbreak along the California coastline and grown in forest plantations for timber in South Africa, New Zealand, and Australia.

64, 445 Sargent Cypress
Cupressus sargentii Jeps.

Description: Evergreen tree with narrow crown, becoming very broad where exposed.
Height: 30–50' (9–15 m).
Diameter: 3' (0.9 m).
Leaves: opposite in 4 rows; more than $\frac{1}{16}$" (1.5 mm) long. Scalelike; *dull green,* often with gland-dot.
Bark: gray, dark brown, or blackish; rough and furrowed, thick and fibrous.
Twigs: 4-angled, stout, stiff, branching in all directions.
Cones: mostly ¾–1" (2–2.5 cm) long; variable in shape, *round or longer than broad,* dull brown or gray; usually 6 or 8 rounded, often pointed, rough, hard cone-scales; remaining closed and attached; many angular, dark brown seeds.

Habitat: Exposed and protected slopes of foothills and mountains; with Digger Pine, oaks, and chaparral.

Range: Coast Ranges of California; to 3000' (914 m).

The second most widely distributed cypress in California, Sargent Cypress was named for Charles Sprague Sargent (1841–1927), the founder and director of Harvard University's Arnold Arboretum and author of the 14-volume *Silva of North America.* Several stands grow within Mount Tamalpais State Park north of San Francisco.

72, 456 California Juniper
Juniperus californica Carr.

Description: Evergreen shrub or small tree with broad, irregular crown.
Height: 40' (12 m).
Diameter: 1–2' (0.3–0.6 m).
Leaves: *usually in 3's;* ¹⁄₁₆–¹⁄₈" (1.5–3 mm) long. *Scalelike,* blunt, forming stout, stiff, rounded twigs; *yellow-green,* with gland-dot.
Bark: gray, fibrous, furrowed, shreddy.
Cones: ½–¾" (12–19 mm) long; berrylike, *longer than broad, bluish with a bloom,* becoming brown, hard, and dry; *mealy* and sweetish; 1–2 seeds.

Habitat: Dry slopes and flats of foothills and lower mountain zones; with pinyons in woodland and with Joshua-trees in semidesert zone.

Range: Mountains of California, extreme S. Nevada, and W. Arizona; also N. Baja California; at 1000–5000' (305–1524 m).

Able to withstand heat and drought, this species extends farther down into the semidesert zone than other junipers and is important in erosion control on dry slopes. Indians used to gather the "berries" to eat fresh and to grind into meal for baking.

74, 457 Alligator Juniper
Juniperus deppeana Steud.

Description: Evergreen tree with short, stout trunk and rounded, spreading crown, becoming irregular and with branches partly dead in vertical strips.
Height: 20–50' (6–15 m).
Diameter: 2–4' (0.6–1.2 m).
Leaves: *opposite; in 4 rows,* forming slender, 4-angled twigs; ¹⁄₁₆–¹⁄₈" (1.5–3 mm) long. Scalelike, *sharp-pointed; blue-green,* with gland-dot and often *whitish resin drop.*

Bark: *blackish* or gray; thick and rough, deeply furrowed into *checkered plates,* suggesting an alligator's back.
Cones: ½" (12 mm) in diameter; berrylike, brownish with *whitish* bloom, hard and dry, *mealy;* 3–5 seeds; maturing second year.

Habitat: Rocky hillsides and mountains; with pinyons, other junipers, oaks, and Ponderosa Pine.

Range: Trans-Pecos Texas northwest to N. Arizona; also Mexico; at 4500–8000' (1372–2438 m).

Alligator Juniper is easily recognized by its distinctive bark. One of the largest junipers, it is used for fuel and fenceposts. New sprouts often appear at the base of cut stumps. The large "berries" are consumed by birds and mammals. Large trees often have a partially dead crown of grotesque appearance with some branches that die and turn light gray instead of falling; other branches die only in a vertical strip and continue to grow on the other side.

76, 482 Redberry Juniper
Juniperus erythrocarpa Cory

Description:  Evergreen shrub or small tree with open, irregular crown of spreading branches and bright red "berries."
Height: 16' (5 m).
Diameter: 8" (20 cm).
Leaves: *opposite;* in *4 rows,* forming 4-angled twigs; ¹⁄₁₆" (1.5 mm) long; scalelike; *gray-green.*
Bark: reddish-brown or gray; furrowed and shreddy.
Cones: ⁵⁄₁₆" (8 mm) in diameter; berrylike, bright *red, juicy,* 1-seeded.

Habitat: Dry foothills and mountains; with Mexican Pinyon and other junipers.

Range: Trans-Pecos Texas west to S. Arizona

and south to N. Mexico; at 3000–
5000′ (914–1524 m).

Though named in 1936, this juniper
was not accepted as a species until
1975; previously it was not separated
from Pinchot Juniper, which has cones
of similar reddish color. The scientific
name is from the Greek words "red"
and "fruit."

71 **Drooping Juniper**
"Weeping Juniper"
"Mexican Drooping Juniper"
Juniperus flaccida Schlecht.

Description: Small evergreen tree or shrub with
rounded crown of weeping branches
unlike any other juniper.
Height: 30′ (9 m).
Diameter: 1–2′ (0.3–0.6 m).
Leaves: *opposite;* in *4 rows* on long,
slender, *drooping twigs;* ¹⁄₁₆–⅛″ (1.5–
3 mm) long. *Scalelike,* long-pointed,
spreading at tip; *yellow-green,* often with
gland-dot.
Bark: reddish-brown, furrowed into
long, narrow, fibrous, shreddy ridges.
Cones: ⅜–½″ (10–12 mm) in
diameter; berrylike, *reddish-brown* with
a bloom, slightly rough, hard, and *dry,*
resinous; 4–12 seeds. Male or pollen
cones on separate trees.

Habitat: Rocky slopes and mountain canyons;
with other junipers, pinyons, and oaks.
Range: Chisos Mountains of Trans-Pecos Texas;
also Mexico; at 4500–7000′ (1372–
2134 m).

In the United States this species is
confined to Big Bend National Park. It
is the rarest native juniper, although
widespread and common in Mexico.
The scientific name, meaning "flaccid"
or "relaxed," describes the odd
drooping habit.

75, 461 Oneseed Juniper
Juniperus monosperma (Engelm.) Sarg.

Description: Evergreen shrub or small tree with several branches curving up from ground, sometimes with short trunk, and much-branched, spreading, and often scraggly crown.
Height: 10–25′ (3–7.6 m).
Diameter: 1′ (0.3 m).
Leaves: *opposite; in 4 rows* on short, stout, crowded twigs; ¹⁄₁₆″ (1.5 mm) long. *Scalelike; yellow-green,* usually with gland-dot.
Bark: gray, fibrous, shreddy.
Cones: ¼″ (6 mm) in diameter; berrylike, *dark blue* with a bloom, soft, *juicy,* sweetish and resinous, 1-seeded. Male or pollen cones on separate trees.

Habitat: Dry plains, plateaus, hills, and mountains, mostly on rocky soils; often in pure, orchardlike stands.

Range: Central Colorado south to NW. and Trans-Pecos Texas and west to central Arizona; also N. Mexico; at 3000–7000′ (914–2134 m).

This abundant juniper is one of the most common small trees in New Mexico. The wood is important for fenceposts and fuel, and Indians used to make mats and cloth from the fibrous bark. Birds and mammals consume the juicy "berries," and goats browse the foliage.

73, 459 Western Juniper
"Sierra Juniper"
Juniperus occidentalis Hook.

Description: Evergreen tree with short trunk and broad crown of stout, spreading branches, becoming ragged and gnarled with age; or a shrub.
Height: 15–30′ (4.6–9 m).
Diameter: 1′ (0.3 m), sometimes much larger.

Leaves: mostly *in 3's;* ¹⁄₁₆″ (1.5 mm) long. *Scalelike,* forming stout, rounded twigs; *gray-green,* with gland-dot. Bark: reddish-brown, furrowed, shreddy. Cones: ¼–³⁄₈″ (6–10 mm) in diameter; *short, elliptical,* berrylike, *blue-black* with a bloom, soft, *juicy,* resinous; maturing second year; 2–3 seeds.

Habitat: Mountain slopes and plateaus, mostly on shallow, rocky soils.

Range: Central and SE. Washington south to S. California; to 10,000′ (3048 m).

Western Juniper is common at high altitudes in the Sierra Nevada. Giants reach a trunk diameter of 16′ (5 m) and an estimated age of more than 2000 years. This species may develop thick, long roots that entwine rock outcrops, mimicking the shape of the branches.

78, 458 Utah Juniper
Juniperus osteosperma (Torr.) Little

Description: Tree with short, upright trunk, low, spreading branches, and rounded or conical, open crown.
Height: 15–40′ (4.6–12 m).
Diameter: 1–3′ (0.3–0.9 m).
Leaves: *generally opposite in 4 rows,* forming stout, stiff twigs; ¹⁄₁₆″ (1.5 mm) long. Scalelike, short-pointed; *yellow-green,* usually without gland-dot.
Bark: gray, fibrous, furrowed, shreddy.
Cones: ¼–⅝″ (6–15 mm) in diameter; berrylike, *bluish with a bloom,* becoming brown, hard and dry; *mealy* and sweetish; 1–2 seeds.

Habitat: Dry plains, plateaus, hills, and mountains, mostly on rocky soils; often in pure stands or with pinyons.

Range: Nevada east to Wyoming, south to W. New Mexico, and west to S. California; local in S. Montana; at 3000–8000′ (914–2438 m).

The most common juniper in Arizona, it is conspicuous at the south rim of the Grand Canyon and on higher canyon walls. Utah Juniper grows slowly, becoming craggier and more contorted with age. American Indians used the bark for cordage, sandals, woven bags, thatching, and matting. They also ate the "berries" fresh or in cakes. Birds and small mammals also consume quantities of juniper "berries." Junipers are also called cedars, and Cedar Breaks National Monument and nearby Cedar City in southwestern Utah are named for this tree. Scattered tufts of yellowish twigs with whitish berries found on the trees are a parasitic mistletoe, which is characteristic of this tree.

77 Pinchot Juniper
"Redberry Juniper"
Juniperus pinchotii Sudw.

Description: Evergreen shrub or small tree, usually with several branches from ground, sometimes with single trunk and broad, irregular crown.
Height: 20' (6 m).
Diameter: 1' (0.3 m).
Leaves: *evergreen; mostly in 3's* on slender twigs; ¹⁄₁₆" (1.5 mm) long. *Scalelike; yellow-green,* with gland-dot.
Bark: light brown or gray; thin, furrowed into scaly ridges.
Cones: ⅜" (10 mm) in diameter; berrylike, *reddish,* hard and *dry,* mealy; 1–2 long, light brown, pointed, prominently angled seeds. Pollen cones on separate trees.

Habitat: Plains, including hills and canyons on gravelly and rocky soils on limestone and gypsum; in open juniper woodlands.
Range: SW. Oklahoma, Texas, and SE. New Mexico; also NE. Mexico; at 1000–5000' (305–1524 m).

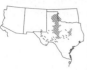

This juniper was named in honor of
Gifford Pinchot (1865–1946), first
chief of the United States Forest
Service. It is well displayed at Palo
Duro State Park near Canyon, Texas,
where it was discovered. The hardy
plants will sprout from stumps after
cutting or burning. The wood is used
principally for fenceposts and fuel.

70, 455 Rocky Mountain Juniper
"Rocky Mountain Redcedar"
"River Juniper"
Juniperus scopulorum Sarg.

Description:  Evergreen tree with straight trunk,
narrow, pointed crown, and slender
branches of aromatic, gray-green foliage
often drooping at ends.
Height: 20–50' (6–15 m).
Diameter: 1½' (0.5 m).
Leaves: *opposite in 4 rows,* forming
slender, 4-angled twigs; $\frac{1}{16}$" (1.5 mm)
long. *Scalelike,* pointed, *gray-green.*
Bark: reddish-brown, thin, fibrous,
shreddy.
Cones: ¼" (6 mm) in diameter;
berrylike, bright *blue* with *whitish* coat,
soft, juicy, sweetish, resinous; usually
2-seeded; maturing second year. Male
or pollen cones on separate trees.

Habitat: Rocky soils, especially on limestone and
lava outcrops; in open woodlands at
lower border of trees to the north; in
foothills with pinyons to the south.

Range: Generally in mountains, from central
British Columbia, east to W. North
Dakota, and south to Trans-Pecos
Texas; also N. Mexico; at 5000–9000'
(1524–2743 m); almost to sea level in
north.

A graceful ornamental, often with
narrow crown of drooping foliage,
several varieties differ in form and in
leaf color. The aromatic wood is
especially suited for cedar chests and is

also used for lumber, fenceposts, and fuel. Wildlife eat the "berries." This species is closely related to Eastern Redcedar (*Juniperus virginiana* L.), which has dark green foliage and "berries" that mature in one year.

58, 453 Incense-cedar
Libocedrus decurrens Torr.

Description: Large, resinous, aromatic tree with tapering, irregularly *angled trunk* and narrow, columnar crown, becoming open and irregular.
Height: 60–150' (18–46 m).
Diameter: 3–5' (0.9–1.5 m).
Leaves: evergreen; *opposite* in 4 rows; ⅛–½" (3–12 mm) long. *Scalelike;* side pair keeled, long-pointed, overlapping next pair, extending down twig; very aromatic when crushed; *shiny green.*
Bark: light or *reddish-brown;* thick, *deeply and irregularly furrowed* into *shreddy ridges.*
Twigs: much-branched and *flattish;* with *wedge-shaped joints* longer than broad; composed of scalelike leaves.
Cones: ¾–1" (2–2.5 cm) long; *oblong; hanging down* at end of slender, leafy stalk; reddish-brown; composed of 6 *paired,* hard, *flattened, pointed cone-scales.* Seeds 4 or fewer in cone, paired, with 2 unequal wings.
Habitat: Mountain soils; in mixed coniferous forests, seldom in pure stands.
Range: W. Oregon south to S. California and extreme W. Nevada; also N. Baja California; at 1200–7000' (366–2134 m).

An important timber species, Incense-cedar is also the leading wood for the manufacture of pencils, because it is soft but not splintery, and can be sharpened in any direction with ease. The aromatic wood is used for cedar chests and closets. Although stands of young trees are killed by fire, the very thick bark protects mature trees.

59 Oriental Arborvitae
"Chinese Arborvitae"
Thuja orientalis L.

Description: Ornamental, cone-bearing, resinous and aromatic evergreen shrub or small tree branching near the base with few trunks and compact, narrow, conical to rounded or irregular crown of many upright branches.
Height: 25′ (7.6 m).
Diameter: 6″ (15 cm).
Leaves: *evergreen;* opposite in 4 rows; ¹⁄₁₆–⅛″ (1.5–3 mm) long. *Scalelike,* short-pointed; side pair keeled, flat pair with gland-dot; *yellow-green.*
Bark: dark reddish-brown, thin, fibrous, finely fissured and shreddy.
Twigs: branching in *vertical plane, flattish, jointed* in fernlike sprays.
Cones: ⅝″ (15 mm) long; *egg-shaped,* whitish or bluish but becoming dark brown, *upright* at end of short twigs; opening widely at maturity and remaining attached; with usually 6 *paired cone-scales,* thick and ending in curved or *hooked point;* 2 tiny, oblong, wingless seeds at base of lower cone-scales.

Habitat: Moist soils in humid, warm temperate and subtropical regions.

Range: Native of China. Planted across the United States, especially in Southeast; sometimes escaped and naturalized in Florida.

Many horticultural varieties grown as shrubs around houses and in parks and gardens have different shapes; some have golden foliage. This species is often trimmed into hedges. Chinese use the fragrant evergreen branches for good luck at New Year celebrations.

60, 452 Western Redcedar
"Giant Arborvitae" "Canoe-cedar"
Thuja plicata Donn ex D. Don

Description: Large to very large tree with tapering trunk, buttressed at base, and with a narrow, conical crown of short, spreading branches *drooping at ends;* foliage is *resinous and aromatic.*

Height: 100–175' (30–53 m) or more.

Diameter: 2–8' (0.6–2.4 m) or more.

Leaves: evergreen; *opposite* in 4 rows; ¹⁄₁₆–¹⁄₈" (1.5–3 mm) long. *Scalelike,* short-pointed; side pair keeled, flat pair usually without gland-dot; *shiny dark green,* usually with whitish marks beneath.

Bark: *reddish-brown,* thin, *fibrous, and shreddy.*

Twigs: much branched in horizontal plane, slightly *flattened in fanlike sprays, jointed.*

Cones: ½" (12 mm) long; clustered and *upright* from short, curved stalk; *elliptical,* brown; with *10–12* paired, thin, leathery, *sharp-pointed cone-scales;* 6 usually bearing 2–3 seeds with 2 wings.

Habitat: Moist, slightly acid soils; forming widespread forests with Western Hemlock, also with other conifers.

Range: SE. Alaska southeast along coast to NW. California; also SE. British Columbia south in Rocky Mountains to W. Montana; to 3000' (914 m) in north; to 7000' (2134 m) in south.

Particularly resistant to rot, Western Redcedar is the chief wood for shingles and one of the most important for siding, utility poles, fenceposts, paneling, outdoor-patio construction, and boatbuilding. Indians of the Northwest Coast carved their famous totem poles and split lumber for their lodges from this durable softwood. The name "Canoe-cedar" refers to the special war canoes hollowed out of giant trunks. Indians also used the wood for boxes, batons, and helmets and the fibrous inner bark for rope, roof thatching, blankets, and cloaks. The largest Western Redcedar measures 21' (6.4 m) in diameter, ranking second only to the Giant Sequoia among native trees; however, this species is not among the tallest.

PALM FAMILY
(Palmae)

Evergreen trees and shrubs and sometimes vines. Stout or sometimes slender, unbranched trunk, not divided into bark and wood and not increasing in diameter. About 2500 species worldwide in tropical and subtropical regions; 11 native and 1 naturalized tree species and 2 native shrub species in North America; many others southward.

Leaves: large, spreading, alternate and crowded at tip of trunk; 2 types: pinnately compound, with many narrow leaflets with fine parallel veins along the axis; and fanlike or palmate-veined; thick and leathery.

Flowers: small, stalkless or short-stalked, generally whitish, commonly male and female on the same plant or bisexual, regular, with calyx of 3 sepals or lobes and corolla of 3 petals or lobes, mostly 6 stamens, and 1 pistil. In large, branched clusters developing from a large bract among leaf bases or below.

Fruit: usually a 1-seeded berry or drupe.

---

315 **Canary Island Date**
"Canary Island Date-palm"
Phoenix canariensis Hort. ex Chabaud.

Description: Ornamental palm with *massive, unbranched trunk* covered by *bases of old leaves* and with very large, *pinnately compound leaves*.
Height: 50′ (15 m).
Diameter: 2–3′ (0.6–0.9 m).
Leaves: *evergreen;* numerous, upright and spreading around top; 12–20′ (3.7–6.1 m) long. Leafstalks stout, bearing *long, green spines* at edges. Many leaflets 1½′ (0.5 m) long, 1″ (2.5 cm) wide; *very narrow,* long-pointed, *folded* into short stalk at base, with *edges turned*

upward; leathery, straight, and not drooping. *Dull green;* dead leaves light brown, hanging down.

Trunk: stout, often wider near crown, with masses of air roots forming near enlarged base; rough, light brown leaf bases often bearing air plants and ferns.

Flowers: male and female on separate trees; *small, whitish;* numerous, along much-branched orange flowerstalks 3–6′ (0.9–1.8 m) long; among leaves.

Fruit: ¾″ (2 cm) long; *egg-shaped, yellow,* edible *dates;* 1-seeded; in heavy clusters.

Habitat: Subtropical regions.

Range: Native of Canary Islands. Introduced along southern border of the United States in California, Arizona, and the Gulf states east to Florida.

This popular street tree is hardy in warm regions and seldom injured by cold. It is sometimes called "Pineapple-palm" from the resemblance of its trunk to that fruit. The fruit is less palatable than that of the Date (*Phoenix dactylifera* L.), which is cultivated as a fruit tree in southeastern California and other hot, dry regions.

313 California Washingtonia
"California-palm" "Fanpalm"
Washingtonia filifera (Linden) H. Wendl.

Description: Tall palm with massive, *unbranched trunk* and very large, *fan-shaped leaves.*

Height: 20–60′ (6–18 m).

Diameter: 2–3′ (0.6–0.9 m).

Leaves: *evergreen;* numerous, spreading around top; if not burned or cut, old dead leaves hang down against trunk in thick thatch. Leafstalks 3–5′ (0.9–1.5 m) long; stout, with *hooked spines* along edges. Leaf blades 3–5′ (0.9–1.5 m) in diameter; *gray-green,* split into *many narrow,* folded, leathery *segments,* with *edges frayed* into many threadlike fibers.

Trunk: gray, smooth, with horizontal lines and vertical fissures.

Flowers: ⅜″ (10 mm) long; with funnel-shaped, deeply *3-lobed white corolla:* short-stalked, slightly fragrant; many together in much-branched clusters 6–12″ (1.8–3.7 m) long; drooping from leaf bases.

Fruit: ⅜″ (10mm) in diameter; *elliptical black berry,* with thin, sweetish, edible pulp, 1 elliptical brown seed.

Habitat: Moist soils along alkaline streams and in canyons of mountains in Colorado and Mojave deserts.

Range: SE. California (San Bernardino County to San Diego County), SW. Arizona (Kofa Mountains, Yuma County; also S. Yavapai County where perhaps introduced) and N. Baja California; at 500–3000′ (152–914 m).

The largest native palm of the continental United States as well as the only western species, it is also known as "Desert-palm." Another name is "Petticoat-palm" from the shaggy mass of dead leaves hanging against the trunk. Groves are in Palm Canyon near Palm Springs and in Joshua Tree National Monument. It is cultivated widely as an ornamental along streets and avenues in southern California, southern Arizona, the Gulf States east to Florida, and in subtropical regions around the world. Indians ate the berries, both fresh and dry, and ground the seeds into meal. This genus honors the first president of the United States.

314 Mexican Washingtonia
"Mexican Fanpalm"
Washingtonia robusta H. Wendl.

Description: Tall, introduced, ornamental palm with
slender, unbranched trunk and very large,
fan-shaped leaves.
Height: 50–70' (15–21 m).
Diameter: 1–1½' (0.3–0.5 m).
Leaves: *evergreen;* numerous, spreading
around top; old, dead, light brown
leaves hang down against trunk in thick
thatch, if not burned or cut. Leafstalks
2½–4' (0.8–1.2 m) long; reddish-
brown, stout, with *hooked spines* along
edges. Leaf blades 3–5' (0.9–1.5 m) in
diameter; *bright green,* split into *many
narrow,* folded, leathery *segments;* edges
with threadlike fibers when young.
Trunk: light brown, nearly *smooth* or
becoming finely fissured, with
horizontal *rings;* enlarged at base.
Flowers: ⅜" (10 mm) long; with
funnel-shaped, deeply *3-lobed, white
corolla;* short-stalked, slightly fragrant;
numerous, in much-branched clusters
6–12' (1.8–3.7 m) long; spreading or
drooping from leaf bases.

Fruit: ⅜" (10 mm) in diameter;
elliptical black berry, with thin,
sweetish, edible pulp and 1 seed.
Habitat: Subtropical regions.
Range: Native or canyons in N. Baja California
and Sonora, Mexico. Introduced along
southern border of the United States in
California, S. Arizona, and the Gulf
states east to Florida.

The Latin species name, meaning
"robust," is more appropriate to the
other species of this genus, California
Washingtonia, which has a stout,
massive trunk. Mexican Washingtonia
is better adapted to planting near the
coast, but it is less cold-hardy than its
native relative.

LILY FAMILY
(Liliaceae)

Mostly perennial herbs, often with bulbs or tubers; sometimes shrubs or small trees with stout, unbranched trunk or with a few branches ending in crowded, narrowed, pointed leaves. A very large family, 4000–5000 species worldwide; 12 native tree species mostly in the yucca genus (*Yucca*), about 30 native shrub, 10 woody vine, and numerous herb species in North America.

Leaves: generally alternate, narrow, not toothed, parallel-veined.

Flowers: mostly clustered and showy, often large and white, bisexual, regular, with 3 sepals and 3 petals nearly equal and mostly separate, usually 6 stamens separate or attached to corolla, and 1 pistil.

Fruit: a capsule or berry with many seeds.

---

305 **Giant Dracaena**
"Green Dracaena" "Dracaena-palm"
Cordyline australis (G. Forst.) Hook. f.

Description: Introduced, evergreen, palmlike tree with stout, unbranched trunk or with irregular, flat-topped crown of a *few large branches,* bearing many crowded, swordlike leaves at the tip.
Height: 30' (9 m).
Diameter: 1' (0.3 m).
Leaves: *evergreen; crowded;* 1–3' (0.3–0.9 m) long, 1½–2½" (4–6 cm) wide. *Very narrow,* very long-pointed, *flat, leathery,* without teeth on edges; green on both surfaces; stalkless.
Trunk: gray; smooth or slightly fissured; branches stout and similar to trunk.
Flowers: ¼" (6 mm) wide; *cup-shaped,* with 6 *white calyx lobes;* fragrant; numerous, crowded in upright or

spreading, branched clusters 2′ (0.6 m)
long among leaves.
Fruit: ¼″ (6 mm) in diameter;
numerous whitish or *bluish-white
berries.*

Habitat: Moist soils in subtropical regions.

Range: Native of New Zealand. Introduced
along Pacific Coast in California; also
Hawaii and other subtropical regions.

Giant Dracaena is quite hardy,
enduring heat, drought, and ocean
winds, but not prolonged frost. It is
propagated by seeds, cuttings, and
roots and grown in tubs northward.
The fibrous leaves can be used for
string. Cultivated varieties have
purplish or striped leaves.

310 Joshua-tree
"Tree Yucca" "Yucca-palm"
Yucca brevifolia Engelm.

Description: A picturesque or grotesque, narrow-leaf
evergreen tree with short, stout trunk;
open, broad crown of many, stout,
widely forking, spreading, and
sometimes drooping branches; and
spiny, daggerlike leaves.
Height: 15–30′ (4.6–9 m).
Diameter: 1–3′ (0.3–0.9 m),
sometimes larger.
Leaves: *evergreen;* numerous, clustered
and spreading at ends of branches; 8–
14″ (20–36 cm) long, ¼–½″ (6–12
mm) wide. *Daggerlike,* stalkless, stiff,
flattish but keeled on outer surface,
smooth or slightly rough; ending in
short, sharp *spine. Blue-green,* the
yellowish edges with many *tiny, sharp
teeth.*
Trunk: small trunks and branches
covered with dead, stiff leaves pressed
downward; larger trunks brown or
gray, rough, corky, deeply furrowed
and cracked into plates.
Flowers: 1¼–1½″ (3–4 cm) long; *bell-*

329

shaped, with 6 *greenish-yellow, leathery* sepals; crowded in upright, much-branched clusters 1–1½' (0.3–0.5 m) long; with unpleasant odor of mushrooms; mostly in early spring, heavily in some years, at irregular intervals.

Fruit: 2½–4" (6–10 cm) long, 2" (5 cm) in diameter; *elliptical,* green to *brown,* 6-celled, slightly fleshy becoming *dry;* falling soon after maturity in late spring, but *not splitting open;* many flat seeds.

Habitat: Dry soils on plains, slopes, and mesas; often common in groves.

Range: Mohave Desert of extreme SW. Utah, Nevada, California, and Arizona; at 2000–6000' (610–1829 m).

Joshua-tree, the largest of the yuccas, is the characteristic tree of the Mohave Desert and has come to symbolize the area. The Mormon pioneers named this species Joshua, because its shape mimics a person praying with uplifted arms or gesturing wildly, referring to the Biblical leader pointing the way to a Promised Land. It is abundant at Joshua Tree National Monument in southern California and Joshua Forest Parkway in western Arizona. Indians made meal from the seeds and a dye for decorating baskets from the reddish rootlets. Red-shafted flickers drill holes in the branches to make nests, which are later occupied by other birds. The desert night-lizard lives in the dead leaves and branches, and woodrats gnaw off the spiny leaves for their nests. The foliage was the primary staple in the diet of the extinct giant sloth.

309   Soaptree Yucca
"Soapweed" "Palmilla"
Yucca elata Engelm.

Description: Evergreen, palmlike shrub or small tree
with single trunk or several clustered
trunks; unbranched or with few upright
branches; and very long, narrow leaves.
Height: 10–17' (3–5 m), rarely to 25'
(7.6 m).
Diameter: 6–10" (15–25 cm).
Leaves: *evergreen;* numerous, spreading,
grasslike; 1–2½' (0.3–0.8 m) long, ⅛–
⅜" (3–10 mm) wide. *Linear, flat,
leathery,* and flexible; ending in sharp
spine. Yellow-green, with fine whitish
threads along edges.
Trunk: gray and slightly furrowed in
lower part, covered by dead leaves
above.
Flowers: 1½–2" (4–5 cm) long; *bell-
shaped,* with 6 *white,* broad, pointed
sepals; crowded on upright branches, in
clusters 3–10' (0.9–3 m) or more in
height including *long stalk;* in spring.
Fruit: 1½–3" (4–7.5 cm) long; a
cylindrical capsule, light brown, dry, 3-
celled; maturing in early summer,
splitting open in 3 parts and remaining
attached; many small, flat, thin, rough,
dull black seeds.
Habitat: Dry, sandy plains, mesas, and washes;
in desert grassland and desert, often in
pure stands with grasses.
Range: Trans-Pecos Texas west to central New
Mexico and central Arizona and local in
SW. Utah; also N. Mexico; at 1500–
6000' (457–1829 m).

Soapy material in the roots and trunks
of this abundant species has been used
as a soap substitute. The leaves are a
source of coarse fiber and were used by
Indians in making baskets. Cattle relish
the young flower stalks, and chopped
trunks and leaves serve as emergency
food during droughts. Indians ate the
flower buds, flowers, and young flower
stalks of this and other yuccas, either

raw or boiled. Growth is extremely
slow, about 1″ (2.5 cm) in height a
year. The local name "Palmilla,"
Spanish for "small palm," refers to the
resemblance of this species to a palm.

307 Faxon Yucca
"Spanish-bayonet" "Spanish-dagger"
Yucca faxoniana Sarg.

Description: Small evergreen tree with 1 or
sometimes more trunks, enlarged at
base and either unbranched or with 1 or
2 upright branches and with long
bayonetlike leaves.
Height: 20′ (6 m).
Diameter: 16″ (0.4 m).
Leaves: *evergreen;* numerous, crowded
and spreading; 20–32″ (0.5–0.8 m)
long and 1¾–2¾″ (4.5–7 cm) wide
near middle. *Bayonetlike* or lance-
shaped, *flattish,* thick, and stiff, ending
in short, stout *spine. Dark green,* with
many gray or brown fibers or *threads
along edges.*
Trunk: dark reddish-brown, scaly;
upper part covered by dead leaves.
Flowers: 2½–3¼″ (6–8 cm) long; bell-
shaped, with short tube and 6 *narrow,
white calyx lobes;* crowded on long,
drooping stalks, in upright clusters
(partly within foliage) 3–4′ (0.9–
1.2 m) long; in spring.

Fruit: 1¼–3½″ (3–9 cm) long, 1–1¼″
(2.5–3 cm) wide; a *cylindrical berry*
ending in long, narrow, curved *point;*
orange turning *black,* with thick, bitter
flesh; maturing in early summer. Many
flat, rough, dull black seeds.
Habitat: Dry, rocky and gravelly mountain
slopes in semidesert; with other yuccas
or in nearly pure stands.
Range: Trans-Pecos Texas and NE. Mexico; at
4000–5000′ (1219–1524 m).

It is named for Charles Edward Faxon
(1846–1918), the illustrator of Charles

Sprague Sargent's 14-volume *Silva of North America*. Birds and mammals eat the fleshy fruit. The name *Yucca* is from *yuca*, the Carib Indian name of the root of cassava (*Manihot*), misapplied to this genus.

308, 381 Mohave Yucca
"Spanish-dagger"
Yucca schidigera Roezl ex Ortgies

Description: Evergreen shrub or small tree, usually with several clustered trunks, often with few upright branches, and with bayonetlike leaves.
Height: 16′ (5 m).
Diameter: 6–12″ (15–30 cm).
Leaves: *evergreen;* numerous, crowded and spreading at top; usually 18–24″ (0.5–0.6 m) long, sometimes to 4′ (1.2 m), and 1–1½″ (2.5–4 cm) wide. *Bayonetlike* or lance-shaped, thick and stiff, *grooved,* broadest at middle; ending in sharp *spine. Yellow-green,* with coarse fibers or *threads on edges.*
Trunk: gray-brown and furrowed at base.
Flowers: 1¼–2″ (3–5 cm) long; *bell-shaped,* with 6 *white* and purplish-tinged *sepals;* on drooping stalks, crowded in upright, branched clusters usually 1½–3′ (0.5–0.9 m) long; raised slightly above longest leaves, bearing flowers almost to the base; in early spring.

Fruit: 2–4″ (5–10 cm) long, 1–1¼″ (2.5–3 cm) wide; a *cylindrical berry,* often curved, blunt-pointed, dull *dark brown* or blackish, with thick, sweetish, edible flesh; maturing in late summer, falling before winter. Many small, flat, rough, dull black seeds.
Habitat: Creosotebush desert and chaparral on dry, gravelly mountain and valley slopes.
Range: Mohave Desert in NW. Arizona, S. Nevada, S. California, and N. Baja

California; at 1000–6000' (305–1829 m), rarely higher.

Indians ate the fleshy fruit, either fresh or roasted, and used the fibrous leaves for ropes and coarse blankets. A soap substitute can be obtained from the roots and trunks of this and other yuccas. Flowers in the *Yucca* genus depend upon the small, white pronuba moth (*Tegeticula*) for pollination. The female moth gathers pollen and works it into a tiny ball before pushing it against the stigma of another flower, where she deposits her eggs in the ovary. The larvae feed on the developing fruit capsule but leave some seeds to mature.

304 Schott Yucca
"Spanish-bayonet" "Spanish-dagger"
Yucca schottii Engelm.

Description: Evergreen shrub or small tree usually with 2 or 3 slightly leaning trunks, either unbranched or with 1 or 2 upright branches, and with bayonetlike leaves.
Height: 16' (5 m).
Diameter: 8–10" (20–25 cm).
Leaves: *evergreen;* numerous, spreading; usually 16–32" (0.4–0.8 m) long and 1¼–2" (3–5 cm) wide. *Bayonetlike* or lance-shaped, *flat,* leathery, and flexible; ending in sharp, reddish *spine.* Dark *blue-green,* edges reddish and *without teeth or fibers.*
Trunk: dark brown, rough and scaly at base; mostly covered by dead leaves above.

Flowers: 1¼–1¾" (3–4.5 cm) long; *bell-shaped,* with 6 *white,* broad, pointed *sepals;* short-stalked and crowded in upright clusters 1–2½' (0.3–0.8 m) long, with finely *hairy branches;* mainly within foliage; in summer.
Fruit: 3–5" (7.5–13 cm) long, 1–1½"

(2.5–4 cm) wide; a *cylindrical berry,*
blunt-pointed, *green turning blackish,*
with thin, sweetish pulp; maturing in
autumn and falling before winter.
Many small, flat, rough, dull black
seeds.

Habitat: Dry slopes and canyons of oak
woodland and rarely upper semidesert
grassland.

Range: Mountains and foothills of extreme
SW. New Mexico, SE. Arizona, and
adjacent N. Mexico; at 4000–7000′
(1219–2134 m).

Schott Yucca extends to a higher
altitude than other tree yuccas. It is
named for Arthur Carl Victor Schott
(1814–75), a German-born naturalist
with the United States and Mexican
Boundary Survey, who discovered this
species in 1855.

306 **Torrey Yucca**
"Spanish-bayonet" "Spanish-dagger"
Yucca torreyi Shafer

Description: Evergreen shrub or small tree with
irregular or ragged shape, 1 or 2 and
sometimes several unbranched trunks,
with long, bayonetlike leaves.
Height: 13′ (4 m).
Diameter: 4–12″ (10–30 cm).
Leaves: *evergreen;* usually 2–3½′ (0.6–
1.1 m) long and 1¼–2″ (3–5 cm)
wide. *Bayonetlike* or lance-shaped,
tapering from englarged base to sharp
spine at end; slightly *grooved,* thick and
stiff, rough. *Yellow-green,* with many
coarse, whitish fibers or *threads along
edges.*
Trunk: dark brown and scaly at base,
covered by dead leaves above.
Flowers: 1½–3″ (4–7.5 cm) long; *bell-
shaped,* with short tube and 6 *narrow,
white* or purplish-tinged *calyx lobes;*
long-stalked and showy; crowded in
upright clusters (partly within foliage)

usually 16–24″ (0.4–0.6 m) long; in
spring.
Fruit: 3–4″ (7.5–10 cm) long, 1–1¼″
(2.5–3 cm) wide; a *cylindrical or egg-
shaped berry,* brown to *black,* fleshy;
falling before winter; many small, flat,
rough, dull black seeds.

Habitat: Dry soils of plains, mesas, and foothill
slopes; in desert grassland and shrub
thickets.

Range: SW. Texas including Trans-Pecos
Texas, S. New Mexico, and NE.
Mexico; at 2000–5000′ (610–
1524 m).

This species was named for John Torrey
(1796–1873), the Columbia University
botanist, who designated this yucca as a
new variety in 1859. Indians ate the
pulpy fruits of this and related shrubby
species either raw or roasted; they also
dried and ground them into meal for
winter use. The coarse fibers of the long
leaves were made into ropes, mats,
sandals, baskets, and cloth.

CASUARINA FAMILY
(Casuarinaceae)

Trees and shrubs; about 60 species native from southeastern Asia to northeastern Australia and Polynesia. Cultivated throughout tropical regions worldwide.

Leaves: tiny and scalelike; in rings, or whorls, of 4–16 at each joint.

Twigs: green; very slender and drooping, like pine needles or wires, with joints and long, fine lines.

Flowers: tiny, dark red, crowded, regular, without calyx and corolla; male and female on same or separate trees. Male composed of 2 tiny scales and 1 stamen; in rings, or whorls, along an axis. Female composed of 2 tiny scales and 1 pistil with superior 1- or 2-celled ovary, 2 ovules, and 2 long styles; many clustered in small balls.

Fruit: multiple; ball-like or conelike, hard and woody, with many tiny, winged seeds.

1 **River-oak Casuarina**
"Cunningham Casuarina" "Beefwood"
Casuarina cunninghamiana Miq.

Description: Evergreen, introduced, ornamental tree with open, irregular or conical crown of gray-green, *drooping, jointed twigs* resembling pine needles.

Height: 50′ (15 m).

Diameter: 1′ (0.3 m).

Leaves: rings, or whorls, of 8–10 tiny brown scales less than ¹⁄₆₄″ (0.4 mm) long, larger on main twigs.

Bark: gray-brown, smooth but becoming rough, thick, and furrowed into narrow ridges.

Twigs: 3–7″ (7.5–18 cm) long, less than ¹⁄₃₂″ (1 mm) wide; *wiry, gray-green,* hairless, with 8–10 *long, fine lines* or ridges ending in scale leaves; at joints or nodes less than ¼″ (6 mm) apart;

evergreen, shedding gradually. Main twigs pale green and finely hairy when young, becoming brownish and rough. Flowers: *tiny, light brown, crowded;* male and female on separate trees. Male in *narrow catkins* ¼–¾" (6–19 mm) long; at end of twig. Female in *balls* more than ¼" (6 mm) in diameter; at leaf bases.

Fruit: a brown or gray, *hard, warty ball* ⅜–½" (10–12 mm) in diameter (multiple); individual fruits nearly ¼" (6 mm) long, pointed, splitting in 2 parts, containing 1 winged seed.

Habitat: Salty, coastal soils, as well as alkaline valley soils.

Range: Native of Australia. Planted in subtropical regions, including Florida, California, S. Arizona, and Hawaii.

This fast-growing, subtropical tree is one of the most cold-hardy of its genus. The plants can be pruned as hedges or fences; they have also been utilized in reforestation. Another common name, "Australian-pine," alludes to the appearance of the needle-like twigs. The green twigs manufacture food, replacing leaves.

WILLOW FAMILY
(Salicaceae)

Deciduous, often aromatic trees and
shrubs. About 350 species in the genera
willow (*Salix*) and poplar (*Populus*);
nearly worldwide, mostly in north
temperate and arctic regions. 35 native
and 5 naturalized tree species and about
60 native shrub species in North
America.
Leaves: alternate, simple, mostly
toothed, with paired stipules.
Flowers: tiny male and female on
separate plants; regular, each above a
scale, crowded in narrow catkins. Male
flowers with cuplike disk or 1–2 glands
and 1–40 stamens separate or united at
base. Female flower with 1 pistil.
Fruit: a capsule opening in 2–4 parts,
containing many tiny seeds with
cottony hairs.

The large genus of willows (*Salix*),
characteristic of wet soils, includes
shrubs and mostly small trees, often
with several stems or trunks from base
and forming thickets. Leaves are narrow
and commonly long-pointed and finely
toothed, with distinct odor when
crushed, turning yellow in autumn;
leafstalks are very short with paired and
often large stipules. Bark is gray or
brown, smooth or becoming rough,
scaly or furrowed, bitter, and aromatic.
The slender or wiry twigs are tough,
flexible, often shedding or easily
detached at forks. The many tiny
yellowish or greenish flowers usually
appear in early spring before leaves;
male and female are on separate plants,
many crowded in mostly erect catkins.
Each flower is above a hairy scale and
has a glandlike disk, without calyx or
corolla. Male flower has 1–2
(sometimes to 12) stamens; female
has a narrow, pointed pistil. The
many conical, 1-celled, long-pointed
capsules along a slender stalk are mostly

light brown and mature in late spring
or early summer, splitting into 2 parts.
The numerous tiny seeds have tufts of
white, cottony hairs.

213, 236 White Poplar
"Silver Poplar"
Populus alba L.

Description: Large, much-branched, introduced tree
with leaves that toss in the slightest
breeze to reveal silvery-white lower
surfaces.
Height: 80' (24 m).
Diameter: 2' (0.6 m).
Leaves: 2½–4" (6–10 cm) long and
nearly as wide. Ovate, *3- or 5-lobed and
maplelike,* blunt-tipped with scattered
small teeth. Dark green above, *densely
white hairy and feltlike beneath;* turning
reddish in autumn. Long leafstalks
covered with white hairs.
Bark: *whitish-gray,* smooth, becoming
rough and furrowed at base.
Twigs: densely covered with white
hairs.
Flowers: catkins 1½–3" (4–7.5 cm)
long; densely covered with white hairs;
male and female on separate trees; in
early spring before leaves.
Fruit: ³⁄₁₆" (5 mm) long; *egg-shaped*
capsules with many tiny, cottony seeds.

Habitat: Moist soils, especially along roadsides
and borders of fields.

Range: Native of Europe and Asia. Planted and
widely naturalized in S. Canada and
across the United States.

Probably introduced in colonial times,
White Poplar is hardy in cities and in
dry areas. Handsome varieties, some
with silvery leaves and one with a
columnar form, are planted for shade
and ornament. It grows rapidly and
spreads readily by root sprouts,
sometimes becoming an undesirable
weed.

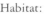

135 Narrowleaf Cottonwood
"Mountain Cottonwood"
"Black Cottonwood"
Populus angustifolia James

Description: Tree with narrow, conical crown of slender, upright branches and with resinous, balsam-scented buds.
Height: 50′ (15 m).
Diameter: 1½′ (0.5 m).
Leaves: 2–5″ (5–13 cm) long, ½–1″ (1.2–2.5 cm) wide. *Lance-shaped, long-pointed* at tip, rounded at base; *finely saw-toothed,* hairless or nearly so, *short-stalked. Shiny green* above, *paler beneath;* turning dull yellow in autumn.
Bark: yellow-green, smooth; becoming gray-brown and furrowed into flat ridges at base.
Twigs: yellow-green, slender, hairless.
Flowers: catkins 1½–3″ (4–7.5 cm) long; reddish; male and female on separate trees; in early spring before leaves.
Fruit: about ¼″ (6 mm) long; broadly *egg-shaped* capsules, light brown, hairless; maturing in spring; splitting into 2 parts; many cottony seeds.

Habitat: Moist soils along streams in mountains, with willows and alders in coniferous forests.

Range: Mountains from S. Alberta and extreme SW. Saskatchewan south to Trans-Pecos Texas and California; also N. Mexico; at 3000–8000′ (914–2438 m).

Discovered in 1805 by Lewis and Clark on their expedition to the Northwest, this is the common cottonwood of the northern Rocky Mountains. It is easily distinguishable from related species by the narrow, short-stalked, willowlike leaves. Its root system makes it suitable for erosion control.

210 Balsam Poplar
"Tacamahac" "Balm"
Populus balsamifera L.

Description: Large tree with narrow, open crown of upright branches and fragrant, resinous buds with strong balsam odor.
Height: 60–80' (18–24 m).
Diameter: 1–3' (0.3–0.9 m).
Leaves: 3–5" (7.5–13 cm) long, 1½–3" (4–7.5 cm) wide. Ovate, pointed at tip, rounded or slightly notched at base; *finely wavy-toothed,* slightly thickened, hairless or nearly so. Shiny dark green above, *whitish, often with rust-colored veins beneath.* Leafstalks slender, *round,* hairy.
Bark: light brown, smooth; becoming gray and furrowed into flat, scaly ridges.

Twigs: brownish, stout, with large, gummy or sticky buds producing *fragrant, yellowish resin.*
Flowers: catkins 2–3½" (5–9 cm) long; brownish; male and female on separate trees; in early spring.
Fruit: ⁵⁄₁₆" (8 mm) long; *egg-shaped* capsules, *pointed,* light brown, *hairless;* maturing in spring and *splitting into 2 parts;* many tiny, cottony seeds.

Habitat: Moist soils of valleys, mainly stream banks, sandbars, and flood plains, also lower slopes; often in pure stands.

Range: Across N. North America along northern limit of trees from NW. Alaska, south to SE. British Columbia, and east to Newfoundland, south to Pennsylvania and west to Iowa; local south to Colorado and in eastern mountains to West Virginia; to 5500' (1676 m) in Rocky Mountains.

The northernmost New World hardwood, Balsam Poplar extends in scattered groves to Alaska's Arctic Slope. Balm-of-Gilead Poplar, an ornamental with broad, open crown and larger, heart-shaped leaves, is a clone or hybrid. Balm-of-Gilead, derived from

the resinous buds, has been used in
home remedies.

209, 540 Fremont Cottonwood
"Rio Grande Cottonwood"
"Meseta Cottonwood"
Populus fremontii Wats.

Description: Tree with broad, flattened, open crown
of large, widely spreading branches.
Height: 40–80' (12–24 m).
Diameter: 2–4' (0.6–1.2 m).
Leaves: 2–3" (5–7.5 cm) long and
wide. *Broadly triangular, often broader
than long,* short-pointed, nearly straight
at base; with *coarse, irregular, curved
teeth;* thick, hairless; leafstalks long,
flattened. *Shiny yellow-green;* turning
bright yellow in autumn.
Bark: gray, thick, rough, deeply
furrowed.
Twigs: light green, stout, hairless.
Flowers: catkins 2–3½" (5–9 cm) long;
reddish; male and female on separate
trees; in early spring.
Fruit: about ½" (12 mm) long; *egg-
shaped* capsules, light brown, hairless;
maturing in spring, splitting into 3
parts; many cottony seeds.

Habitat: Wet soils along streams, often with
sycamores, willows, and alders; in
deserts, grasslands, and woodlands.

Range: S. and W. Colorado west to N.
California and southeast to Trans-Pecos
Texas; also N. Mexico; to 6500'
(1981 m).

This species, including varieties, is the
common cottonwood at low altitudes
along the Rio Grande and Colorado
River and in the rest of the Southwest,
as well as in California. Fremont
Cottonwood grows only on wet soil and
is an indicator of permanent water and
shade. Easily propagated from cuttings,
it is extensively planted in its range
along irrigation ditches, and although

it grows rapidly, it is short-lived. To this day, Hopi Indians of the Southwest carve cottonwood roots into kachina dolls, the representations of supernatural beings, that have become valuable collectors' items. Horses gnaw the sweetish bark of this species; beavers also feed on the bark and build dams with the branches. Greenish clumps of parasitic mistletoes are often scattered on the branches. Fremont Cottonwood is named for its discoverer, General John Charles Frémont (1813–90), politician, soldier, and explorer.

212 Lombardy Poplar
Populus nigra L. 'Italica'

Description: Medium-sized, introduced tree with straight, stout trunk, often enlarged at base, and *narrow, columnar crown* of upright, short, brittle branches.
Height: 30–60' (9–18 m).
Diameter: 1–2' (0.3–0.6 m).
Leaves: 1½–3" (4–7.5 cm) long and wide. *Triangular,* long-pointed at tip, straight at base; *wavy saw-toothed,* hairless or nearly so. Green above, light green beneath. Slender, flattened leafstalks 1–2" (2.5–5 cm) long.
Bark: gray, thick, deeply furrowed.
Twigs: orange, turning gray; stout, hairless, with sticky, scaly buds.
Flowers: catkins 2" (5 cm) long; narrow, drooping, with many male flowers; in early spring before leaves.
Habitat: Moist soils in temperate regions.
Range: Widely planted nearly throughout the United States.

Lombardy Poplar is a clone or cultivated variety (designated as 'Italica') of Black Poplar, which is native to Europe and western Asia. Apparently this clone originated in northern Italy before 1750. The trees are male and bear no seeds; female

flowers and fruit are not produced and
propagation is by cuttings and root
sprouts. They are generally grown in
rows for shelterbelts or windbreaks,
screens, roadsides, and formal gardens.
It is especially common around rural
homes in treeless, irrigated areas and
conspicuous in the landscape of the
Great Basin. The trees grow rapidly but
are short-lived. They are subject to
European canker disease in the trunk
and upper branches, which destroys the
columnar form.

211, 339   Quaking Aspen
"Trembling Aspen" "Golden Aspen"
Populus tremuloides Michx.

Description:  The most widely distributed tree in
North America, with a narrow,
rounded crown of thin foliage.
Height: 40–70' (12–21 m).
Diameter: 1–1½' (0.3–0.5 m).
Leaves: 1¼–3" (4–7.5 cm) long. *Nearly
round,* abruptly short-pointed, rounded
at base; *finely saw-toothed,* thin. Shiny
green above, dull green beneath;
turning *golden-yellow* in autumn before
shedding. Leafstalks slender, *flattened.*
Bark: *whitish, smooth,* thin; on very
large trunks becoming dark gray,
furrowed, and thick.
Twigs: shiny brown, slender, hairless.
Flowers: catkins 1–2½" (2.5–6 cm)
long; brownish; male and female on
separate trees; in spring before leaves.
 Fruit: ¼" (6 mm) long; narrowly
conical, light green capsules in drooping
catkins to 4" (10 cm) long; maturing in
late spring and splitting in 2 parts.
Many tiny, cottony seeds; rarely
produced in West, where propagation
is by root sprouts.
Habitat: Many soil types, especially sandy and
gravelly slopes; often in pure stands and
in western mountains in an altitudinal
zone below spruce-fir forest.

Range: Across N. North America from Alaska to Newfoundland, south to Virginia, and in Rocky Mountains south to S. Arizona and N. Mexico; from near sea level northward; to 6500–10,000′ (1981–3048 m) southward.

The names refer to the leaves, which tremble in the slightest breeze on their flattened leafstalks. The soft, smooth bark is sometimes decorated with carved initials and marked by bear claws. A pioneer tree after fires and logging and on abandoned fields, it is short-lived and replaced by conifers. Principal uses of the wood include pulpwood, boxes, furniture parts, matches, excelsior, and particle-board. The twigs and foliage are browsed by deer, elk, and moose, as well as by sheep and goats. Beavers, rabbits, and other mammals eat the bark, foliage, and buds, and grouse and quail feed on the winter buds.

---

207 Black Cottonwood
"Western Balsam Poplar"
"California Poplar"
Populus trichocarpa Torr. & Gray

Description: The tallest native cottonwood, with open crown of erect branches and sticky, *resinous buds* with balsam odor. Height: 60–120′ (18–37 m). Diameter: 1–3′ (0.3–0.9 m), sometimes much larger. Leaves: 3–6″ (7.5–15 cm) long, 2–4″ (5–10 cm) wide, larger on young twigs. *Broadly ovate;* short- or long-pointed at tip, rounded or slightly notched at base; *finely wavy-toothed,* slightly thickened, hairless or nearly so. *Shiny dark green* above, *whitish* and often with *rusty veins beneath;* turning yellow in autumn. Leafstalks slender, round, hairy.

Bark: gray, smooth, becoming thick

and deeply furrowed into flat, scaly ridges.

Twigs: brownish, stout, often hairy when young.

Flowers: catkins 1½–3¼" (4–8 cm) long; reddish-purple; male and female on separate trees; in early spring.

Fruit: ¼" (6 mm) in diameter; *round* capsules, light brown, *hairy;* maturing in spring, *splitting into 3 parts;* many cottony seeds.

Habitat: Moist to wet soils of valleys, mainly on stream banks and flood plains, also on upland slopes; often in pure stands and with willows and Red Alder.

Range: S. Alaska south to S. California and east in mountains to extreme SW. Alberta and Montana; also local in SW. North Dakota and N. Baja California; to 2000' (610 m) in north; to 9000' (2743 m) in south.

Black Cottonwood is the tallest native western hardwood. The current champion in Yamhill County, Oregon, measures 147' (44.8 m) in height, 30.2' (9.2 m) in trunk circumference, and 97' (29.6 m) in crown spread. The wood is used for boxes and crates, pulpwood, and excelsior. Black Cottonwood intergrades with its northern relative, Balsam Poplar, where both meet in southern Alaska and elsewhere.

101 Feltleaf Willow
Salix alaxensis (Anderss.) Cov.

Description: Shrub or small tree, occasionally dwarfed where exposed, the leaves distinctively whitish and hairy on lower surface.

Height: 20–30' (6–9 m).
Diameter: 6" (15 cm).
Leaves: 2–4" (5–10 cm) long, ½–1½" (1.2–4 cm) wide. *Elliptical* or reverse lance-shaped, short-pointed, *often widest*

beyond middle and tapering to base, *mostly without teeth*. Dull green and mostly hairless above, *covered with dense, feltlike mat of whitish hairs beneath*.

Bark: gray, smooth, becoming rough and furrowed into scaly plates.

Twigs: *stout, with white, woolly hairs*.

Flowers: catkins 2–3½" (5–9 cm) long; scales blackish with white hairs; stalkless; in early spring before leaves.

Fruit: ¼" (6 mm) long; white, woolly capsules; maturing in early summer.

Habitat: Wet soils, especially in valleys.

Range: Almost throughout Alaska, east in NW. Canada to Mackenzie, and south to SW. Alberta and central British Columbia; also local east to Keewatin and extreme N. Quebec; not in contiguous United States; to tree line and beyond to 6000′ (1829 m). Also in NE. Asia.

This is the only tree willow that extends beyond the limits of the spruce-birch forest in Alaska. In many parts of northern Alaska it is the only available fuelwood. The scientific name, meaning "of Alaxa," is from an old Russian word for Alaska. Moose browse this species by pulling down and breaking the smaller branches. People have reportedly survived on the inner bark in emergencies. Feltleaf Willow is one of several species sometimes producing "diamond willow" caused by fungi. The resultant diamond-shaped patterns on the trunks are made into canes and novelties. Except for a few shrubby willow species rarely reaching tree size, Feltleaf Willow and Littletree Willow are the only Alaskan tree species absent from the contiguous 48 states.

130 White Willow
"European White Willow"
Salix alba L.

Description: Naturalized tree with 1–4 trunks and open crown of spreading branches.
Height: 50–80' (15–24 m).
Diameter: 2' (0.6 m) or more.
Leaves: 2–4½" (5–11 cm) long, ⅜–1¼" (1–3 cm) wide. *Lance-shaped* to elliptical, *finely saw-toothed*, firm. *Shiny dark green above, whitish and silky beneath;* turning yellow in autumn.
Bark: gray; rough, furrowed into narrow ridges.
Twigs: *yellow to brown;* flexible and often slightly drooping, *silky when young.*
Flowers: catkins 1¼–2¼" (3–6 cm) long; with yellow, hairy scales; at end of short, leafy twigs; in early spring.
Fruit: ³⁄₁₆" (5 mm) long; hairless, light brown capsules, maturing in late spring or early summer.

Habitat: Wet soils of stream banks and valleys near cities.

Range: Native from Europe and N. Africa to central Asia. Naturalized in SE. Canada and E. United States.

Introduced in colonial times, this handsome willow is planted as a shade and ornamental tree and for shelterbelts, fenceposts, and fuel. Some cultivated varieties have golden-yellow or reddish twigs that have been used in basketmaking. One variety in England is prized for making cricket bats.

134 Peachleaf Willow
"Peach Willow" "Almond Willow"
Salix amygdaloides Anderss.

Description: Tree with 1 or sometimes several straight trunks, upright branches, and spreading crown.
Height: 60' (18 m).

Diameter: 2' (0.6 m).

Leaves: 2–4½" (5–11 cm) long, ½–1¼" (1.2–3 cm) wide. *Lance-shaped,* often slightly curved to one side, tapering to *long, narrow point; finely saw-toothed,* with long, slender leafstalks, becoming hairless. Shiny green above, *whitish beneath.*

Bark: dark brown, rough, furrowed into flat, scaly ridges.

Twigs: shiny orange or brown, hairless.

Flowers: catkins 1½–3" (4–7.5 cm) long; with yellow, hairy scales; on short, leafy twigs; in spring with leaves.

Fruit: ½" (6 mm) long; long-stalked, reddish-yellow, hairless capsules; maturing in late spring or early summer.

Habitat: Wet soils of valleys, often bordering stream banks with cottonwoods.

Range: SE. British Columbia east to extreme S. Quebec and New York, south to NW. Pennsylvania, and west to W. Texas; also N. Mexico; at 500–7000' (152–2134 m).

This is the common willow across the northern plains, where it is important in protecting riverbanks from erosion. Both common and scientific names refer to the leaf shape, which suggests that of Peach.

---

143 **Littletree Willow**
Salix arbusculoides Anderss.

Description: Upright shrub or commonly a small tree, often forming dense thickets, with shiny reddish-brown twigs.

Height: 10–30' (3–9 m).

Diameter: 6" (15 cm).

Leaves: 1–3" (2.5–7.5 cm) long, ⅜–¾" (1–2 cm) wide. *Lance-shaped,* pointed at ends, *finely but shallowly toothed.* Dark green above, *whitish and with silvery hairs beneath.*

Bark: gray to reddish-brown, smooth.

Twigs: much branched, slender; yellowish-brown and slightly hairy when young.

Flowers: catkins 1–1½" (2.5–4 cm) long; with hairy blackish or brownish scales; in early spring before or with leaves.

Fruit: ³⁄₁₆" (5 mm) long; slightly hairy capsules, light brown; maturing in early summer.

Habitat: Along streams, also in upland sites; with Black Spruce after fires and with White Spruce and Paper Birch.

Range: NW. Alaska east across NW. Canada to Kewatin, south to central Manitoba, and west to central British Columbia. Not in contiguous United States. To tree line and beyond to 4000' (1219 m).

Littletree Willow is widespread in interior Alaska and one of the most common willows there, forming dense thickets. The scientific name means "like a little tree." Except for a few shrubby willow species rarely attaining tree size, this and Feltleaf Willow are the only tree species native in Alaska that are absent from the contiguous 48 states. It is one of several species sometimes producing "diamond willow"; these trunks with diamond-shaped patterns caused by fungi are prized for carving into canes.

---

128 Weeping Willow
"Babylon Weeping Willow"
Salix babylonica L.

Description: A handsome, naturalized tree with short trunk and broad, open, irregular crown of *drooping branches.*

Height: 30–40' (9–12 m).

Diameter: 2' (0.6 m), sometimes much larger.

Leaves: 2½–5" (6–12 cm) long, ¼–½" (6–12 mm) wide. *Narrowly lance-*

shaped, with long-pointed tips, finely saw-toothed; hanging from short leafstalks. Dark green above, whitish or gray beneath.

Bark: gray, rough, thick, deeply furrowed in long, branching ridges.

Twigs: yellowish-green to brownish; very slender, unbranched, drooping vertically.

Flowers: catkins ⅜–1" (1–2.5 cm) long; greenish, at end of short leafy twigs; in early spring; plants mostly female.

Fruit: 1/16" (1.5 mm) long; light brown capsules; maturing in late spring or early summer.

Habitat: Parks, gardens, and cemeteries, especially near water.

Range: Native of China. Naturalized locally from extreme S. Quebec and S. Ontario, south to Georgia, and west to Missouri. Also planted in the West.

This willow is well known for its distinctive weeping foliage. It is among the first willows to bear leaves in spring and among the last to shed them in autumn. China, not Babylon, was its native home; when named, it was confused with Euphrates Poplar (*Populus euphratica* Olivier).

102, 537 Bebb Willow
"Beak Willow" "Diamond Willow"
Salix bebbiana Sarg.

Description: Much-branched shrub or small tree with broad, rounded crown.
Height: 10–25' (3–7.6 m).
Diameter: 6" (15 cm).
Leaves: 1–3½" (2.5–9 cm) long, ⅜–1" (1–2.5 cm) wide. *Elliptical*, often broadest beyond middle, *short-pointed* at ends; slightly saw-toothed or wavy; firm, slightly hairy. *Dull green above*, gray or whitish and *net-veined* beneath.

Bark: gray, smooth, becoming rough and furrowed.

Twigs: *reddish-purple,* slender, widely forking; with pressed hairs when young.

Flowers: catkins ¾–1½" (2–4 cm) long; with yellow or brown scales; on short, leafy stalks; before or with leaves.

Fruit: ⅜" (10 mm) long; very slender capsules; hairy, light brown, *ending in long point,* long-stalked; maturing in early summer.

Habitat: Moist, open uplands and borders of streams, lakes, and swamps.

Range: Central and SW. Alaska south to British Columbia and east to Newfoundland, south to Maryland, west to Iowa, and south in Rocky Mountains to S. New Mexico; to 11,000' (3353 m) southward. Also in NE. Asia.

Bebb Willow is the most important "diamond willow," a term applied to several species which sometimes have diamond-shaped patterns on their trunks. These are caused by fungi, usually in shade or poor sites. The contrasting whitish and brownish stems are carved into canes, lamps, posts, furniture, and candleholders. This species forms willow thickets as a weed on uplands after forest fires. It is named for Michael Shuck Bebb (1833–95), U.S. specialist on willows.

---

133 Bonpland Willow
"Red Willow" "Polished Willow"
Salix bonplandiana H.B.K.

Description: Tree with broad, rounded crown of spreading branches.
Height: 20–50' (6–15 m).
Diameter: 1–2' (0.3–0.6 m).
Leaves: 3–6" (7.5–15 cm) long, ½–1¼" (1.2–3 cm) wide. *Lance-shaped* to

oblong; long-pointed at tip, short-pointed at base; *inconspicuously finely toothed,* slightly *thickened,* becoming *hairless. Shiny green* above, *paler and whitish beneath;* shedding irregularly in winter and evergreen in far south.
Bark: dark brown or blackish; rough, furrowed into irregular, flat, scaly ridges.

Twigs: slender, yellow to brown, hairless.
Flowers: catkins 1½–3″ (4–7.5 cm) long; with long, hairy scales; on short, leafy twigs; in early spring.
Fruit: nearly ¼″ (6 mm) long; yellowish, hairless capsules; maturing in late spring and early summer.

Habitat: Wet soils along streams mostly in mountains, also in deserts; with oaks, pinyons, and junipers.

Range: N. California east to SW. Utah, and south to SE. Arizona; also south to Mexico and Guatemala; to 5000′ (1524 m).

This willow becomes evergreen southward in Mexico. A cultivated form of this species with a narrow, columnar crown borders the floating flower gardens at Xochimilco near Mexico City. It is named for its discoverer, Aimée Bonpland (1773–1858), a French botanist.

150, 319 **Pussy Willow**
Salix discolor Muhl.

Description: Many-stemmed shrub or small tree with open, rounded crown; silky, furry catkins appear in late winter and early spring.

Height: 20′ (6 m).
Diameter: 8″ (0.2 m).
Leaves: 1½–4″ (4–10 cm) long, ⅜–1¼″ (1–3 cm) wide. *Lance-shaped* or narrowly elliptical, *irregularly wavy-toothed,* stiff; hairy when young;

slender-stalked. *Shiny green above, whitish beneath.*

Bark: gray, fissured, scaly.

Twigs: brown, stout, hairy when young.

Flowers: catkins 1–2½" (2.5–6 cm) long; cylindrical, thick with blackish scales, covered with *silky, whitish hairs; in late winter and early spring* long before leaves.

Fruit: ⁵⁄₁₆–½" (8–12 mm) long; narrow, light brown, finely hairy capsules; in early spring before leaves.

Habitat: Wet meadow soils and borders of streams and lakes; usually in coniferous forests.

Range: N. British Columbia to Labrador, south to Delaware, west to NE. Missouri, and north to N. Wyoming and North Dakota; to 4000' (1219 m).

The large flower buds burst and expose their soft, silky hairs, or "pussy fur," early in the year. In winter, cut Pussy Willow twigs can be put in warm water and the flowers forced at warm temperatures. Some twigs will produce beautiful golden stamens, while others will bear slender greenish pistils. The Latin species name refers to the contrasting colors of the leaf surfaces, which aid in recognition.

---

127 Sandbar Willow
"Coyote Willow" "Narrowleaf Willow"
Salix exigua Nutt.

Description: Thicket-forming shrub with clustered stems or rarely a tree, with *very narrow leaves.*

Height: 3–10' (1–3 m), sometimes to 20' (6 m).

Diameter: 5" (13 cm).

Leaves: 1½–4" (4–10 cm) long, ¼" (6 mm) wide. Linear, very long-pointed at ends; *few tiny, scattered teeth* or none; varying from *hairless to densely hairy*

with pressed, silky hairs; almost
stalkless. Yellow-green to gray-green
on both surfaces.
Bark: gray, smooth or becoming
fissured.
Twigs: reddish- or yellowish-brown;
slender, upright, hairless or with gray
hairs.
Flowers: catkins 1–2½" (2.5–6 cm)
long; with hairy yellow scales; at end of
leafy twigs; in spring.
Fruit: ¼" (6 mm) long; light brown
capsules, usually hairy; maturing in
early summer.

Habitat: Wet soils, especially riverbanks,
sandbars, and silt flats.

Range: Central Alaska east to Ontario and New
York, southwest to Mississippi, and
west to S. California; also local east to
Quebec and Virginia and in N. Mexico;
to 8000' (2438 m).

This hardy species has perhaps the
greatest range of all tree willows: from
the Yukon River in central Alaska to
the Mississippi River in southern
Louisiana. A common and characteristic
shrub along streams throughout the
interior, especially the Great Plains and
Southwest, it is drought-resistant and
suitable for planting on stream bottoms
to prevent surface erosion. Livestock
browse the foliage; Indians made
baskets from the twigs and bark.

129 **River Willow**
"Sandbar Willow"
Salix fluviatilis Nutt.

Description: Shrub or small tree with *silvery hairs on
young twigs and young leaves.*
Height: 7–20' (2–6 m).
Diameter: 4" (10 cm).
Leaves: 2–6" (5–15 cm) long, ¼–⅝"
(6–15 mm) wide. *Narrowly lance-
shaped,* long-pointed, with *scattered
small teeth or none,* becoming nearly

hairless; almost stalkless; green above, *whitish beneath.*

Bark: light brown, thick, furrowed into irregular scaly ridges.

Twigs: slender, greenish becoming brownish and nearly hairless.

Flowers: catkins 1½–3″ (4–7.5 cm) long; with yellow, hairy scales; on leafy twigs; in *early summer.*

Fruit: ³⁄₁₆″ (5 mm) long; light brown, hairy capsules; maturing in summer.

Habitat: Banks of Columbia River from the mouth of the Deschutes River to the lower Willamette River.

Range: SW. Washington and NW. Oregon only; near sea level.

This species has the most restricted natural range of all native tree willows. The Latin species name, meaning "of rivers," and the common name refer to its habitat. It was discovered on the Columbia River by Thomas Nuttall (1786–1859), the British-American botanist, who named it in 1843.

89, 338 Hinds Willow
"Sandbar Willow" "Valley Willow"
Salix hindsiana Benth.

Description: Shrub or small tree with many trunks and very narrow leaves covered with silky hairs; forming thickets.

Height: 7–23′ (2–7 m).

Diameter: 10″ (25 cm).

Leaves: 1½–3¼″ (4–8 cm), usually ⅛–¼″ (3–6 mm) wide. *Very narrow* or linear, tapering at both ends; *usually without teeth;* covered with *gray, silky hairs* on both surfaces; almost stalkless.

Bark: gray and furrowed.

Twigs: *gray or silvery,* covered with *woolly hairs when young.*

Flowers: catkins ¾–1½″ (2–4 cm) long; covered with yellow, densely hairy scales; on leafy twigs; after leaves appear in spring.

Fruit: ¼″ (6 mm) long; densely hairy capsules, light brown; almost stalkless; maturing in late spring or early summer.

Habitat: Moist soils of ditches, sandbars, and stream banks.

Range: SW. Oregon to S. California and NW. Baja California; to 3000′ (914 m).

Hinds Willow is named for Richard Brinsley Hinds (1812–47), a British botanist who collected plant specimens along the West Coast on a surveying expedition with the ship *Sulphur* in 1836–42. This willow holds soil banks but is considered a weed when it clogs irrigation ditches.

---

104 Hooker Willow
"Bigleaf Willow" "Coast Willow"
Salix hookerana Barratt

Description: Shrub or small tree with many stems, broad, rounded crown, and leaves nearly half as wide as long.
Height: 10–30′ (3–9 m).
Diameter: 1′ (0.3 m).
Leaves: 1½–4½″ (4–11 cm) long, ¾–2″ (2–5 cm) wide. *Elliptical, blunt or rounded at tip,* mostly rounded at base, often broadest beyond middle; *generally without teeth* or sparsely wavy-toothed; with *stout,* hairy *leafstalk* ⅜″ (10 mm) long. *Shiny yellow-green* and nearly hairless above, *whitish and usually hairy beneath.*
Bark: gray, smooth, thin, becoming rough and scaly.
Twigs: *dark brown, densely covered with white or gray, woolly hairs; stout,* brittle.
Flowers: catkins 3–4″ (7.5–10 cm) long; blackish scales with long, whitish hairs; densely crowded, on short, leafy stalks; before or with leaves in spring.
Fruit: ¼″ (6 mm) long; light brown capsules, hairless or nearly so, crowded;

maturing in spring and early summer.

Habitat: Wet soils, including beach ridges, sand dunes, coastal meadows, and edges of streams and salt marshes; bordering coniferous forests.

Range: Narrow zone along Pacific Coast from extreme SW. British Columbia south to NW. California; also local north to S. Alaska (Yakutat Bay); to 500′ (152 m). Reported also in E. Siberia.

Hooker Willow's relatively broad leaves aid in recognition. It is named after William Jackson Hooker (1785–1865) a British botanist, in whose book the original description of this species was published. The isolated Alaskan plants were formerly regarded as a different species, Yakutat Willow (*S. amplifolia* Cov.).

132 Pacific Willow
"Western Black Willow"
"Yellow Willow"
Salix lasiandra Benth.

Description: Tree with open, irregular crown; sometimes a thicket-forming shrub.
Height: 20–50′ (6–15 m).
Diameter: 2′ (0.6 m).
Leaves: 2–5″ (5–13 cm) long, ½–1″ (1.2–2.5 cm) wide. *Narrowly lance-shaped, very long-pointed,* mostly rounded at base; *finely saw-toothed,* becoming almost hairless; leafstalks slender, with *glands* at upper end. *Shiny green above, whitish beneath.*
Bark: gray or dark brown; becoming rough and deeply furrowed into flat, scaly ridges.
Twigs: shiny reddish to brownish or yellow, hairless.
Flowers: catkins 1½–4″ (4–10 cm) long; with hairy, yellow or brown scales; at ends of leafy twigs; with leaves in spring.
Fruit: ¼″ (6 mm) long; light reddish-

brown, hairless capsules; maturing in early summer.

Habitat: Wet soils along streams, lakes, and roadsides; in valleys and on mountains.

Range: Central and SE. Alaska east to Saskatchewan and south mostly in mountains to S. New Mexico and S. California; to 8000′ (2438 m).

As the common name suggests, Pacific Willow is familiar along riverbanks and valleys through the Pacific states. The scientific name alludes to the shaggy-haired flower stamens.

98, 142 **Arroyo Willow**
"White Willow"
Salix lasiolepis Benth.

Description: Usually a thicket-forming shrub with clustered stems; sometimes a small tree with slender, erect branches forming narrow, irregular crown.

Height: 30′ (9 m).
Diameter: 6″ (15 cm).
Leaves: 2½–4″ (6–10 cm) long, ⅜–¾″ (1–2 cm) wide. *Narrow, reverse lance-shaped, short- or blunt-pointed* at ends, *broadest beyond middle; without teeth* or slightly wavy with few small teeth, *thick* and leathery. *Dark green* and hairless above, *whitish* and usually hairy beneath.
Bark: pale gray-brown with whitish areas; smooth, becoming darker, rough, and furrowed into broad ridges.
Twigs: yellow to brown, finely hairy.
Flowers: catkins 1–2″ (2.5–5 cm) long; black or brown scales with dense, long, white hairs; almost stalkless; in early spring before or with leaves.

Fruit: ¼″ (6 mm) long; light reddish-brown, hairless capsules, crowded; maturing in late spring.

Habitat: Wet soils along streams and arroyos, or gullies, in valleys, foothills, and mountains.

Range: Washington and Idaho south to S. California and New Mexico; also in N. Mexico; to 7500' (2286 m).

The name "White Willow" may come from the light-colored bark and leaves with whitish lower surfaces. The scientific name, meaning "shaggy scale," refers to the white hairs on the scales of the flowers.

136 Mackenzie Willow
Salix mackenzieana (Hook.) Anderss.

Description: Shrub or small tree with slender trunk, upright branches, and narrow crown.
Height: 7–20' (2–6 m).
Diameter: 4" (10 cm).
Leaves: 2½–4" (6–10 cm) long, ⅝–1½" (1.5–4 cm) wide. *Lance-shaped, long-pointed,* with rounded to notched base, *finely saw-toothed, hairless;* slender-stalked; dark green above, *whitish beneath.*
Bark: gray, smooth, with scattered, raised dots or warts.

Twigs: yellowish or brown, slender, shiny, mostly hairless.
Flowers: catkins 1–2" (2.5–5 cm) long; with dark brown, hairy scales; at ends of leafy twigs; after leaves appear in spring.
Fruit: ³⁄₁₆" (5 mm) long; light brown, hairless capsules on stalks of ⅛" (3 mm); maturing in late spring.
Habitat: Wet soils of stream borders and swamps; in coniferous forests.
Range: S. Mackenzie and SE. Yukon south in mountains to Sierra Nevada of central California and N. Utah; from near sea level in north; to 7000' (2134 m) in south.

This species was first found along the Mackenzie River, which was discovered by Alexander Mackenzie (1755?–

1820), the Scottish fur trader and explorer.

---

131 Black Willow
"Swamp Willow" "Goodding Willow"
Salix nigra Marsh.

Description: Large tree with 1 or more straight and usually leaning trunks, upright branches, and narrow or irregular crown.
Height: 60–100′ (18–30 m).
Diameter: 1½–2½′ (0.5–0.8 m).
Leaves: 3–5″ (7.5–13 cm) long, ⅜–¾″ (10–19 mm) wide. *Narrowly lance-shaped, often slightly curved to one side, long-pointed, finely saw-toothed,* hairless or nearly so; *shiny green above, paler beneath.*
Bark: dark brown or blackish; deeply furrowed into scaly, forking ridges.

Twigs: brownish; very slender, easily detached at base.
Flowers: catkins 1–3″ (2.5–7.5 cm) long; with yellow hairy scales; at end of leafy twigs; in spring.
Fruit: ³⁄₁₆″ (5 mm) long; reddish-brown, hairless capsules; maturing in late spring.

Habitat: Wet soils of banks of streams and lakes, especially flood plains; often in pure stands and with cottonwoods.
Range: S. New Brunswick and Maine south to NW. Florida, west to S. Texas, and north to SE. Minnesota; also from W. Texas west to N. California; local in N. Mexico; to 5000′ (1524 m).

This is the largest and most important New World willow, with one of the most extensive ranges across the country. In the lower Mississippi Valley it attains commercial timber size, reaching 100–140′ (30–42 m) in height and 4′ (1.2 m) in diameter. The uses of the wood include millwork, furniture, doors, cabinetwork, boxes,

barrels, toys, and pulpwood. In pioneer
times the wood of this and other
willows was a source of charcoal for
gunpowder. Large trees are valuable in
binding soil banks, thus preventing soil
erosion and flood damage. Mats and
poles made from Black Willow trunks
and branches provide further protection
of riverbanks and levees. This species is
also a shade tree and a honey plant.

---

144 **Balsam Willow**
"Bog Willow"
Salix pyrifolia Anderss.

Description: Usually a shrub, sometimes a small
tree, with clumps of slender stems
branched near the top and a fragrance of
balsam.
Height: 20' (6 m).
Diameter: 4" (10 cm).
Leaves: 2–3½" (5–9 cm) long, 1–1½"
(2.5–4 cm) wide. *Ovate or elliptical,
short-pointed,* base rounded and usually
notched; finely saw-toothed, becoming
hairless, aromatic. Dark green above,
paler and whitish with yellow midvein
and conspicuous *network of small veins*
beneath.
Bark: gray, smooth, thin.

Twigs: *shiny reddish-brown,* slightly
stout, *hairless,* with *shiny, bright red
winter buds.*
Flowers: catkins 1–1½" (2.5–4 cm)
long; yellowish; on short, leafy twigs;
in late spring.
Fruit: ¼" (6 mm) long; dark orange,
hairless capsules; maturing in early
summer.
Habitat: Cold, wet bogs.
Range: Yukon, south to E. British Columbia
and across Canada to Labrador, south to
Maine, and west to Minnesota; to
2000' (610 m).

The common name refers to the
aromatic, gland-toothed young leaves,

while the Latin species name means "pear leaf." In winter, Balsam Willow is easily recognized by the shiny reddish buds and twigs.

---

103 Scouler Willow
"Black Willow" "Fire Willow"
Salix scoulerana Barratt ex Hook.

Description: Shrub or small tree with erect trunk and compact, rounded crown; sometimes medium-sized. Freshly stripped bark of twigs usually has *skunklike odor.*
Height: 15–50' (4.6–15 m).
Diameter: 1½' (0.5 m).
Leaves: spreading fanlike; 2–5" (5–13 cm) long, ½–1½" (1.2–4 cm) wide. Variable in shape; mostly obovate to *elliptical* and *broadest beyond middle, short-pointed* at tip and tapering to base; *without teeth* to sparsely wavy-toothed. *Dark green* and nearly hairless above, *whitish with gray hairs* or with few *reddish* hairs beneath.
Bark: gray, smooth, thin; becoming dark brown and fissured into broad, flat ridges.
Twigs: yellow to reddish-brown, stout; densely hairy when young, with reddish buds.
Flowers: catkins 1–2" (2.5–5 cm) long; stout, stalkless or nearly so, with black, long-haired scales; abundant in early spring before leaves.
Fruit: ⅜" (10 mm) long; narrow, stalkless, light brown capsules, with gray, woolly hairs; maturing in early summer.

Habitat: Upland coniferous forests under larger trees, in cutover areas, and in clearings, including dry sites.
Range: Central Alaska east to Manitoba and southwest to Idaho and California; also in Black Hills of South Dakota and in mountains to S. New Mexico; to 10,000' (3048 m) in mountains.

This species is sometimes called "Fire Willow" because it rapidly occupies burned areas, forming blue-green thickets. A pussy willow and one of th earliest flowering species, it is an important browse plant for moose in Alaska and for sheep and cattle elsewhere. It is one of several species sometimes forming "diamond willow"; these stems with diamond-shaped patterns caused by fungi are in demand for canes, novelties, and furniture posts. It is named for its discoverer, John Scouler (1804–71), the Scottish naturalist and physician.

141 Northwest Willow
"Velvet Willow" "Soft-leaf Willow"
Salix sessilifolia Nutt.

Description: Handsome shrub or small tree with distinctive, soft, whitish, velvety-hairy twigs, foliage, and flowers.
Height: 7–20′ (2–6 m).
Diameter: 4″ (10 cm).
Leaves: 1–2″ (2.5–5 cm) long, ½–¾″ (12–19 mm) wide. *Lance-shaped,* short-pointed, base blunt-pointed; with *few short teeth mainly in upper half;* almost stalkless. *Blue-green, densely covered with velvety or silky white hairs,* especially on midvein beneath.
Bark: gray, smooth, with scattered, raised dots or warts.
Twigs: slender, green, becoming brown.
Flowers: catkins 1½–2½″ (4–6 cm) long; *densely covered with whitish hairs;* a ends of short, leafy twigs; after leaves appear in spring.
Fruit: ¼″ (6 mm) long; light brown capsules covered with *whitish hairs;* maturing in early summer.
Habitat: Wet soils, riverbanks.
Range: Extreme S. British Columbia, Washington, and W. Oregon; to 1000 (305 m).

Northwest Willow is very showy in spring when the whitish blossoms are abundant. The scientific name means "stalkless leaves." It was discovered near the mouth of the Willamette River and named by Thomas Nuttall (1786–1859), the British-American botanist.

---

99 Sitka Willow
"Silky Willow" "Coulter Willow"
Salix sitchensis Sanson ex Bong.

Description:
Usually a large shrub or small tree, much-branched with rounded crown; a low shrub in exposed places.
Height: 10–30′ (3–9 m).
Diameter: 4–12″ (10–30 cm).
Leaves: 2–4″ (5–10 cm) long, ¾–1½″ (2–4 cm) wide. *Reverse lance-shaped* or elliptical, *blunt-pointed, broadest beyond middle,* tapering to narrow base; mostly *without teeth* or wavy-edged. Shiny dark green and with sparse, short hairs when young above, paler and with short, *silvery, silky hairs* beneath.
Bark: gray, smooth, becoming slightly furrowed and scaly.
Twigs: reddish-brown, slender, brittle; hairy when young.

Flowers: catkins 1½–4″ (4–10 cm) long; slender, with densely *hairy, black* or brown scales; on short, leafy stalks; with leaves in spring.
Fruit: ¼″ (6 mm) long; light brown capsules covered with silvery hairs; maturing in late spring.

Habitat:
Moist soils, along beaches and streams and in open areas through coastal coniferous forests.

Range:
Pacific Coast from SW. Alaska southeast to central California and east in mountains to W. Montana and central Alberta; mostly near sea level; inland locally to 5000′ (1524 m).

Sitka Willow is easily recognized by the satiny sheen on the lower leaf surfaces.

Indians used the flexible twigs for basketmaking and for stretching skins, and the pounded bark to heal wounds. As the smoke from Sitka Willow fires does not have a bad odor, the wood is used in drying fish. This species is named for Sitka, in southeastern Alaska, where it was first collected.

---

81 **Yewleaf Willow**
"Yew Willow"
Salix taxifolia H.B.K.

Description: Large shrub or small to medium-sized tree with dense, rounded crown of erect and spreading branches, *very small narrow leaves,* and very short catkins.
Height: 20–40' (6–12 m).
Diameter: 2' (0.6 m).
Leaves: *densely crowded;* ½–1¼" (1.2–3 cm) long, less than ⅛" (3 mm) wide. Linear, short-pointed, *usually without teeth,* densely covered with *silvery hairs when young,* remaining hairy but hairs *turning gray;* almost stalkless.
Bark: light gray-brown, rough, fissured into scaly ridges.
Twigs: covered with whitish or silvery hairs, slender, much-branched.
Flowers: catkins ⅜–¾" (10–19 mm) long; with yellowish, hairy scales; at end of short leafy twigs; in early spring and sometimes again in autumn.

Fruit: ¼" (6 mm) long; very narrow capsules, reddish-brown, hairy, stalkless; maturing in spring.
Habitat: Along streams in foothills and mountains; in oak woodlands, sometimes in deserts and desert grasslands.
Range: Mexican border region from Trans-Pecos Texas, west to SE. Arizona, and south to Guatemala; at 3500–6000' (1067–1829 m).

Yewleaf Willow is easily recognized by the needlelike leaves, resembling those of yews (although not evergreen). This species has the smallest leaves of any tree willow. A good soil binder, it is fairly drought-resistant and is an excellent browse for livestock.

---

97 Tracy Willow
Salix tracyi Ball

Description: Slender shrub or small, rounded tree with relatively broad leaves, more than ⅓ as wide as long; distributed locally on the Pacific Coast.
Height: 7–20′ (2–6 m).
Diameter: 3″ (7.5 cm).
Leaves: 1¼–2″ (3–5 cm) long, ⅝–¾″ (15–19 mm) wide. Reverse lance-shaped and *broadest beyond middle,* long-pointed, *without teeth, hairless;* green above, *whitish beneath.*
Bark: grayish-green, smooth.
Twigs: slender, hairless, yellow to brownish.
Flowers: catkins ¾–2″ (2–5 cm) long; yellow; with or before leaves in early spring.
Fruit: 3/16–¼″ (5–6 mm) long; narrow, long-pointed, hairless capsules, light brown; maturing in late spring.
Habitat: Sand and gravel bars of streams; in Redwood and mixed evergreen forests.
Range: Pacific Coast region of extreme SW. Oregon and NW. California only; to 500′ (152 m).

This willow of restricted range is confined to a narrow coastal zone about 200 miles (322 km) long. Not distinguished until 1934, it was named for Joseph Prince Tracy (1879–1953), an amateur botanist of Eureka, California, who made detailed studies of the plants of that area.

BAYBERRY (WAXMYRTLE) FAMILY
(Myricaceae)

Nearly worldwide; about 40 species of small trees and shrubs, mostly in the bayberry genus (*Myrica*); 5 native tree species and 3 shrub species in North America.

Leaves: alternate, simple, often oblanceolate, toothed, and leathery, with orange or yellow resinous dots, very aromatic when crushed, mostly without stipules.

Flowers: tiny, greenish or yellowish, male and female usually on the same plant or on separate plants in short lateral clusters, regular, without calyx or corolla, each above a scale. Male flower usually with 4–8 (2–20) stamens, sometimes united. Female with 1 pistil.

Fruit: a small, rounded, whitish drupe covered with wax, 1-seeded.

---

139 Pacific Bayberry
"Pacific Waxmyrtle"
"Western Waxmyrtle"
Myrica californica Cham.

Description: Evergreen, much-branched shrub or small tree with a narrow, rounded crown and waxy brownish berries.
Height: 30' (9 m).
Diameter: 1' (0.3 m).
Leaves: *evergreen;* 2–4½" (5–11 cm) long, ½–¾" (12–19 mm) wide. *Reverse lance-shaped,* usually broadest near short-pointed tip; *saw-toothed* except near long-pointed base, *aromatic,* slightly thickened, hairless. *Shiny dark green* above, yellow green with tiny *black gland-dots* beneath.
Bark: gray or brown, smooth, thin.
Twigs: green or brown, slender, hairy when young.
Flowers: *tiny yellowish male* flowers in

almost stalkless clusters ⅜–¾″ (10–19 mm) long; at base of lower leaves. *Tiny reddish-green female* flowers in clusters ⁵⁄₁₆–½″ (8–12 mm) long; at base of upper leaves of same plant. Both in early spring.

Fruit: ¼–⁵⁄₁₆″ (6–8 mm) in diameter; *brownish-purple,* warty with *whitish wax coat,* 1-seeded; several along stalk at leaf base; maturing in early autumn.

Habitat: Moist sand dunes, hillsides, and canyons; forming thickets with coastal scrub, Redwood, and Shore Pine.

Range: SW. Washington south near Pacific Coast to S. California; to 500′ (152 m).

Pacific Bayberry is sometimes planted as an ornamental shrub for the showy berries and dense, shiny evergreen foliage. The fruit is eaten in small quantities by myrtle warblers and many other birds. The waxy covering of the fruit apparently is not used; colonists extracted the wax from related eastern bayberries or waxmyrtles in boiling water and made fragrant-burning candles.

WALNUT FAMILY
(Juglandaceae)

Deciduous, aromatic trees, including hickories and Pecan (*Carya*) and Butternut and walnuts (*Juglans*). About 50 species worldwide in north temperate and tropical regions; 17 in North America, others southward.
Leaves: mostly alternate, odd pinnately compound, without stipules; leaflets with toothed border and resin dots beneath.
Flowers: male and female on same tree; tiny, greenish. Male, usually many in long, narrow catkins, composed of 3 bracts, 4 or fewer sepals, no corolla, and 3–40 or more stamens. Female, few or only 1–2, composed of 3 bracts and 1 pistil in short, erect clusters.
Fruit: a nut with hard shell, often splitting open or sometimes winged, or a drupe with large, oily, edible seed.

---

294 **Southern California Walnut**
"California Walnut"
"California Black Walnut"
Juglans californica Wats.

Description: Small to medium-sized tree, usually forked near base, with rounded crown.
Height: 40' (12 m).
Diameter: 1' (0.3 m).
Leaves: pinnately compound; 6–9" (15–23 cm) long; axis with hairy glands. *11–15 leaflets* 1–2½" (2.5–6 cm) long; *oblong lance-shaped, short- or long-pointed,* finely *saw-toothed;* stalkless. Shiny green above, paler, with tufts of hairs along veins beneath; turning yellow or brown in autumn.
Bark: dark brown or blackish, rough and furrowed into broad ridges.
Twigs: brown, stout, with chambered pith.
Flowers: small, greenish; in early spring. Male with 30–40 stamens,

numerous, in catkins 2–3″ (5–7.5 cm) long. Female with 2-lobed style, few, at tip of same twig.

Fruit: 1–1¼″ (2.5–3 cm) in diameter; *walnut* with thin, dark brown husk, thick, *grooved shell,* and edible seed; maturing in early autumn.

Habitat: Moist soils along streams in canyons and foothills; with cottonwoods, willows, and sycamores.

Range: Coastal S. California only; to 2500′ (762 m).

First found north of Los Angeles in 1850 by the survey of the Mexican Boundary Commission, this species is now established by planting beyond its local range. It is often small and shrublike; in contrast, Northern California Walnut becomes a tall tree with larger nuts to 2″ (5 cm) in diameter.

---

291 Northern California Walnut
"Hinds Walnut"
Juglans hindsii Jeps. ex R.E. Smith

Description: Tree with single, tall trunk and narrow, rounded, dense crown.
Height: 30–70′ (9–21 m).
Diameter: 1–2′ (0.3–0.6 m).
Leaves: pinnately compound; 7–12″ (18–30 cm) long; axis covered with soft hairs. *Usually 15–19 leaflets* 2½–4″ (6–10 cm) long; *lance-shaped, long-pointed, saw-toothed,* stalkless. Shiny green above, paler and hairy with tufts of hairs along veins beneath.
Bark: gray-brown, furrowed into narrow ridges.
Twigs: brown, stout, with chambered pith.
Flowers: small, greenish; in early spring. Male with 30–40 stamens, numerous, in catkins 3–5″ (7.5–13 cm) long. Female with 2-lobed style, few at tip of same twig.

Fruit: 1⅜–2″ (3.5–5 cm) in diameter; *walnut* with dark brown, thin husk; thick, *nearly smooth shell* and edible seed; maturing in early autumn.

Habitat: Moist soils of stream borders and valleys; with cottonwoods, willows, and sycamores.

Range: Local in central California; to 500′ (152 m).

The original range of this tree is uncertain. It is now naturalized along streams and perhaps was spread earlier by Indians around their camp sites. It was discovered along the lower Sacramento River in 1837 by Richard Brimsley Hinds (1812–47), a British botanist. Grown as a street tree in central California, it is also an important hardy stock for grafting English Walnuts.

292 Arizona Walnut

"Arizona Black Walnut" "Nogal"
Juglans major (Torr.) Heller

Description: Tree often with forked trunk and rounded crown of widely spreading branches and with distinct walnut odor.

Height: 30–50′ (9–15 m).

Diameter: 1–2′ (0.3–0.6 m).

Leaves: pinnately compound; 7–14″ (18–36 cm) long. *Usually 9–13 leaflets* 2–4″ (5–10 cm) long; *broadly lance-shaped,* often slightly curved, *coarsely saw-toothed;* covered with scurfy hairs when young, becoming hairless or nearly so; yellow-green.

Bark: gray-brown, smoothish, becoming thick and deeply furrowed into ridges.

Twigs: brown, stout, with chambered pith.

Flowers: small, greenish; in early spring. Male with 30–40 stamens, numerous, in drooping catkins. Female with 2-lobed style, few at tip of same twig.

Fruit: 1–1½" (2.5–4 cm) in diameter; *walnut* with thin, densely hairy, brown husk; thick, grooved, hard shell; small, edible seed; maturing in early autumn.

Habitat: Moist soils along streams and canyons, mostly in mountains, in desert, desert grassland, and oak woodland zones.

Range: Central Texas west to Arizona; also Mexico; at 2000–7000' (610–2134 m).

The small walnuts, known in Spanish as *nogales,* are gathered locally. The wood, like that of Black Walnut, is used for furniture and gunstocks, but the supply is limited. The valuable enlarged burls and bases of the trunks make beautifully patterned tabletops and veneer.

293, 340 Little Walnut
"Texas Walnut" "Nogal"
Juglans microcarpa Berland.

Description: Large shrub or small tree, usually branching near ground, with broad rounded crown, small nuts, and characteristic walnut odor.
Height: 10–20' (3–6 m).
Diameter: ½–1½' (0.15–0.5 m).
Leaves: pinnately compound; 8–13" (20–33 cm) long. *Usually 7–13 leaflets 2–3" (5–7.5 cm) long; narrowly lance-shaped,* long-pointed, usually *slightly curved, finely saw-toothed or almost without teeth,* becoming hairless or nearly so; very short-stalked. Yellow-green; turning yellow in autumn.
Bark: gray, smooth to deeply furrowed.
Twigs: gray, slender, with brown *chambered pith.*
Flowers: small; greenish; in early spring. Male with about 20 stamens, many in catkins. Female with 2-lobed style, few at tip of same twig.

Fruit: ½–¾" (12–19 mm) in diameter; thin, hairy *husk becoming brown;* nut with hard, grooved, thick shell; and *very small,* edible seed; maturing in early autumn.

Habitat: Moist soils along streams in plains and foothills, grasslands, and deserts.

Range: SW. Kansas west to New Mexico and south to S. Texas; also NE. Mexico; at 1500–4000′ (457–1219 m).

Squirrels and other rodents consume these nuts, which are mostly shell. The common and scientific names describe the tiny marblelike fruit, the smallest of the walnuts.

BIRCH FAMILY
(Betulaceae)

Trees, often large, and some shrubs, including alders (*Alnus*), hornbeams (*Carpinus*), and hophornbeams (*Ostrya*), as well as birches (*Betula*). About 135 species worldwide; about 20 native and 1 naturalized tree species and 8 shrub species in North America.

Leaves: deciduous, alternate, often spreading in 2 rows, simple, mostly ovate or elliptical, doubly saw-toothed with several, nearly straight side veins; paired stipules shedding early.

Bark: mostly smooth (peeling in papery layers in birch, *Betula*).

Flowers: male and female on same plant, usually in early spring before or with leaves; tiny, greenish, with 4 to no sepals and no petals. Male in long narrow catkins, with 1–20 stamens; female in short conelike or headlike clusters, with 1 pistil.

Fruit: usually many in conelike cluster, small nuts or nutlets; often short-winged; 1-seeded.

---

190 **Arizona Alder**
"New Mexican Alder" "Mexican Alder"
Alnus oblongifolia Torr.

Description: Tall, straight-trunked tree with open, rounded crown.
Height: 80' (24 m).
Diameter: 2' (0.6 m).
Leaves: *in 3 rows;* 1½–3¼" (4–8 cm) long, 1–1½" (2.5–4 cm) wide. *Ovate or elliptical,* usually *doubly saw-toothed* but not lobed, with *7–10* nearly straight *parallel veins* on each side. *Dark green* and almost hairless above, paler and often finely hairy with tufts of rust-colored hairs in vein-angles beneath.
Bark: *dark gray, smooth,* thin, becoming fissured and scaly.
Twigs: slender, brown, finely hairy

when young, with 3-angled pith.
Flowers: tiny; in early spring before
leaves. Male yellowish, in 3–4
drooping, narrowly cylindrical catkins
2–3″ (5–7.5 cm) long. Female reddish,
in narrow cones ¼″ (6 mm) long.
Cones: ½–¾″ (12–19 mm) long; 3–8
clustered on short stalks, elliptical,
with many hard black scales; remaining
attached; tiny, elliptical, flat nutlets;
maturing in late summer.

Habitat: Wet canyon soils in mountains and
along streams.

Range: SW. New Mexico and Arizona; local in
N. New Mexico; also in N. Mexico; at
4500–7500′ (1372–2286 m).

This is one of the largest native alders
and a handsome tree of rocky canyon
bottoms. It is recognized by its
smooth, dark gray bark. The soft wood
is used as fuel.

191, 439 White Alder
"Sierra Alder"
Alnus rhombifolia Nutt.

Description: Medium-sized to large tree with tall,
straight trunk and open, rounded
crown; showy in winter with long,
golden-colored male catkins hanging
from slender, leafless twigs.
Height: 70′ (21 m).
Diameter: 2′ (0.6 m).
Leaves: *in 3 rows;* 2–3½″ (5–9 cm)
long, 1½–2″ (4–5 cm) wide. *Ovate* or
elliptical, *finely saw-toothed* but not
lobed, slightly thickened, with *9–12*
nearly straight, *parallel veins* on each
side. Dull *dark green,* hairless or nearly
so, and often with tiny gland-dots
above, light yellow-green and slightly
hairy beneath.
Bark: light or dark brown, fissured into
flat, scaly ridges.
Twigs: slender, light green, finely hairy
when young, with 3-angled pith.

Flowers: tiny; in winter and early spring before leaves. Male yellowish in drooping, narrowly cylindrical catkins 1½–5″ (4–13 cm) long. Female reddish in narrow cones ⅜″ (10 mm) long.

Cones: ⅜–¾″ (10–19 mm) long; 3–7 clustered on short stalks, elliptical, with many hard, black scales; remaining closed until early spring; tiny, elliptical, flat nutlets; maturing in late summer.

Habitat: With chaparral and Ponderosa Pine in foothill woodlands.

Range: W. Idaho and Washington south in mountains to W. Nevada and S. California; at 100–8000′ (30–2438 m); generally below 5000′ (1524 m).

White Alder, named for its pale green foliage, is the only alder native in southern California. Limited to permanent streams, it is a good indicator of water. It is sometimes planted as an ornamental in wet sites.

187 Red Alder
"Oregon Alder" "Western Alder"
Alnus rubra Bong.

Description: Graceful tree with straight trunk, pointed or rounded crown, and mottled, light gray to whitish, smooth bark.
Height: 40–100′ (12–30 m).
Diameter: 2½′ (0.8 m), sometimes larger.
Leaves: *in 3 rows;* 3–5″ (7.5–13 cm) long, 1¾–3″ (4.5–7.5 cm) wide. *Ovate* to elliptical, short-pointed at both ends, slightly thickened, *wavy-lobed* and *doubly saw-toothed,* edges slightly turned under, with *10–15* nearly straight *parallel veins* on each side. *Dark green* and usually hairless above, gray-green with *rust-colored hairs* beneath.
Bark: mottled light gray to whitish, smooth or becoming slightly scaly,

thin; inner bark reddish-brown.

Twigs: slender, light green, covered with gray hairs when young, with 3-angled pith.

Flowers: tiny; in spring before leaves. Male yellowish, in drooping, narrowly cylindrical catkins 4–6″ (10–15 cm) long, ¼″ (6 mm) wide. Female reddish, in narrow cones ⅜–½″ (10–12 mm) long.

Cones: ½–1″ (1.2–2.5 cm) long; 4–8 on short stalks, elliptical, with many hard black scales; remaining attached; tiny, rounded, flat nutlets with 2 narrow wings; maturing in late summer.

Habitat: Moist soils including loam, gravel, sand, and clay, along streams and lower slopes; often in nearly pure stands.

Range: SE. Alaska southeast to central California; also local in N. Idaho; to 2500′ (762 m).

The leading hardwood in the Pacific Northwest, Red Alder is used for pulpwood, furniture, cabinetwork, and tool handles. It is planted as an ornamental in wet soils and is a pioneer on landslides, roadsides, and moist sites after logging or fire. Red Alder thickets are short-lived and serve as a cover for seedlings of the next coniferous forest. Alder roots, like those of legumes, often have swellings or root nodules containing nitrogen-fixing bacteria, which enrich the soil by converting nitrogen from the air into chemicals like fertilizers that the plants can use. The common name describes the reddish-brown inner bark and heartwood.

189, 320   Speckled Alder
"Tag Alder" "Gray Alder"
Alnus rugosa (Du Roi) Spreng.

Description: A low and clump-forming shrub;
sometimes a small tree.
Height: 20' (6 m).
Diameter: 4" (10 cm).
Leaves: *in 3 rows;* 2–4" (5–10 cm) long,
1¼–3" (4–7.5 cm) wide. *Elliptical* or
ovate, broadest near or below middle,
doubly and irregularly saw-toothed and
wavy-lobed, with *9–12* nearly straight
parallel veins on each side; short, hairy
stalks. *Dull dark green* with network of
sunken veins above, whitish-green and
often with soft hairs and with
prominent veins and veinlets arranged
in rows like a ladder beneath.
Bark: gray, smooth.
Twigs: gray-brown, slender, slightly
hairy when young; with 3-angled pith.
Flowers: tiny; in early spring before
leaves. Male yellowish, in drooping
catkins 1½–3" (4–7.5 cm) long.
Female reddish, in cones ¼" (6 mm)
long.

Cones: ½–⅝" (12–15 mm) long;
elliptical, blackish, hard, short-stalked;
maturing in autumn; with tiny,
rounded, flat nutlets.

Habitat: Wet soils along streams and lakes and
in swamps.

Range: Widespread across Canada from
Yukon and British Columbia to
Newfoundland, south to West
Virginia, west to NE. Iowa, and north
to NE. North Dakota; almost to
northern limit of trees; in south to
2600' (792 m).

The Latin species name, meaning
"rugose" or "wrinkled," refers to the
network of sunken veins prominent on
the lower leaf surfaces. It is planted as
an ornamental at water edges. Alder
thickets provide cover for wildlife,
browse for deer and moose, and seeds
for birds.

192, 441 Sitka Alder
"Mountain Alder" "Wavyleaf Alder"
Alnus sinuata (Regel) Rydb.

Description: Thicket-forming shrub or small tree,
often with several trunks, and with
shiny yellow-green leaves, gummy
when young.
Height: 30' (9 m).
Diameter: 8" (20 cm).
Leaves: 2½–5" (6–13 cm) long, 1½–3"
(4–7.5 cm) wide. *Ovate, shallowly
wavy-lobed* and *doubly saw-toothed* with
long-pointed teeth and 6–10 nearly
straight *parallel veins* on each side;
gummy or sticky when young. *Shiny,
speckled yellow-green on both surfaces,*
paler and often slightly hairy beneath.
Bark: gray to light gray, smooth and
thin; inner bark red.
Twigs: *gummy,* finely hairy, and orange-
brown when young; becoming light
gray, slender, and slightly zigzag.
Flowers: tiny; in spring with or after
leaves. Male flowers yellowish,
drooping, narrowly cylindrical; in
catkins 3–5" (7.5–13 cm) long, ⅜"
(10 mm) wide. Female flowers reddish,
in narrow cones ⅜" (10 mm) long.
Cones: ½–¾" (12–19 mm) long; 3–6
clustered on slender, spreading, long
stalks; elliptical, with many hard, black
scales; remaining attached. Tiny,
elliptical, flat nutlets with *2 broad
wings;* maturing in summer.
Habitat: Along streams and lakes and in valleys.
Range: SW. and central Alaska and Yukon
southeast to NW. California and
central Montana; in Alaska to alpine
zone above timberline; in NW.
California to 7000' (2134 m).

In Alaska, Sitka Alder is a pioneer in
disturbed areas, following landslides,
logging, and glacial retreat. Adapted to
soils too barren for other trees, this
species improves soil conditions by
adding organic matter and nitrogen
from bacteria in its root nodules. It acts

as a short-lived nurse tree for Sitka
Spruce, later dying when shaded by the
larger conifer.

188, 440 Mountain Alder
"Thinleaf Alder" "River Alder"
Alnus tenuifolia Nutt.

Description: Shrub with spreading, slender branches
or sometimes a small tree with several
trunks and a rounded crown; often
forming thickets.
Height: 30′ (9 m).
Diameter: 6″ (15 cm).
Leaves: *in 3 rows;* 1½–4″ (4–10 cm)
long, 1–2½″ (2.5–6 cm) wide. *Ovate* or
elliptical, *wavy-lobed* and *doubly saw-
toothed, rounded at base,* with 6–9 nearly
straight *parallel veins* on each side. *Dull
dark green* above, light yellow-green and
sometimes finely hairy beneath.
Bark: gray, thin, smooth, becoming
reddish-gray and scaly.
Twigs: slender, reddish and hairy when
young, becoming gray, with 3-angled
pith.
Flowers: tiny; in early spring before
leaves. Male yellowish, in catkins 1–
2¾″ (2.5–7 cm) long. Female
brownish, in narrow cones ¼″ (6 mm)
long.
Cones: ⅜–⅝″ (10–15 mm) long; 3–9
clustered on short stalks; elliptical,
with many hard black scales; maturing
in late summer and remaining attached.
Tiny, elliptical, flat nutlets.
Habitat: Banks of streams, swamps, and
mountain canyons in moist soils.
Range: Central Alaska, Yukon, and Mackenzie
southeast mostly in mountains to New
Mexico and central California; near sea
level in north; to 9000′ (2743 m) in
south.

This is the common alder throughout
the Rockies. The Navajo Indians
made a red dye from the powdered bark.

205 Water Birch
"Red Birch" "Black Birch"
Betula occidentalis Hook.

Description: Shrub or small tree with rounded crown of spreading and drooping branches; usually forming clumps and often in thickets.

Height: 25' (7.6 m).

Diameter: 6–12" (15–30 cm).

Leaves: 1–2" (2.5–5 cm) long, ¾–1" (2–2.5 cm) wide. *Ovate,* sharply and often doubly saw-toothed, usually with *4–5 veins on each side.* Dark green above, pale yellow-green with tiny gland-dots beneath; turning dull yellow in autumn.

Bark: *shiny, dark reddish-brown, smooth,* with horizontal lines, not peeling.

Twigs: greenish, slender, with gland-dots.

Flowers: tiny; in early spring. Male yellowish, with 2 stamens, many in long, drooping catkins near tip of twigs. Female greenish in short, upright catkins back of tip of same twig.

Cones: 1–1¼" (2.5–3 cm) long; cylindrical, brownish, upright or spreading on slender stalk; with many 2-winged nutlets; maturing in late summer.

Habitat: Moist soils along streams in mountain canyons; usually in coniferous forests and with cottonwoods and willows.

Range: NE. British Columbia east to S. Manitoba and south to N. New Mexico and California; at 2000–8000' (610–2438 m).

 This uncommon but widespread species is the only native birch in the Southwest and the southern Rocky Mountains. Sheep and goats browse the foliage.

206 Paper Birch
"Canoe Birch" "White Birch"
Betula papyrifera Marsh.

Description: One of the most beautiful native trees, with narrow, open crown of slightly drooping to nearly horizontal branches; sometimes a shrub.
Height: 50–70′ (15–21 m).
Diameter: 1–2′ (0.3–0.6 m).
Leaves: 2–4″ (5–10 cm) long, 1½–2″ (4–5 cm) wide. *Ovate, long-pointed,* coarsely and doubly saw-toothed, usually with *5–9 veins on each side.* Dull dark green above, light yellow-green and nearly hairless beneath; turning light yellow in autumn.
Bark: *chalky to creamy white, smooth, thin,* with long horizontal lines; separating into *papery strips* to reveal orange inner bark; becoming brown, furrowed, and scaly at base; bronze to purplish in varieties.
Twigs: reddish-brown, slender, mostly hairless.
Flowers: tiny; in early spring. Male yellowish, with 2 stamens, many in long, drooping catkins near tip of twigs. Female greenish in short, upright catkins back of tip of same twig.

Cones: 1½–2″ (4–5 cm); narrowly cylindrical, brownish, hanging on slender stalk, with many 2-winged nutlets; maturing in autumn.
Habitat: Moist upland soils and cutover lands; often in nearly pure stands.
Range: Transcontinental across North America near northern limit of trees from NW. Alaska, east to Labrador, south to New York, and west to Oregon; local south to N. Colorado and W. North Carolina; to 4000′ (1219 m), higher in southern mountains.

Paper Birch is used for specialty products such as ice cream sticks, toothpicks, spools, and toys, as well as pulpwood. Indians made their

lightweight, birchbark canoes by stretching the stripped bark over frames of Northern White-cedar, sewing it with thread from Tamarack roots, and caulking the seams with pine or Balsam Fir resin. Souvenirs of birch bark should always be from a fallen log, since stripping bark from living trees leaves permanent, ugly black scars.

208 European White Birch
"European Birch"
Betula pendula Roth

Description:

Ornamental, planted tree with open, pyramid-shaped or spreading crown of long, *drooping branches.*
Height: 50′ (15 m).
Diameter: 1′ (0.3 m).
Leaves: 1¼–2¾″ (3–7 cm) long, 1–1½″ (2.5–4 cm) wide. *Ovate* or nearly *triangular,* very long-pointed at tip, blunt or almost straight at base; *doubly saw-toothed,* with 6–9 veins on each side, sticky when young; long-stalked. Dull green above, paler beneath; turning yellow in autumn.
Bark: *white, smooth,* flaky, peeling in *papery* strips.
Twigs: slender, *drooping,* with tiny resin gland-dots.
Flowers: tiny; in early spring. Male yellowish, with 2 stamens, many in long, drooping catkins near tip of twigs. Female greenish, in short, upright catkins back of tip of same twig.

Cones: ¾–1¼″ (2–3 cm) long; cylindrical, hanging on slender stalks, composed of many small, *2-winged nutlets* and 3-lobed bracts; maturing in autumn.

Habitat: Moist soils on lawns and in parks and cemeteries. Sometimes an escape in thickets and open forest areas.

Range: Native of Europe and Asia Minor. Planted across the United States.

European White Birch is a graceful,
short-lived ornamental, grown for its
white, papery bark and drooping
branches. Commonly cultivated
varieties have very long branches and
finely divided or lobed leaves.

---

199 Chisos Hophornbeam
"Big Bend Hophornbeam"
Ostrya chisosensis Correll

Description: Tree with cylindrical crown of few
slender branches and with fruit like
hops; occurring only in the Chisos
Mountains of Trans-Pecos Texas.
Height: 40' (12 m).
Diameter: 10" (25 cm).
Leaves: 1½–2½" (4–6 cm) long, ¾–
1¼" (2–3 cm) wide. *Elliptical,* often
widest beyond middle, blunt or short-
pointed at tip, rounded and slightly
notched at base; finely *doubly saw-
toothed,* slightly hairy, nearly stalkless.
Dark green above, paler beneath.
Bark: gray, finely fissured into *long
narrow scaly ridges,* shreddy, thin,
forming blocks at base.
Twigs: green, slender, willowy, hairy.
Flowers: in spring before leaves. Male
tiny, greenish; clustered in drooping,
narrowly cylindrical catkins 1¼–1½"
(3–4 cm) long. Female tiny, greenish;
in short catkins.
Fruit: ¾" (19 mm) long; *conelike*
clusters, densely hairy; composed of
(usually) 6 small, light brown nutlets
each within a swollen, light brown,
egg-shaped cover that is *papery* and
sacklike; maturing in summer.
Habitat: Moist, rocky soils of canyons.
Range: Local in Chisos Mountains, Trans-Pecos
Texas; at about 5000' (1524 m).

This rare tree was only distinguished
and named in 1965. Confined to the
Big Bend National Park, it is protected
as an endangered species.

200, 533　Knowlton Hophornbeam
"Western Hophornbeam" "Ironwood"
Ostrya knowltonii Cov.

Description:　Shrub or small tree with 1 or more trunks, a narrow, rounded crown of slender branches, and fruit like hops.
Height: 30′ (9 m).
Diameter: 6″ (15 cm).
Leaves: 1–2½″ (2.5–6 cm) long, ⅝–1½″ (1.5–4 cm) wide. *Elliptical,* sharply *doubly saw-toothed,* with 5–8 slightly curved veins. *Yellow-green* and slightly hairy above, paler and with soft hairs beneath; turning yellow in autumn. Hairy leafstalks less than ¼″ (6 mm) long.
Bark: gray, finely fissured into *long, narrow, scaly ridges;* shreddy, thin.
Twigs: slender, tough, green, densely hairy when young.
Flowers: tiny, greenish; in spring before leaves. Male clustered in drooping, *narrowly cylindrical* catkins ¾–1¼″ (2–3 cm) long. Female in *narrowly cylindrical* catkins ¼–⅜″ (6–10 mm) long; at end of twig.
Fruit: 1–1½″ (2.5–4 cm) long, ¾″ (19 mm) wide; *conelike* clusters composed of many small nutlets each within a swollen, light brown, egg-shaped, *papery,* and *sacklike* cover; maturing in late spring.

Habitat:　Moist canyons and mountains; with oaks, pinyons, junipers, and Ponderosa Pine.

Range:　SE. Utah, N. Arizona, SE. New Mexico, and Trans-Pecos Texas; at 4200–7000′ (1280–2134 m).

This local tree is found below both rims at Grand Canyon National Park, where is was discovered in 1889 by Frank H. Knowlton (1869–1926), a U.S. botanist and paleobotanist.

BEECH FAMILY
(Fagaceae)

Trees, often large, and some shrubs including chestnuts and chinkapins (*Castanea*) and oaks (*Quercus*), as well as beeches (*Fagus*). About 700–900 species nearly worldwide except tropical South America and tropical Africa. About 65 native and 1 naturalized tree species and 10 of shrubs in North America.

Leaves: usually deciduous or in warm climates evergreen; alternate, simple, mostly toothed or lobed, with narrow paired stipules shedding early.

Flowers: male and female on same plant, usually in early spring before or at same time as the leaves; tiny, without petals. Male mostly in long, narrow catkins composed of 4- to 7-lobed calyx and 4–40 stamens; 1–3 female flowers, often in spikes, consisting of cup of scales and 1 pistil.

Fruit: a 1-seeded nut or 2–3 nuts within a cup; often edible acorns, chestnuts, beechnuts, etc.

The oak genus (*Quercus*), including the most important native hardwoods, is mostly deciduous or in warmer climates evergreen. Leaves are alternate in 5 rows, often variable in shape, with lobed, toothed, or straight edges, often prominently veined, and short-stalked. Bark is light gray and scaly or blackish and furrowed. Twigs are slender, often slightly 5-angled and hairy, with star-shaped pith. Small, greenish flowers appear in early spring, male and female on same twig. Male flowers are composed of a bell-shaped calyx and usually 6 stamens; many are clustered in slender, drooping catkins; 1 or a few female flowers at leaf bases have a cup of many overlapping scales and a pistil with 3 styles protruding. The hard-shelled acorns, containing 1 large bitter seed, are borne within a cup of many overlapping scales. Two groups (subgenera) of oaks are distinguishable. The leaves and lobes of red (or black) oaks are bristle-tipped; bark is usually blackish and furrowed; the bitter acorns mature in autumn of the second year, with two sizes usually present. The leaves of white oaks are not bristle-tipped; bark is usually light gray and scaly; the less bitter and sometimes edible acorns mature the first year.

92, 501 Giant Chinkapin
"Golden Chinkapin"
"Goldenleaf Chestnut"
Castanopsis chrysophylla (Dougl.) A. DC.

Description: Evergreen tree with straight trunk, becoming grooved or fluted, and with stout, spreading branches and broad, rounded crown; also a shrub variety.
Height: 40–80' (12–24 m).
Diameter: 1–3' (0.3–0.9 m).
Leaves: *evergreen;* 2–5" (5–13 cm) long, ⅝–1½" (1.5–4 cm) wide. *Lance-shaped* or oblong; *thick* and leathery; edges slightly turned under, *without teeth. Shiny dark green* with scattered scales above, covered with tiny *golden-yellow scales* beneath; turning yellow before falling. Shrub variety has leaves folded upward.
Bark: gray and smooth when young, becoming reddish-brown, thick, deeply furrowed into plates; inner bark bright red.
Twigs: stiff, scurfy with tiny, golden-yellow scales when young, turning dark reddish-brown.
Flowers: about ⅛" (3 mm) long; *whitish,* stalkless; many in catkins 2–2½" (5–6 cm) long; upright near ends of twigs; mostly male; few female at base or in separate clusters; mostly in early summer.
Fruit: 1–1½" (2.5–4 cm) in diameter; a nearly stalkless, *spiny bur;* maturing in

autumn of second year, splitting irregularly into 4 parts. 1–2 nuts ⁵⁄₁₆–½" (8–12 mm) long; broadly egg-shaped or rounded, light brown, hard-shelled, edible.

Habitat: Gravelly and rocky soils in mountain slopes and canyons in Redwood and evergreen forests. Shrub variety on dry ridges in chaparral and Knobcone Pine forests.

Range: Pacific Coast region from SW. Washington south in Coast Ranges to central California; also local in the Sierra Nevada of central California; to 1500' (457 m); shrub variety to 6000' (1829 m).

A handsome tree with a massive trunk, this species attains a large size. The showy, whitish blossoms have a strong odor. The edible nuts, like small chestnuts, borne in small quantities, are usually consumed by chipmunks and ground squirrels.

196, 337 European Beech
Fagus sylvatica L.

Description: Cultivated tree with stout trunk and dense, rounded crown of spreading branches extending almost to ground; producing edible beechnuts.
Height: 70' (21 m).
Diameter: 2½' (0.8 m).
Leaves: spreading in *2 rows:* 2–4" (5–10 cm) long, 1½–3" (4–7.5 cm) wide. *Broadly elliptical* or ovate, edges with small teeth, with 5–9 straight parallel veins on each side, hairy when young, short-stalked. *Shiny dark green* above, light green beneath; turning reddish-brown or bronze in autumn.
Bark: dark gray, smooth.
Twigs: gray or brown, finely hairy when young, slender, ending in long, narrow, brown, scaly buds; with short side twigs or spurs.

Fruit: 1″ (2.5 cm) long; a light brown *prickly bur;* maturing in autumn, splitting into 4 parts; usually containing 2 triangular, shiny, brown, edible seeds, known as *beechnuts.*

Habitat: Tolerates most soils; best in calcareous or deep sandy loam; hardy in cool, moist temperate regions.

Range: Native in central and S. Europe to high altitudes. Planted in NE. United States and Pacific states.

European Beech is one of the most popular, large shade trees. Numerous horticultural varieties include purple, copper, fernleaf, cutleaf, oakleaf, and roundleaf foliage and columnar and weeping habits. It can be pruned and clipped into arbors and hedges. An important hardwood in its native range, it forms extensive forests. Beechnuts serve as food for people and animals. The words "beech" and "book" come from the same root; ancient Saxons and Germans wrote on pieces of beech board.

171, 505   Tanoak
"Tanbark-oak"
Lithocarpus densiflorus (Hook. & Arn.) Rehd.

Description: Evergreen tree with a great central trunk and crown varying from narrow and conical to broad and rounded; sometimes a shrub.
Height: 50–80′ (15–24 m).
Diameter: 1–2½′ (0.3–0.8 m).
Leaves: *evergreen;* 2½–5″ (6–13 cm) long, ¾–2¼″ (2–6 cm) wide. *Oblong, thick* and leathery, with many *straight parallel* sunken side *veins;* with *wavy-toothed* border sometimes turning under stout, hairy leafstalks. *Shiny light green* and becoming hairless or nearly so above; with *whitish* or yellowish *hairs,* woolly when young, beneath.
Bark: brown, thick, deeply furrowed into ridges and plates.

Twigs: stout, with dull yellow hairs.
Flowers: numerous, tiny, stalkless,
whitish flowers in catkins 2–4″ (5–10
cm) long; with unpleasant odor;
upright from base of leaf; in early
spring, sometimes also in autumn;
usually all male, sometimes also 1–2
tiny, greenish female flowers at base.
Acorns: ¾–1¼″ (2–3 cm) long; *egg-
shaped;* 1–2 on stout, long stalk;
yellow-brown, with shallow, *saucer-
shaped cup* covered by long, *slender,
spreading scales;* maturing second year.

Habitat: Moist valleys and mountain slopes; in
oak forests and sometimes in nearly
pure stands.

Range: Pacific Coast from SW. Oregon south
to S. California and in Sierra Nevada to
central California; to 5000′ (1524 m).

Tanoak is placed in a separate genus
with more than 100 species native to
southeast Asia and Indomalaysia. While
the acorns resemble those of true oaks,
the flowers are like those of chinkapins
and chestnuts. Tanoak bark was once
the main commercial western source of
tannin. Indians ground flour from the
large acorns after removing the shells
and washing the seeds in hot water to
remove the bitter taste.

175 **Coast Live Oak**
"California Live Oak" "Encina"
Quercus agrifolia Née

Description: Evergreen tree with short, stout trunk;

many large, crooked, spreading
branches; and broad, rounded crown;
sometimes shrubby.
Height: 30–80′ (9–24 m).
Diameter: 1–3′ (0.3–0.9 m) or more.
Leaves: *evergreen;* ¾–2½″ (2–6 cm)
long, ½–1½″ (1.2–4 cm) wide. *Oblong*
or elliptical, short-pointed or rounded
at both ends; with edges turned under
and bearing *spiny teeth;* thick and

leathery. *Shiny dark green* above, yellow-green and often hairy beneath.
Bark: dark brown, thick, deeply furrowed.
Acorns: 1–1½" (2.5–4 cm) long; *narrowly egg-shaped*, ⅓ enclosed by *deep, thin cup* with many brownish, finely hairy scales outside and silky hairs inside; 1 or few together; stalkless; maturing first year.

Habitat: In valleys and on slopes, usually in open groves; often with Canyon Live Oak and California Black Oak.

Range: Coast Ranges mostly, central to S. California, including Santa Cruz and Santa Rosa Islands; also N. Baja California; to about 3000' (914 m).

This is the common oak of the California coast and foothills, forming parklike groves that often appear in the scenery of motion pictures made in Hollywood. The acorns were among those preferred by Indians; after removing the shells, they ground the seeds into meal, which was washed to remove the bitter taste, and boiled into mush or baked in ashes as bread.

173, 508   Arizona White Oak
"Arizona Oak"
Quercus arizonica Sarg.

Description: Medium-sized evergreen tree with irregular, spreading crown of stout branches.
Height: 30–60' (9–18 m).
Diameter: 1–2' (0.3–0.6 m).
Leaves: *evergreen;* 1½–3" (4–7.5 cm) long, ¾–1½" (2–4 cm) wide. *Oblong* or obovate; *slightly wavy-lobed* and toothed toward tip, base notched or rounded; thick and stiff. *Dull blue-green* and nearly hairless with sunken veins above, paler and densely hairy with raised veins beneath; shedding gradually in spring as new leaves unfold.

Bark: light gray, furrowed into narrow,
scaly plates and ridges.
Acorns: ¾–1" (2–2.5 cm) long; *oblong,*
about ⅓ enclosed by *deep cup* of finely
hairy scales thickened at base; 1–2 on
short stalk or stalkless; maturing first
year.

Habitat: Mountain slopes and canyons; in oak
woodland with other evergreen oaks.

Range: Trans-Pecos Texas west to Arizona; also
N. Mexico; at 5000–7500' (1524–
2286 m).

One of the largest southwestern oaks,
this handsome tree reaches its greatest
size in canyons and other moist sites.
Although a good fuel, the hard wood is
difficult to cut and split.

177, 504 **Canyon Live Oak**
"Canyon Oak" "Goldcup Oak"
Quercus chrysolepis Liebm.

Description: Evergreen tree with short trunk, large,
spreading, horizontal branches, and
broad, rounded crown; sometimes
shrubby.
Height: 20–100' (6–30 m).
Diameter: 1–3' (0.3–0.9 m) or
more.
Leaves: *evergreen;* 1–3" (2.5–7.5 cm)
long, ½–1½" (1.2–4 cm) wide.
Elliptical to oblong; short-pointed at
tip, rounded or blunt at base; with
edges turned under and *often with spiny
teeth* (especially on young twigs); *thick*
and leathery. *Shiny green* above; with
yellow hairs or *becoming gray* and nearly
hairless beneath.
Bark: light gray, nearly smooth or
scaly.
Acorns: ¾–2" (2–5 cm) long; variable
in size and shape, *egg-shaped.* turbanlike
with *shallow, thick cup* of scales densely
covered with *yellowish* hairs; stalkless or
short-stalked; maturing second year.

Habitat: In canyons and on sandy, gravelly, and

rocky slopes; in pure stands and mixed forests.

Range: SW. Oregon south through Coast Ranges and Sierra Nevada to S. California; local in W. Nevada and in W. and central Arizona; at 1000–6500' (305–1981 m); in Arizona at 5500–7500' (1676–2286 m).

Many consider this to be the most beautiful of the California oaks. The species name, meaning "golden-scale," refers to the yellowish acorn cups. The hard, heavy wood was used locally for farm implements and wagon axles and wheels. Another name, "Maul Oak," refers to the early use for heads of mauls or wedges for splitting Redwood ties.

242, 513 Blue Oak
"Mountain White Oak" "Iron Oak"
Quercus douglasii Hook. & Arn.

Description: Tree with short, leaning trunk; short, stout branches; broad, rounded crown; and brittle, hairy twigs; sometimes shrubby.
Height: 20–60' (6–18 m).
Diameter: 1' (0.3 m).
Leaves: 1¼–4" (3–10 cm) long, ¾–1¾" (2–4.5 cm) wide. *Oblong* or elliptical, rounded or blunt at both ends; *shallowly 4- or 5-lobed,* coarsely toothed or without teeth; thin but stiff. *Pale blue-green* and nearly hairless above, paler and slightly hairy beneath.
Bark: light gray, thin, scaly.
Acorns: ¾–1¼" (2–3 cm) long; *elliptical,* broad or narrow, with *shallow cup* of *warty scales;* stalkless or nearly so; maturing first season.
Habitat: Dry, loamy, gravelly, and rocky slopes; with other oaks and Digger Pine.
Range: N. to S. California mostly in foothills of Coast Ranges and Sierra Nevada; at 300–3500' (91–1067 m).

Recognized from a distance by the bluish foliage, this handsome California oak was named for its discoverer, David Douglas (1798–1834), the Scottish botanical explorer. It is used principally for fuel. The acorns, often abundant, are eaten by livestock as well as by wildlife.

244 California Scrub Oak
"Scrub Oak"
Quercus dumosa Nutt.

Description:

Evergreen, much-branched, thicket-forming shrub or small tree with rounded crown.
Height: 3–10' (0.9–3 m), sometimes to 25' (7.6 m).
Diameter: 6" (15 cm).
Leaves: *evergreen; small;* ⅝–1" (1.5–2.5 cm) long. *Oblong or elliptical,* short-pointed at ends; often 3- to 9-lobed, usually with short, *sharp or spiny teeth; thick and stiff.* Shiny green above, paler and hairy beneath.
Bark: gray or brown, scaly.
Twigs: stiff, slender, brown and hairy when young; becoming stout and gray.
Acorns: mostly ½–1" (1.2–2.5 cm) long; egg-shaped, ⅓ enclosed by a light brown *half-round, thick-walled cup* covered with many hairy, *warty,* overlapping *scales;* stalkless or nearly so; maturing first season.

Habitat: Dry, barren slopes, forming dense thickets; mainly in chaparral and foothill woodland.
Range: Coast Ranges and western base of Sierra Nevada in N. California south to N. Baja California; to 5000' (1524 m).

This common and variable shrubby oak hybridizes with tree species found nearby. Some plants classed as trees may be hybrids with size inherited from the larger parent. The scientific name means "bushy" or "shrubby."

178, 503 Dunn Oak
"Palmer Oak"
Quercus dunnii Kellogg

Description: Evergreen, thicket-forming shrub or
small tree with several small trunks.
Height: 6–20' (1.8–6 m).
Diameter: 4" (10 cm).
Leaves: *evergreen;* ½–1¼" (12–32 mm)
long, ½–1" (12–25 mm) wide. *Broadly
elliptical* to nearly round, curved and
wavy rather than flat, rounded to
notched at base; several *spiny teeth* at tip
and edges, with few spreading side
veins; *thick, stiff. Gray-green* above,
with white hairs beneath.
Bark: gray-brown, scaly.
Twigs: very stiff, widely forking, hairy.
Acorns: ¾–1¼" (2–3 cm) long; *egg-
shaped,* ¼ enclosed by *shallow,* thin,
scaly *cup;* short-stalked; maturing
second year.

Habitat: Dry slopes and canyons; in chaparral
and pinyon-juniper woodland.

Range: Extreme SW. New Mexico, Arizona,
and S. California and adjacent Mexico;
at 3000–6000' (914–1829 m).

Dunn Oak is closely related to Canyon
Live Oak and was once regarded as a
variety of that species.

172, 511 Emory Oak
"Black Oak" "Blackjack Oak"
Quercus emoryi Torr.

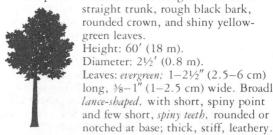

Description: Medium-sized evergreen tree with
straight trunk, rough black bark,
rounded crown, and shiny yellow-
green leaves.
Height: 60' (18 m).
Diameter: 2½' (0.8 m).
Leaves: *evergreen;* 1–2½" (2.5–6 cm)
long, ⅜–1" (1–2.5 cm) wide. Broadly
lance-shaped, with short, spiny point
and few short, *spiny teeth,* rounded or
notched at base; thick, stiff, leathery.

Both surfaces *shiny yellow-green* and nearly hairless; shedding gradually in spring as new leaves unfold.
Bark: black, thick, deeply furrowed into scaly plates.
Twigs: slender, stiff, finely hairy.
Acorns: ½–¾" (12–19 mm) long; *oblong,* ⅓–½ enclosed by *deep,* scaly *cup;* almost stalkless; slightly bitter but edible; maturing first year.

Habitat: On slopes of foothills and mountains and in canyons; common to abundant in oak woodland with other evergreen oaks.

Range: Trans-Pecos Texas west to central Arizona; also NW. Mexico; at 4000–7000' (1219–2134 m); rarely to 8000' (2438 m).

Emory Oak is the most characteristic tree of the oak woodland in mountains along the Mexican border. The acorns (*bellotas* in Spanish) are only slightly bitter and are gathered and eaten locally. They are also consumed in quantities by quail, wild turkeys, squirrels, and other wildlife. The foliage is browsed by deer and, to a lesser extent, by livestock. This species was named for its discoverer, William Hemsley Emory (1811–87), leader of two southwestern expeditions.

166 Engelmann Oak
"Evergreen White Oak" "Mesa Oak"
Quercus engelmannii Greene

Description: Tree with stout, spreading branches and broad, irregular crown, evergreen or nearly so.
Height: 20–60' (6–18 m).
Diameter: 1–2½' (0.3–0.8 m).
Leaves: *evergreen* or nearly so; 1–2¾" (2.5–7 cm) long, ½–1¼" (1.2–3.2 cm) wide. *Oblong* or obovate, blunt or rounded at ends; *often wavy-toothed; thick* and leathery. *Blue-green* or gray-green

and nearly hairless above, light yellow-green and usually hairy beneath; *shedding in spring* when new leaves appear.
Bark: light gray, thin, scaly.
Twigs: stiff, brown, densely hairy.
Acorns: ⅝–1″ (1.5–2.5 cm) long; *elliptical,* nearly ½ enclosed by *deep,* scaly *cup;* stalkless or on slender stalks; maturing first year.

Habitat: Dry slopes in foothills; with other oaks.
Range: SW. California and Santa Catalina Island; to 4000′ (1219 m).

This oak of limited distribution was named for George Engelmann (1809–84), German-born botanist of St. Louis.

246, 507   **Gambel Oak**
"Rocky Mountain White Oak"
"Utah White Oak"
Quercus gambelii Nutt.

Description: Tree with rounded crown, often in dense groves; or a thicket-forming shrub.
Height: 20–70′ (6–21 m).
Diameter: 1–2½′ (0.3–0.8 m).
Leaves: 2–6″ (5–15 cm) long, 1¼–3¼″ (3–8 cm) wide. *Elliptical* or oblong, rounded at tip, short-pointed at base; *deeply 7- to 11-lobed* halfway or more to middle, edges straight or wavy; varying in size, lobing, and hairiness. *Shiny dark green* and usually hairless above, paler and with soft hairs below; turning yellow and reddish in autumn.
Bark: gray, rough, thick, deeply furrowed or scaly.
Acorns: ½–¾″ (12–19 mm) long; *egg-shaped,* about ⅓ enclosed by *deep,* thick, scaly *cup;* 1–2 on short stalk or nearly stalkless; maturing first year.
Habitat: Slopes and valleys, in mountains, foothills, plateaus; scattered with Ponderosa Pine.
Range: N. Utah east to extreme S. Wyoming,

south to Trans-Pecos Texas, and west to S. Arizona; local in extreme NW. Oklahoma and S. Nevada; also N. Mexico; at 5000–8000' (1524–2438 m).

Gambel Oak is the common oak of the Rocky Mountains, abundant in Grand Canyon National Park. It is closely related to White Oak (*Quercus alba* L.) of the eastern United States. The foliage is browsed by deer and sometimes by livestock. Wild turkeys, squirrels, and other wildlife, as well as hogs and other domestic animals eat the sweetish acorns. The wood is used mainly for fenceposts and fuel. This species is named for William Gambel (1821–49), a naturalist from Philadelphia.

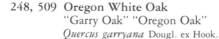

248, 509 **Oregon White Oak**
"Garry Oak" "Oregon Oak"
Quercus garryana Dougl. ex Hook.

Description: Tree with dense, rounded, spreading crown of stout branches; sometimes shrubby.
Height: 30–70' (9–21 m).
Diameter: 1–2½' (0.3–0.8 m).
Leaves: 3–6" (7.5–15 cm) long, 2–4" (5–10 cm) wide. *Elliptical,* blunt or rounded at both ends; *deeply lobed* halfway or more to midvein, with blunt or slightly toothed lobes; slightly thickened. *Shiny dark green* above, light green and usually hairy beneath; sometimes turning reddish in autumn.
Bark: light gray or whitish; thin; scaly or furrowed into broad ridges.
Acorns: 1–1¼" (2.5–3 cm) long; *elliptical,* ¼–⅓ enclosed by *shallow,* thin, scaly *cup;* stalkless or short-stalked; sweetish and edible.
Habitat: In valleys and on mountain slopes; often in pure stands and with other oaks.

Range: SW. British Columbia south to central
California in Coast Ranges and Sierra
Nevada; to 3000' (914 m) in north and
at 1000–5000' (305–1524 m) in
south.

The oak of greatest commercial
importance in the West, this species is
used for furniture, shipbuilding,
construction, cabinetwork, interior
finish, and fuel. It is the only native
oak in Washington and British
Columbia. The sweetish acorns, often
common in alternate years, are relished
by livestock and wildlife and were eaten
by Indians. Planted for shade and
ornament, it resembles the eastern
White Oak (*Quercus alba* L.).

240 Chisos Oak
Quercus graciliformis C. H. Muller

Description: Small tree with long, slender, arching
branches and drooping foliage; confined
to Chisos Mountains, Texas.
Height: 25' (7.6 m).
Diameter: 1' (0.3 m).
Leaves: nearly evergreen; 3–4" (7.5–10
cm) long, ¾–1¼" (2–3 cm) wide.
Narrowly *lance-shaped,* very long-
pointed and bristle-tipped, short-
pointed at base; with *8–10 deep lobes*
ending in *bristle-tipped teeth;* with
slender, flexible leafstalk. Yellow-green
when young, becoming dull green
above, slightly coppery-colored
beneath.
Bark: greenish-brown, smooth, slightly
warty.
Twigs: shiny red or brown; very
slender, angled, becoming hairless,
ending in small, shiny brown buds.

Acorns: ⅝" (15 mm) long; *narrowly egg-
shaped,* with small, *shallow cup,* brown;
1 or 2 grouped together and nearly
stalkless; maturing second year.
Habitat: Dry, rocky canyons.

Range: Local in Chisos Mountains, Trans-Pecos Texas; also Coahuila, Mexico; at 5500′ (1676 m).

This rare species is protected within Big Bend National Park. The scientific name, meaning "slender form," refers to the long, flexible branches. Chisos Oak is closely related to Canby Oak (*Quercus canbyi* Trel.), of Mexico, which has leaves with a greater number of lobes and acorns with a very shallow cup and is sometimes regarded as a northern variation of that species.

---

247 Graves Oak
"Chisos Red Oak"
Quercus gravesii Sudw.

Description: Small to medium-sized tree with very rough, blackish bark and with leaves deeply lobed and bristle-tipped; confined to Trans-Pecos Texas and adjacent Mexico.
Height: 40′ (12 m).
Diameter: 2′ (0.6 m).
Leaves: 2–4″ (5–10 cm) long and 1½–3¼″ (4–8 cm) wide. *Ovate,* long-pointed at tip, rounded or blunt at base; *deeply divided* into *3–7 short-pointed lobes,* with *few bristle-tipped teeth. Shiny dark green* and becoming nearly hairless above, dull light green and often with tufts of hairs in vein angles beneath; turning red in fall.
Bark: blackish or gray; very rough, furrowed into narrow, scaly ridges; gray and smooth on branches.

Acorns: ½–⅝″ (12–15 mm) long; *egg-shaped,* ¼–½ enclosed by *deep,* scaly *cup;* 1 or 2 on short stalk; maturing second year.
Habitat: Mountain woodlands, especially in moist canyons.
Range: Mountains of Trans-Pecos Texas and Coahuila, Mexico; at 4000–7000′ (1219–2134 m).

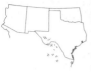

Common locally within its restricted range, it is easily seen in Big Bend National Park. It was named after Henry Solon Graves (1871–1950), chief of the United States Forest Service and later dean of Yale University School of Forestry.

107 Gray Oak
"Scrub Oak" "Shin Oak"
Quercus grisea Liebm.

Description: Low clump-forming shrub or small tree, sometimes medium-sized, with grayish foliage.
Height: 60' (18 m).
Diameter: 1' (0.3 m).
Leaves: deciduous or nearly evergreen; ¾–2" (2–5 cm) long, ⅜–¾" (10–19 mm) wide. *Elliptical* to ovate, base notched or rounded; sometimes with few short teeth toward tip; slightly thickened and rigid. *Gray-green* or blue-green, *shiny* and sparsely hairy above, dull and *finely hairy* with slightly raised veins beneath.

Bark: light gray, fissured into shaggy plates.
Acorns: ½–⅝" (12–15 mm) long; *egg-shaped,* ⅓–½ enclosed by *deep,* scaly *cup;* 1–2 on stalk or nearly stalkless; maturing first year.

Habitat: Dry, rocky slopes of mountains and foothills and in canyons; with other oaks, pinyon, and junipers.

Range: SW. Texas west to Arizona; also N. Mexico; at 5000–7000' (1524–2134 m).

Of greatest size in moist canyons, Gray Oak is most common as a shrub in New Mexico. It is easily seen in the Chisos Mountains in Big Bend National Park. It is closely related to Arizona White Oak, a larger tree with larger leaves and sunken veins. Common and scientific names describe the color of the foliage.

90 Silverleaf Oak
"White-leaf Oak"
Quercus hypoleucoides A. Camus

Description: Evergreen tree with rounded, spreading crown of foliage that is shiny above and silvery-white beneath; sometimes a clump-forming shrub.
Height: 30–60' (9–18 m).
Diameter: 1–2½' (0.3–0.8 m).
Leaves: *evergreen;* 2–4" (5–10 cm) long, ½–1" (1.2–2.5 cm) wide. *Lance-shaped,* sharp-pointed at tip, rounded at base; *edges rolled under* and usually without teeth or with a few small, spiny teeth; *very thick* and leathery. *Shiny yellow-green* above, densely covered with *white, woolly hairs* beneath.
Bark: blackish, deeply furrowed into ridges and plates.

Acorns: ½–⅝" (12–15 mm) long; *egg-shaped,* ⅓ enclosed by *deep,* thick, scaly *cup;* 1 or 2 on short stalk; maturing in 1 or sometimes 2 years.
Habitat: Mountain slopes and canyons; in oak woodland with other evergreen oaks.
Range: Trans-Pecos Texas west to SE. Arizona; also N. Mexico; at 5000–7000' (1524–2134 m).

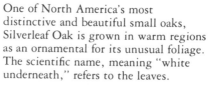

One of North America's most distinctive and beautiful small oaks, Silverleaf Oak is grown in warm regions as an ornamental for its unusual foliage. The scientific name, meaning "white underneath," refers to the leaves.

249 California Black Oak
"Black Oak" "Kellogg Oak"
Quercus kelloggii Newb.

Description: Tree with large branches and irregular, broad, rounded crown of stout, spreading branches.
Height: 30–80' (9–24 m).
Diameter: 1–3' (0.3–0.9 m).
Leaves: 3–8" (7.5–20 cm) long, 2–5"

(5–13 cm) wide. *Elliptical,* usually 7-*lobed* about halfway to midvein, each lobe with few *bristle-pointed teeth;* slightly thick. *Shiny dark green* above, light yellow-green and often hairy beneath; turning yellow or brown in autumn.

Bark: dark brown, thick, becoming furrowed into irregular plates and ridges; on small trunks, smooth, light brown.

Acorns: 1–1½″ (2.5–4 cm) long; *elliptical,* ⅓–⅔ enclosed by *deep,* thin, *scaly cup;* 1 or few on short stalk; maturing second year.

Habitat: Sandy, gravelly, and rocky soils of foothills and mountains; often in nearly pure stands and in mixed coniferous forests.

Range: SW. Oregon south in Coast Ranges and Sierra Nevada to S. California; at 1000–8000′ (305–2438 m).

This is the common oak in valleys of southwestern Oregon and in the Sierra Nevada. The large, deeply lobed leaves with bristle-tipped teeth differ from all other western oaks, but resemble those of Black Oak (*Quercus velutina* Lam.) of the eastern United States. Woodpecker drill holes in the bark and bury acorns there for future use, where they are safe from squirrels which cannot extract them. Slow-growing and long-lived, it is a popular fuelwood and hardy shade tree in dry soils. Deer and livestock browse the foliage.

245, 512 **Valley Oak**
"Valley White Oak"
"California White Oak"
Quercus lobata Née

Description: Large, handsome tree with stout, short trunk and large, *widely spreading branches* drooping at ends, forming broad, open crown.

Height: 40–100′ (12–30 m).
Diameter: 3–4′ (0.9–1.2 m),
sometimes much greater.
Leaves: 2–4″ (5–10 cm) long and 1¼–
2½″ (3–6 cm) wide. *Elliptical,* rounded
or blunt at both ends; *deeply 7- to 11-
lobed* more than halfway to midvein,
larger lobes broadest and notched at
end. *Dark green* and nearly hairless
above, paler and finely hairy beneath.
Bark: light gray or brown; thick,
deeply furrowed and broken
horizontally into thick plates.
Acorns: 1¼–2¼″ (3–6 cm) long;
oblong, pointed, ⅓ enclosed by *deep,*
half-round *cup* with light brown scales,
the lowest ones thick and *warty* or
knobby; sweetish and edible; maturing
first year.

Habitat: Valleys and slopes on rich loam soils;
forming groves in foothill woodland.

Range: N. to S. California, also Santa Cruz and
Santa Catalina islands; to 5000′
(1524 m).

Valley Oak is the largest of the western
deciduous oaks and a handsome,
graceful shade tree. This relative of the
eastern White Oak (*Quercus alba* L.) is
common through California's interior
valleys. Acorn crops, often abundant,
are consumed by many kinds of wildlife
and domestic animals, especially hogs.
California Indians roasted these large
acorns and also ground the edible
portion into meal which they prepared
as bread or mush.

---

241 **McDonald Oak**
"Island Scrub Oak"
Quercus macdonaldii Greene

Description: Small to medium-sized tree with
compact, rounded crown; occurring
only on California islands.
Height: 20–50′ (6–15 m).
Diameter: 1½′ (0.5 m).

Leaves: 1½–2¾" (4–7 cm) long, 1–1¾" (2.5–4.5 cm) wide. *Oblong* to obovate, *blunt* at both ends; with *2–4 blunt or sharp-pointed,* bristle-tipped *lobes* on each side; *thick* and stiff. Shiny *green* above, paler and hairy beneath with prominent network of veins beneath; shedding in winter.

Bark: gray or brown, scaly.

Twigs: gray, densely hairy.

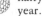

Acorns: ¾–1¼" (2–3 cm) long; *oblong,* nearly ⅓ enclosed by *deep,* half-round, hairy *cup;* stalkless; maturing first year.

Habitat: Ravines and canyons; in chaparral and woodlands.

Range: Santa Rosa, Santa Cruz, and Santa Catalina islands of S. California; absent from mainland; near sea level.

This rare and local oak apparently is a hybrid between California Scrub Oak and Valley Oak. It is named for Capt. James M. McDonald, who financed the publication of *Illustrations of West American Oaks* (1889) by Albert Kellogg and Edward L. Greene.

---

238 Mohr Oak
"Shin Oak" "Scrub Oak"
Quercus mohriana Buckl. ex Rydb.

Description: Thicket-forming shrub or sometimes small tree with rounded crown, usually evergreen.

Height: 20' (6 m).

Diameter: 8" (20 cm).

Leaves: *evergreen* or deciduous; 1–3" (2.5–7.5 cm) long, ½–1" (1.2–2.5 cm) wide. *Oblong* or elliptical, *edges straight or wavy* lobed with few teeth, *thick and leathery;* leafstalks very short. Shiny dark green above, with *dense gray hairs* and prominent veins beneath.

Bark: gray-brown, thin, furrowed.

Acorns: ⅜–⅝" (10–15 mm) long; *broadly elliptical,* green becoming

brown, about ½ enclosed by *deep cup;*
1–2 on short stalk; maturing first year.
Habitat: Plains and hills, especially in limestone
soils; forming thickets in oak brush or
"shinnery."
Range: W. Oklahoma, central, W., and Trans-
Pecos Texas, and NE. New Mexico;
also NE. Mexico; at 2000–4000' (610–
1219 m).

The names "Shin Oak" and "shinnery"
refer to the dense thickets, scarcely
knee-high, of dwarf evergreen oaks of
this and related species on the uplands
of western Texas and borders of
adjacent states.

170, 342, 510 **Chinkapin Oak**
"Chestnut Oak" "Rock Oak"
Quercus muehlenbergii Engelm.

Description: Tree with narrow, rounded crown;
characteristic of limestone uplands.
Height: 50–80' (15–24 m).
Diameter: 2–3' (0.6–0.9 m).
Leaves: 4–6" (10–15 cm) long, 1½–3"
(4–7.5 cm) wide. *Narrowly elliptical* to
obovate, pointed at tip, narrowed to
base; with *many straight, parallel side
veins,* each ending in curved tooth on
wavy edges; slightly thickened. Shiny
green above, *whitish-green* and covered
with tiny hairs beneath; turning brown
or red in fall.
Bark: light gray, thin, fissured and
scaly.
Acorns: ½–1" (1.2–2.5 cm) long; *egg-
shaped,* ⅓ or more enclosed by *deep, thin
cup of many overlapping,* hairy, long-
pointed, gray-brown *scales;* usually
stalkless; maturing first year.
Habitat: Mostly on limestone outcrops in
alkaline soils, including dry bluffs and
rocky river banks; often with other
oaks.
Range: S. Ontario east to W. Vermont, south
to NW. Florida, west to central Texas,

and north to Iowa; local in SE. New Mexico, Trans-Pecos Texas, and NE. Mexico; at 400–3000' (122–914 m).

The common name refers to the resemblance of the foliage to Chinkapins (*Castanea*), while the Latin species name honors Henry Ernst Muehlenberg (1753–1815), a Pennsylvania botanist. The wood is marketed as White Oak.

---

106 Mexican Blue Oak
Quercus oblongifolia Torr.

Description: Small evergreen tree with many branches and a spreading, rounded crown of bluish foliage; or a shrub.
Height: 30' (9 m).
Diameter: 20" (0.5 m).
Leaves: *evergreen;* 1–2" (2.5–5 cm) long, ½–¾" (12–19 mm) wide. *Oblong,* rounded at both ends or notched at base, *without teeth* (rarely wavy-toothed), thin and *stiff. Shiny blue-green* above, paler beneath; turning yellow and shedding in spring as new leaves appear. Short, stout stalks.
Bark: light gray, fissured into small rectangular or square plates.

Acorns: ½–¾" (12–19 mm) long; *egg-shaped,* ⅓ enclosed by *deep,* scaly *cup;* 1 on short stalk or stalkless; maturing first year.
Habitat: Foothills and mountains, sometimes in canyons; in oak woodland, often common, forming open groves at upper edge of desert grassland.
Range: Extreme SW. New Mexico, SE. Arizona, and N. Mexico; at 4500–6000' (1372–1829 m).

This handsome, small oak is limited to the Mexican border region. It is recognized by its light gray, checkered bark and small, blue-green, hairless leaves without teeth. Deer browse the

foliage. Although difficult to split, the
wood of Mexican Blue Oak is
sometimes used for fuel.

---

239 Sandpaper Oak
"Scrub Oak" "Shin Oak"
Quercus pungens Liebm.

Description: Evergreen, thicket-forming shrub or
small tree with small, rough, sharply
toothed leaves.
Height: 20' (6 m).
Diameter: 4" (10 cm).
Leaves: *evergreen;* mostly ¾–2" (2–5
cm) long, ⅜–¾" (10–19 mm) wide.
Oblong to elliptical, tip short-pointed or
blunt, base rounded or short-pointed;
edges wavy-lobed, with several *sharp teeth;*
thick and stiff. *Shiny green* and *rough like*
sandpaper with tiny, stiff hairs above,
densely covered with gray hairs and
with prominent veins beneath.
Bark: gray-brown, deeply furrowed into
scaly ridges.

Acorns: ½–⅝" (12–15 mm) long;
oblong, about ¼ enclosed by *shallow,*
scaly cup; 1 or 2 on very short stalk;
maturing first year.
Habitat: Dry, rocky slopes and hillsides, often
on limestone.
Range: Central Texas (Edwards Plateau) west to
SE. Arizona; also N. Mexico; at 2000–
5000' (610–1524 m).

Sandpaper Oak is one of a few species of
shrubby oaks in the Southwest; most do
not usually attain tree size. The
scientific name, meaning "pungent" or
"sharp-pointed," refers to the spiny-
toothed leaves.

243, 341 English Oak
Quercus robur L.

Description: Introduced tree with short, stout trunk,
widespreading branches, and broad,
rounded, open crown.
Height: 80′ (24 m), taller with age.
Diameter: 2–3′ (0.6–0.9 m).
Leaves: 2–5″ (5–13 cm) long, 1¼–2½″
(3–6 cm) wide. *Oblong,* with 6–14
shallow, rounded lobes, including 2
small, ear-shaped lobes at very *short-
stalked base.* Dark green above, pale
blue-green beneath.
Bark: dark gray, deeply and irregularly
furrowed, becoming thick.

Acorns: ⅝–1″ (1.5–2.5 cm) long; *egg-
shaped,* about ⅓ enclosed by *half-round
cup,* becoming brown; 1–5 on long,
slender stalk; maturing first year.

Habitat: Spreading from cultivation in moist
soils, along roadsides and forest edges.

Range: Native of Europe, N. Africa, and W.
Asia. Naturalized locally in SE. Canada
and NE. United States; also planted in
southeastern and Pacific states.

This noble oak is one of the most
characteristic British trees, attaining
very large size with age. It supplied
timbers for wooden ships of the Royal
Navy and oak paneling for the
Parliament and other famous buildings.
The bark was once a source of tannin.
Many horticultural varieties are
distinguished by crown shape, leaf
shape, and color.

174, 502 Netleaf Oak
Quercus rugosa Née

Description: Evergreen tree with broad, rounded
crown and broad leaves with raised
veins on lower surface; also a shrub.
Height: 40′ (12 m).
Diameter: 1′ (0.3 m).
Leaves: *evergreen;* 1–4″ (2.5–10 cm)

long, ¾–2¾" (2–7 cm) wide. *Broadly obovate,* varying in shape and size, slightly notched at base, *widest beyond middle,* with several *small, spiny teeth* mainly toward rounded tip, *thick and stiff. Dark green* and slightly hairy with *sunken veins above,* paler and covered with yellow hairs and with *network of raised veins beneath.*
Bark: gray, thin, fissured and scaly.
Acorns: ½–¾" (12–19 mm) long; *oblong,* ¼ enclosed by *shallow, scaly cup;* 1–3 on *long, slender stalk* of 1–2½" (2.5–6 cm); maturing first year.

Habitat: Mountain slopes and canyons; in oak woodland; with other evergreen oaks.

Range: Trans-Pecos Texas west to central Arizona; also south to S. Mexico; at 4000–8000' (1219–2438 m).

This uncommon oak of mountains along the Mexican border has acorns borne on a long stalk, as well as unusual leaves. The scientific name, meaning "wrinkled," refers to the veins of the leaves which are sunken above and raised in a network beneath.

169 Island Live Oak
"Island Oak"
Quercus tomentella Engelm.

Description: Evergreen tree with rounded crown, spreading branches, and fine white hairs on twigs, young leaves, and acorn cups.
Height: 20–40' (6–12 m).
Diameter: 1–2' (0.3–0.6 m).
Leaves: *evergreen;* 1¼–3½" (3–9 cm) long, ¾–2" (2–5 cm) wide. *Oblong* or lance-shaped; short-pointed at tip, blunt or rounded at base; edges mostly *wavy-toothed* and often turned under; *thick* and leathery, with many parallel side veins. *Shiny dark green* above, paler and hairy with prominent *network of veins* beneath.
Bark: reddish-brown, thin, scaly.

Twigs: covered with white hairs, turning brown.

Acorns: 1–1½" (2.5–4 cm) long; *egg-shaped,* with *shallow,* thin *cup* of *densely hairy* scales; stalkless or nearly so; maturing second year.

Habitat: Canyons and ravines; in chaparral woodland.

Range: Santa Rosa, Santa Cruz, Anacapa, Santa Catalina, and San Clemente islands of S. California and Guadalupe Island of Baja California; absent from mainland; near sea level.

One of the rarest species of native oaks, it was discovered on Guadalupe Island off the coast of Mexico in 1875 and later found on 5 islands off southern California. The scientific name means "with fine, woolly hairs."

108 Toumey Oak
Quercus toumeyi Sarg.

Description: Evergreen shrub or small tree with short trunk, broad, irregular crown of many branches, and numerous crowded, very *small leaves.*

Height: 30' (9 m).

Diameter: 8" (20 cm).

Leaves: *evergreen;* ½–1" (1.2–2.5 cm) long, ¼–½" (6–12 mm) wide. *Elliptical,* sharp-pointed, rounded at base; *without teeth* or sometimes with *few short, spiny teeth, slightly thick* and rigid. *Shiny yellow-green* above, pale and with fine hairs beneath; shedding in spring as new leaves appear.

Bark: dark brown, thin, scaly or flaky.

Acorns: ½–⅝" (12–15 mm) long; *oblong,* ¼ enclosed by *shallow, scaly cup;* stalkless; maturing first year.

Habitat: Mountain slopes; in oak woodland with other evergreen oaks.

Range: Extreme SW. New Mexico, SE. Arizona, and adjacent N. Mexico; at 4000–7000' (1219–2134 m).

Toumey Oak probably has the smallest leaves of all native tree oaks. While local in distribution, it is not rare. It is named for James William Toumey (1865–1932), the U.S. forester and botanist, who discovered this species in 1894.

179, 506 Turbinella Oak
"Shrub Live Oak" "Scrub Oak"
Quercus turbinella Greene

Description: Evergreen, much-branched, thicket-forming shrub or small tree with a spreading crown.
Height: 5–15' (1.5–4.6 m), sometimes larger.
Diameter: 4" (10 cm).
Leaves: *evergreen; small;* ⅝–1½" (1.5–4 cm) long, ⅜–¾" (10–19 mm) wide. *Elliptical* or oblong, short-pointed at tip, rounded or notched at base, *spiny-toothed, thick* and stiff. *Blue-green* with whitish bloom and nearly hairless above, yellow-green and with fine hairs beneath.
Bark: gray, fissured and scaly.
Acorns: ⅝–1" (1.5–2.5 cm) long; *narrowly oblong,* ¼–⅓ enclosed by *shallow, scaly cup;* 1 or few at end of stalk; maturing first year.

Habitat: On mountain slopes, forming thickets; also with other oaks, pinyons, and junipers.

Range: SW. Colorado south S. New Mexico, west to S. California, and south to Baja California; at 4000–8000' (1219–2438 m).

The name *turbinella,* meaning "like a little top," refers to the acorns. Turbinella Oak is the characteristic shrub in the chaparral vegetation of Arizona mountain slopes. The foliage is browsed by wildlife and occasionally by livestock.

168 Interior Live Oak
"Highland Live Oak" "Sierra Live Oak"
Quercus wislizeni A. DC.

Description: Evergreen tree with short trunk and broad, rounded crown of stout, spreading branches; sometimes a shrub.
Height: 30–70' (9–21 m).
Diameter: 1–3' (0.3–0.9 m).
Leaves: *evergreen;* 1–2" (2.5–5 cm) long, ½–1¼" (1.2–3 cm) wide. *Lance-shaped* to elliptical, short-pointed at tip, blunt or rounded at base, often with short, *spiny teeth, thick* and leathery, hairless. *Shiny dark green above,* light yellow-green with prominent *network of veins* beneath.
Bark: gray, becoming furrowed into narrow, scaly ridges.

Acorns: ¾–1½" (2–4 cm) long; *egg-shaped,* long-pointed, often with *long, dark lines,* about ½ enclosed by *deep, thin, scaly cup;* 1 or 2 on short stalks or stalkless; maturing second year.
Habitat: Valleys and slopes in foothill woodlands; with other oaks and Digger Pine.
Range: N. to S. California, mostly in foothills of Sierra Nevada and inner Coast Ranges, and N. Baja California; at 1000–5000' (305–1524 m).

This species is named for its discoverer, Friedrich Adolph Wislizenus (1810–89), a German-born physician of St. Louis, Missouri. Although slowgrowing, Interior Live Oak is planted as an ornamental. Deer browse the foliage, and the wood is used for fuel.

ELM FAMILY
(Ulmaceae)

Trees and shrubs, and sometimes
woody vines, including hackberries
(*Celtis*) and elms (*Ulmus*). About 200
species nearly worldwide; 14 native and
1 naturalized tree species and 2 native
shrub species in North America.
Leaves: alternate in 2 rows,
asymmetrical or unequal at base, often
with 3 main veins, generally toothed,
with paired stipules.
Flowers: tiny, inconspicuous, greenish;
usually 1 to many, along twigs; male
and female (bisexual in *Ulmus*), with
calyx of 4–8 persistent sepals or lobes,
no corolla, 4–8 stamens opposite
sepals, and 1 pistil with 1-celled ovary.
Fruit: a drupe or winged key (samara).

201, 474 **Netleaf Hackberry**
"Western Hackberry" "Sugarberry"
Celtis reticulata Torr.

Description: Shrub or small tree with short trunk
and open, spreading crown.
Height: 20–30′ (6–9 m).
Diameter: 1′ (0.3 m).
Leaves: in 2 rows; 1–2½″ (2.5–6 cm)
long, ¾–1½″ (2–4 cm) wide. Shape
very variable, mostly *ovate;* short- or
long-pointed; with 3 *main veins* from
unequal-sided, rounded, or slightly
notched base; *without teeth* or sometimes
coarsely saw-toothed; usually *thick.*
Dark green and *rough above,* yellow-
green with *prominent network of raised
veins* and slightly hairy beneath;
shedding in late autumn or winter.
Bark: *gray; smooth* or becoming rough
and fissured, with *large, corky warts.*
Twigs: light brown, slender, slightly
zigzag, hairy.
Flowers: ⅛″ (3 mm) wide; greenish;
male and female at base of young
leaves; in early spring.

Fruit: ¼–⅜" (6–10 mm) in diameter;
orange-red, 1-seeded, sweet berries;
slender-stalked at leaf bases; maturing
in autumn.

Habitat: Moist soils usually along streams, in
canyons, and on hillsides in desert,
grassland, and woodland zones.

Range: Central Kansas south to Texas, west to
S. California, and north to E.
Washington; also N. and central
Mexico; usually at 1500–6000' (457–
1829 m).

This is the native hackberry of the
western United States, mainly in the
Southwest, but extending eastward into
the prairie states. The sweetish fruit is
eaten by wildlife and was a food source
for Indians. The branches often have
deformed bushy growths called
witches'-brooms, produced by mites
and fungi. The leaves bear rounded,
swollen galls caused by tiny, jumping
plant lice. This hackberry is mostly
confined to areas with a constant water
supply.

152, 321, 531 **American Elm**
"White Elm" "Soft Elm"
Ulmus americana L.

Description: Large, handsome, graceful tree, often
 with enlarged buttresses at base,
usually forked into *many spreading
branches, drooping at ends,* forming a very
broad, rounded, flat-topped or vaselike
crown, often wider than high.
Height: 100' (30 m).
Diameter: 4' (1.2 m), sometimes much
larger.
Leaves: in *2 rows;* 3–6" (7.5–15 cm)
long, 1–3" (2.5–7.5 cm) wide.
Elliptical, abruptly long-pointed, base
rounded with sides unequal, doubly
saw-toothed, with many straight,
parallel side veins, thin. *Dark green* and
usually hairless or slightly rough above,

paler and usually with soft hairs
beneath; turning bright yellow in
autumn.
Bark: light gray, deeply furrowed into
broad, forking, scaly ridges.
Twigs: brownish, slender, hairless.
Flowers: ⅛" (3 mm) wide; greenish;
clustered along twigs; in early spring.
Fruit: ⅜–½" (10–12 mm) long;
elliptical, flat, 1-seeded *keys* (samaras);
with *wing hairy on edges,* deeply
notched, with points curved inward,
long-stalked; maturing in early spring.

Habitat: Moist soils, especially in valleys and
flood plains; in mixed hardwood
forests.

Range: SE. Saskatchewan east to Cape Breton
Island, south to central Florida, and
west to central Texas; to 2500'
(762 m). Widely planted in the West
beyond the native eastern range.

This well-known, once abundant
species, familiar on lawns and city
streets, has been ravaged by the Dutch
Elm disease, caused by a fungus
introduced accidentally about 1930 and
spread by European and native elm bark
beetles. The wood is used for
containers, furniture, and paneling.

156 Chinese Elm
Ulmus parvifolia Jacq.

Description: Introduced tree with dense, broad,
rounded crown of spreading branches
and *small leaves.*
Height: 50' (15 m).
Diameter: 1½' (0.5 m).
Leaves: in *2 rows;* ¾–2" (2–5 cm) long,
⅜–¾" (10–19 mm) wide. Elliptical,
*saw-toothed, slightly thickened. Shiny dark
green* above, paler and hairy when young
and in vein angles beneath; turning
reddish or purplish in autumn or
remaining nearly evergreen in warm
climates.

Bark: *mottled brown,* smooth, shedding in irregular, *thin flakes* and exposing reddish-brown inner bark.

Twigs: slender, slightly zigzag, hairy.

Flowers: ⅛" (3 mm) wide; greenish; clustered at base of leaves; in *autumn.*

Fruit: ⅜" (10 mm) long; *elliptical, flat,* 1-seeded *keys* (samaras), with *broad, pale yellow wing;* maturing in *autumn.*

Habitat: Moist soils in humid temperate regions.

Range: Native of China, Korea, and Japan. Planted across the United States, especially in Gulf and Pacific regions.

Fast-growing and hardy, Chinese Elm is a handsome ornamental with showy bark and a compact crown and is also cultivated for shade and shelterbelts. It should not be confused with Siberian Elm (*Ulmus pumila* L.), sometimes erroneously called Chinese Elm.

154 English Elm
Ulmus procera Salisb.

Description: Large, introduced shade tree with tall, straight trunk and dense, broad, rounded crown of spreading and nearly upright branches.

Height: 80' (24 m).

Diameter: 3' (0.9 m).

Leaves: in *2 rows;* 2–3¼" (5–8 cm) long, 1¼–2" (3–5 cm) wide. Broadly elliptical, abruptly long-pointed at tip, base with *very unequal* sides, *doubly saw-toothed,* with 10–12 straight veins on each side. Dark green and *rough* above, paler and covered with *soft hairs* and with tufts in vein angles beneath; remaining green late, turning yellow and shedding late in autumn.

Bark: gray, deeply furrowed into rectangular plates.

Twigs: brown, slender, densely covered with hairs when young; sometimes with corky wings.

Flowers: ⅛" (3 mm) wide; dark red;

clustered along twigs; in early spring.
Fruit: ½″ (12 mm) long; *rounded, flat,*
greenish *keys* (samaras); *hairless;* with
1 seed near narrow notch at tip; short-
stalked; maturing in spring.

Habitat: Scattered in moist soils, in thickets,
and along roadsides and forest borders,
spreading from cultivation.

Range: Native of England and W. Europe.
Widely planted since colonial times and
escaping in northeastern and Pacific
states.

This species is propagated by suckers
from roots, which often encircle the
trunk or appear at a distance. Several
cultivated varieties differ in habit and
leaf color (including white-striped, dark
purple, and yellowish forms). English
Elm is widely planted in England,
where the trees reach great size with
age.

---

151 **Siberian Elm**
"Asiatic Elm" "Dwarf Elm"
Ulmus pumila L.

Description: Small to medium-sized, introduced tree
with open, rounded crown of slender,
spreading branches.
Height: 60′ (18 m).
Diameter: 1½′ (0.5 m), usually
smaller.
Leaves: ¾–2″ (2–5 cm) long, ½–1″
(1.2–2.5 cm) wide. *Narrowly elliptical,*
blunt-based, *saw-toothed,* with many
straight side veins, slightly thickened.
Dark green above, paler and nearly
hairless beneath; turning yellow in
autumn.
Bark: gray or brown, rough, furrowed.
Flowers: ⅛″ (3 mm) wide; greenish; in
clusters; in early spring.

Fruit: ⅜–⅝″ (10–15 mm) long; several
clustered, *rounded, flat,* 1-seeded *keys*
(samaras), bordered with *broad, notched*
wing; in early spring.

Habitat: Dry regions, tolerant of poor soils and city smoke, also scattered in moist soils along streams.

Range: Native from Turkestan to E. Siberia and N. China. Naturalized from Minnesota south to Kansas and west to Utah; at 1000–5000' (305–1524 m).

A fast-growing tree in dry regions such as the Great Plains, it is less suited to moist regions. Siberian Elm is hardy and resistant to the Dutch Elm disease. This species has been known erroneously as Chinese Elm. However, Chinese Elm (*Ulmus parvifolia* Jacq.), a cultivated species in this country, has brown, smoothish, mottled bark, small leaves turning purplish in autumn, and elliptical, flattish fruit borne in autumn.

155 Japanese Zelkova
Zelkova serrata (Thunb.) Mak.

Description: Large introduced tree with short trunk and broad, rounded crown of many spreading branches.
Height: 70' (21 m).
Diameter: 2' (0.6 m).
Leaves: in *2 rows;* 1–3½" (2.5–9 cm) long, ¾–1¾" (2–4.5 cm) wide. *Ovate* or elliptical, *sharply saw-toothed,* with *8–14 straight, parallel veins* on each side of midvein, short-stalked. *Dark green* and *rough* above, paler and usually hairless beneath; turning yellow to reddish-yellow in autumn.
Bark: gray, smooth, becoming scaly.
Twigs: slender, mostly hairless.
Flowers: ⅛" (3 mm) long; greenish; male clustered at base of new lower leaves, 1 or few female at upper leaves; almost stalkless; in early spring.

Fruit: ³⁄₁₆" (5 mm) long and wide; *egg-shaped* drupes; oblique, almost stalkless; maturing in autumn.
Habitat: Moist soils in humid temperate regions.

Range: Native of Japan. Planted across the
United States.

Japanese Zelkova resembles its
relatives, the elms, which have doubly
saw-toothed leaf edges and winged
fruits. It has been suggested as a
substitute for American Elm, since it is
resistant to the Dutch Elm disease.
Propagated by seeds, layers, and grafts,
it grows rapidly. In Japan, the wood is
an important timber and is valued for
making furniture, lacquerware, and
trays; the plants are often used for
bonsai.

MULBERRY FAMILY
(Moraceae)

Trees and shrubs, including mulberries
(*Morus*), sometimes herbs, with white
sap or latex usually present and often
abundant. About 1400 species mostly
in tropical and subtropical areas, a few
in temperate regions; 5 native and 3
naturalized tree species.
Leaves: alternate, often in 2 rows,
simple; without teeth, toothed, or
lobed; pinnate- or palmate-veined.
Twigs: with 1–2 large stipules forming
long-pointed bud, soon falling and
leaving scars or rings at nodes.
Flowers: tiny, often greenish, male and
female on the same plant or separate
plants, usually numerous and crowded,
often in spikes or heads. Calyx usually
with 4 sepals or lobes and no petals;
male with 4–1 opposite stamens and
female with 1 pistil.
Fruit: a drupe or achene, often multiple
and fleshy, sometimes edible.

112, 334, 495 **Osage-orange**
"Bodark" "Hedge-apple"
Maclura pomifera (Raf.) Schneid.

Description: Medium-sized, spiny tree with short,
often crooked trunk; broad, rounded or
irregular crown of spreading branches;
single, *straight, stout spines* at base of
some leaves; and *milky sap.*
Height: 50' (15 m).
Diameter: 2' (0.6 m).
Leaves: 2½–5" (6–13 cm) long, 1½–3"
(4–7.5 cm) wide. Narrowly *ovate,* long-
pointed, *not toothed,* hairless. Shiny dark
green above, paler beneath; turning
yellow in autumn.
Bark: gray or brown, thick, deeply
furrowed into narrow, forking ridges;
inner bark of roots orange, separating into
thin, papery scales.
Twigs: brown, stout, with single spine

¼–1″ (0.6–2.5 cm) long at some nodes and short twigs or spurs.

Flowers: tiny, greenish; crowded in rounded clusters less than 1″ (2.5 cm) in diameter; male and female on separate trees; in early spring.

Fruit: 3½–5″ (9–13 cm) in diameter; a *heavy yellow-green ball,* hard and fleshy, containing many light brown nutlets; maturing in autumn and soon falling.

Habitat: Moist soils of river valleys.

Range: The native range uncertain. SW. Arkansas to E. Oklahoma and Texas; widely planted and naturalized in the eastern and northwestern states.

Rows of these spiny plants served as fences in the grassland plains before the introduction of barbed wire. The name "Bodark" is from the French *bois d'arc,* meaning "bow wood," referring to the Indians' use of the wood for archery bows. It is also used for fenceposts. Early settlers extracted a yellow dye for cloth from the root bark. The fruit is eaten by livestock, which has given rise to yet another common name, "Horse-apple."

204, 252, 463 **White Mulberry**
"Silkworm Mulberry"
"Russian Mulberry"
Morus alba L.

Description: Naturalized small tree with rounded crown of spreading branches, *milky sap,* and edible mulberries.
Height: 40′ (12 m).
Diameter: 1′ (0.3 m).
Leaves: in 2 rows; 2½–7″ (6–18 cm) long, 2–5″ (5–13 cm) wide. Broadly ovate but variable in shape, with 3 main veins from rounded or notched base, *coarsely toothed, often divided into 3 or 5 lobes;* long-stalked. Shiny green above, paler and slightly hairy beneath.
Bark: light brown, smoothish,

becoming furrowed into scaly ridges.

Twigs: light brown, slender.

Flowers: tiny, greenish; crowded in short clusters; male and female on same or separate trees; in spring.

Fruit: ⅜–¾" (10–19 mm) long; a cylindrical *mulberry,* purplish, pinkish, or white; composed of many tiny, beadlike, 1-seeded fruits, sweet and juicy, edible; in late spring.

Habitat: Hardy in cities, drought-resistant, and adapted to dry, warm areas.

Range: Native of China. Widely cultivated across the United States; naturalized in the East and in the Pacific states.

White Mulberry has been cultivated for centuries, the leaves serving as the main food of silkworms. It was introduced long ago in the southeastern United States, where silk production was not successful. It grows rapidly and produces abundant berries that are enjoyed by birds as well as by many people. The trees spread like weeds in cities, where the berries litter sidewalks. Varieties include a hardy one for shelterbelts, another with drooping or weeping foliage, and a fruitless form.

202, 251   **Texas Mulberry**
"Mexican Mulberry"
"Mountain Mulberry"
Morus microphylla Buckl.

Description: Small tree or clump-forming shrub with variable leaf shape, *milky sap,* and edible mulberries.
Height: 20' (6 m).
Diameter: 8" (20 cm).
Leaves: in 2 rows; 1–2½" (2.5–6 cm) long, ¾–1¼" (2–3 cm) wide. *Ovate* but variable in shape, sometimes 3-lobed, pointed at tip, with *3 main veins* from rounded or notched base, coarsely saw-toothed. Dark green and rough above, paler and with soft hairs

beneath; turning yellow in autumn.
Bark: light gray, smooth, becoming
furrowed and scaly.
Twigs: brown, slender, hairy when
young.
Flowers: tiny, greenish; crowded in
short clusters ⅜–¾" (10–19 mm) long;
male and female on separate trees; in
spring with leaves.

Fruit: ½" (12 mm) long; a cylindrical
mulberry; red to purple or black,
composed of many tiny, beadlike, 1-
seeded fruits, sweet and juicy, edible;
in late spring.

Habitat: Moist soils, mostly along streams,
canyons, washes, and rocky slopes; in
foothills and mountains, woodland,
upper desert, and grassland zones.

Range: S. Oklahoma and Texas west to
Arizona; also N. Mexico; at 1000–
6000′ (305–1829 m).

The small fruit is eaten by wildlife and
was a food of southwestern Indians.
This species was introduced at the
bottom of the Grand Canyon by
Havasupai Indians, probably in
prehistoric times, from wild trees
farther south.

203 Red Mulberry
"Moral"
Morus rubra L.

Description: Medium-sized tree with short trunk,
broad, rounded crown, and *milky sap.*
Height: 60′ (18 m).
Diameter: 2′ (0.6 m).
Leaves: in 2 rows; 4–7" (10–18 cm)
long, 2½–5" (6–13 cm) wide. Ovate,
abruptly long-pointed, with *3 main
veins* from often unequal base, coarsely
saw-toothed, often with 2 or 3 lobes on
young twigs. Dull dark green and
rough above, with soft hairs beneath;
turning yellow in autumn.
Bark: brown, fissured into scaly plates.

Twigs: brown, slender.

Flowers: *tiny,* about ⅛" (3 mm) long; crowded in narrow clusters; male and female on same or separate trees; in spring when leaves appear.

Fruit: 1–1¼" (2.5–3 cm) long; a cylindrical *mulberry, red to dark purple,* composed of *many tiny,* beadlike, *1-seeded fruits,* sweet and juicy, edible; in late spring.

Habitat: Moist soils in hardwood forests.

Range: S. Ontario east to Massachusetts, south to S. Florida, west to central Texas, and north to SE. Minnesota; to 2000' (610 m). Cultivated in the West.

The wood is used locally for fenceposts, furniture, interior finish, and agricultural implements. People, domestic animals, and wildlife (especially songbirds) eat the berries. Choctaw Indians wove cloaks from the fibrous inner bark of young mulberry shoots.

PROTEA FAMILY
(Proteaceae)

Shrubs and trees; about 1000 species in tropical regions, especially those with a long dry season, including Africa, Australia, and South America. In the New World, 1 species ranges north to southern Mexico. Others are introduced in subtropical regions and indoors northward.

Leaves: mostly alternate; simple; without teeth or deeply lobed; lacking stipules.

Flowers: mostly stalked on 1 side of axis or in showy heads; bisexual; often irregular; calyx 4-lobed, colored, without corolla, 4 stamens opposite and inserted on calyx, and 1 pistil often stalked; with superior 1-celled ovary, 1 to many ovules, and slender style.

Fruit: podlike (follicle), nut, or stone (drupe).

297, 328 Silk-oak
"Silky-oak"
Grevillea robusta A. Cunn.

Description: Handsome, introduced, subtropical evergreen tree with straight trunk, narrow crown of many branches, and graceful, fernlike foliage.

Height: 70′ (21 m).
Diameter: 1′ (0.3 m).

Leaves: *evergreen;* pinnately compound; 6–12″ (15–30 cm) long. *Fernlike;* deeply divided into many *narrow, long-pointed lobes turned under* at edges. *Dark green* above, *silky* with *whitish* hairs beneath.

Bark: gray, smooth on branches, becoming rough and deeply furrowed.

Twigs: stout, gray, finely hairy.

Flowers: ½″ (12 mm) long; with *4 narrow, yellow or orange sepals* and no petals; crowded on slender stalks on *1 side of unbranched axis* 3–5″ (7.5–13

cm) long on trunk; on previous year's twigs back of leaves and at leaf bases; in spring and early summer.
Fruit: ¾" (19 mm) long; *black, curved, podlike,* with slender stalks and long, threadlike styles; maturing in summer and autumn, splitting open on 1 side. 1 or 2 elliptical, winged brown seeds.

Habitat: Well-drained, sandy soils in subtropical regions.

Range: Native of Australia. Planted and escaping from cultivation in Florida; planted in S. Arizona, California, Hawaii, and Puerto Rico.

Silk-oak is widely cultivated along streets as an attractive, flowering tree. It grows rapidly, is drought-resistant and tolerates cold, but branches break easily and become ragged with age. The fernlike plants are raised as houseplants in the north. The common name refers to the silky lower leaf surfaces and to the wood's similarity to oak. In Australia, the durable wood is used for furniture, cabinets, and barrel staves.

MAGNOLIA FAMILY
(Magnoliaceae)

About 200 species of trees and shrubs
in warm temperate and tropical regions;
11 native tree species in North
America, including anise-trees (*Illicium*)
and Yellow-poplar (*Liriodendron*), as
well as magnolias (*Magnolia*).
Leaves: alternate, simple, not toothed,
mostly with large stipules that form the
bud and leave ring scars at nodes.
Flowers: often large and showy,
frequently solitary, bisexual, regular,
with 3 to many sepals, 6 to many
commonly white petals, many stamens
in spiral on elongated axis, and many
simple pistils with 1-celled ovary.
Fruit: many follicles or berries, often
aggregate and conelike.

234, 336, 534 Yellow-poplar
"Tuliptree" "Tulip-poplar"
Liriodendron tulipifera L.

Description: One of the tallest and most beautiful
eastern hardwoods, with a long,
straight trunk, a narrow crown that
spreads with age, and large, showy
flowers resembling tulips or lilies.
Height: 80–120′ (24–37 m).
Diameter: 2–3′ (0.6–0.9 m),
sometimes much larger.
Leaves: 3–6″ (7.5–15 cm) long and
wide. Blades of unusual shape, with
broad tip and base nearly straight *like a
square* and with *4 or sometimes 6 short-
pointed, paired lobes,* hairless; long-
stalked. Shiny dark green above, paler
beneath; turning yellow in autumn.
Bark: dark gray, becoming thick and
deeply furrowed.
Twigs: brown, stout, hairless, with *ring
scars at nodes.*
Flowers: 1½–2″ (4–5 cm) long and
broad; *cup-shaped,* with 6 rounded,
green petals (orange at base); solitary

and upright at end of leafy twig; in
spring.

Fruit: 2½–3″ (6–7.5 cm) long; *conelike,*
light brown, composed of many
overlapping, 1- or 2-seeded nutlets 1–
1½″ (2.5–4 cm) long (including *narrow
wing*); shedding from upright axis in
autumn, the axis persistent in winter.

Habitat: Moist, well-drained soils, especially
valleys and slopes; often in pure stands.

Range: Extreme S. Ontario east to Vermont
and Rhode Island, south to N. Florida,
west to Louisiana, and north to S.
Michigan; to 1000′ (305 m) in north
and to 4500′ (1372 m) in southern
Appalachians. Planted on the Pacific
Coast.

The earliest Virginia colonists
introduced Yellow-poplar into Europe;
it is grown also on the Pacific
Coast. Very tall trees with massive
trunks existed in the primeval forests
but were cut for the valuable soft wood.
Pioneers hollowed out a single log to
make a long, lightweight canoe. One of
the chief commercial hardwoods,
Yellow-poplar is used for furniture, as
well as for crates, toys, musical
instruments, and pulpwood.

96, 374, 485   **Southern Magnolia**
"Evergreen Magnolia" "Bull-bay"
Magnolia grandiflora L.

Description: A beautiful ornamental tree introduced
from southeastern United States;
evergreen, with straight trunk, conical
crown, and very fragrant, *very large
white flowers.*

Height: 60–80′ (18–24 m).
Diameter: 2–3′ (0.6–0.9 m).
Leaves: *evergreen;* 5–8″ (13–20 cm)
long, 2–3″ (5–7.5 cm) wide. *Oblong or
elliptical, thick and firm,* with edges
slightly turned under. *Shiny bright green
above, pale and with rust-colored hairs*

beneath. Stout leafstalks, with rust-colored hairs.

Bark: dark gray, smooth, becoming furrowed and scaly.

Twigs: covered with *rust-colored hairs* when young, with *ring scars at nodes,* ending in *buds* also covered with *rust-colored hairs.*

Flowers: 6–8″ (15–20 cm) wide; *cup-shaped, with 3 white sepals* and 6 or more petals, very fragrant, solitary at end of twig; in late spring and summer.

Fruit: 3–4″ (7.5–10 cm) long; *conelike,* oblong, pink to brown, covered with rust-colored hairs; composed of many separate, short-pointed, 2-seeded fruits that split open in early autumn.

Habitat: Moist soils of valleys and low uplands; with various other hardwoods.

Range: E. North Carolina to central Florida and west to E. Texas; to 400′ (122 m). Also planted on the Pacific Coast.

Cultivated around the world in warm temperate and subtropical regions, it is a popular ornamental and shade tree. Several horticultural varieties have been developed. Principal uses of the wood are furniture, boxes, cabinetwork, and doors. The dried leaves are used by florists in decorations.

113 Saucer Magnolia
"Chinese Magnolia"
Magnolia ×*soulangiana* Soul.-Bod.

Description: Ornamental shrub or small tree, usually with several trunks, and with widely spreading crown of coarse foliage and abundant, large flowers in early spring before the leaves.

Height: 25′ (7.6 m).

Diameter: 6″ (15 cm).

Leaves: 5–8″ (13–20 cm) long, 2½–4½″ (6–11 cm) wide. Reverse ovate or *elliptical,* broadest toward abruptly short-pointed tip, blunt at base, not

toothed. Dull green above, hairy
beneath.
Bark: light gray, smooth.
Twigs: light gray, stout, hairless, with
ring scars at nodes.
Flowers: 5–10" (13–25 cm) wide; 6
spreading petals, showy, *like large
tulips, bell-shaped* but opening widely
like *saucers,* pink, purple or white;
white within; often fragrant; borne
singly at ends of twigs.

Fruit: 4–5" (10–13 cm) long; *conelike,
cylindrical,* curved, composed of many
separate, pointed fruits, splitting open
to expose red seeds; in early autumn.

Habitat: Humid temperate regions.
Range: Widely planted across the United
States.

This popular magnolia with gorgeous,
early spring flowers is a hybrid of two
Chinese species, Yulan Magnolia
(*Magnolia heptapeta* (Buc'hoz) Dandy) and
Lily Magnolia (*M. quinquepeta* (Buc'hoz)
Dandy). It originated in 1820 as a
chance seedling in the garden of
Étienne Soulange-Bodin, a French
nurseryman. It flowers when still a
small shrub. Several cultivated varieties
differ mainly in color, size, time of
flowering, and overall shape.

LAUREL FAMILY
(Lauraceae)

Mostly large trees and a few shrubs
with aromatic bark, wood, and leaves.
About 2000 species in tropical and
warm temperate regions; 5 native and 2
naturalized trees, 3 native shrubs, and
1 native herbaceous vine species in
North America.

Leaves: mostly alternate, sometimes
opposite or whorled; simple, commonly
elliptical, without teeth; pinnately-
veined with long, curved, side veins;
often leathery, with tiny gland-dots,
without stipules.

Flowers: mostly small, many in
branched clusters along twigs, bisexual
or sometimes male and female on
separate plants, regular, often with
short cup, 3 sepals and 3 similar petals
(or 6 petals), 9–12 stamens (or some
reduced to staminodes), the anthers
opening by 2 or 4 pores with lids, and
1 pistil.

Fruit: a berry or drupe with 1 large
seed, mostly with cup or tube from
calyx or corolla persistent at base.

95 Camphor-tree
Cinnamomum camphora (L.) Karst.

Description: Introduced, aromatic evergreen tree
with rounded, *dense crown, wider than
high;* crushed foliage has *odor of camphor.*
Height: 40′ (12 m).
Diameter: 2′ (0.6 m).
Leaves: *evergreen; partly opposite:* 2½–4″
(6–10 cm) long, 1–2¼″ (2.5–6 cm)
wide. *Elliptical,* pointed, with *3 main
veins* from rounded base, not toothed,
slightly thickened, long-stalked.
Pinkish when young, becoming *shiny
green above, dull whitish beneath.*
Bark: gray, smooth, becoming
rough, thick, and furrowed.
Twigs: green, slender, hairless.

Flowers: ⅛" (3 mm) long; yellowish; in clusters 1½–3" (4–7.5 cm) long; in spring.
Fruit: ⅜" (10 mm) in diameter; a black, 1-seeded berry, with greenish cup and spicy taste of camphor; maturing in autumn.

Habitat: Moist soils, along roadsides and in waste places in humid subtropical regions.

Range: Native of tropical Asia from E. China to Vietnam, Taiwan, and Japan. Extensively cultivated and naturalized locally from Florida to S. Texas; also in S. California.

Popular as a street tree in the deep South, Camphor-tree produces dense shade. In Asia, camphor oil and gum for medicine and industry are obtained by steam distillation of leaf clippings and wood. The insect-repellent wood is used for cabinetry and chests.

93, 333, 498   **California-laurel**
"Oregon-myrtle" "Pepperwood"
Umbellularia californica (Hook. & Arn.) Nutt.

Description: Evergreen tree with short trunk, usually forked into several large, spreading branches, forming a broad, rounded, *dense crown* of *aromatic, peppery foliage;* in exposed situations a low, thicket-forming shrub.
Height: 40–80' (12–24 m).
Diameter: 1½–2½' (0.5–0.8 m).
Leaves: *evergreen;* 2–5" (5–13 cm) long, ½–1½" (1.2–4 cm) wide. *Elliptical* or lance-shaped, short-pointed or rounded at ends, *thick and leathery,* with edges slightly turned under. *Shiny dark green above, dull and paler beneath* with prominent network of veins; turning yellow or orange before shedding gradually after second year.
Bark: dark reddish-brown, thin, with flat scales.

Twigs: stout, hairy; yellow-green when young, becoming reddish-brown.

Flowers: ¼" (6 mm) long; *pale yellow,* numerous; clustered on stalk at leaf bases; in late winter or early spring.

Fruit: ¾–1" (2–2.5 cm) long; an *elliptical or nearly round berry,* greenish to purple, with thin pulp and large brown seed; maturing in late autumn.

Habitat: Moist soils, especially in mountain canyons and valleys; in mixed forests.

Range: SW. Oregon south in Coast Ranges and Sierra Nevada to S. California; to 4000' (1219 m); at southern limit, 2000–6000' (610–1829 m).

A handsome ornamental and street tree on the West Coast, it is also known as "California-bay." When crushed, the foliage, twigs, and other parts are pungently aromatic. The attractive light brown wood with darker streaks takes a beautiful finish and is used for veneer in furniture and paneling, cabinetwork, and interior trim. Prized for novelties and woodenware, it is often marketed as "Oregon-myrtle," though a member of the Laurel Family. California-laurel and Sassafras are the northernmost New World representatives of this tropical family.

WITCH-HAZEL FAMILY
(Hamamelidaceae)

Trees, often large, and shrubs; about 100 species in subtropical and warm temperate regions; 2 native tree and 5 shrub species in North America.

Leaves: deciduous or in warm climates evergreen, alternate, simple, with gland-teeth or palmately lobed; often with star-shaped hairs; with paired stipules.

Flowers: bisexual or male and female; mostly tiny; in heads or racemes; composed of calyx of usually 4–5 sepals united at base and to ovary (sometimes no sepals), corolla of 4–5 petals borne on calyx or none, stamens 2 to many, separate, and 1 pistil.

Fruit: a capsule, often hard with 2-layered wall, splitting open across top into 2 parts; seeds few, sometimes winged.

222, 538   Sweetgum
"Redgum" "Sapgum"
Liquidambar styraciflua L.

Description: Large aromatic tree with straight trunk and conical crown that becomes round and spreading.

Height: 60–100' (18–30 m).
Diameter: 1½–3' (0.5–0.9 m).
Leaves: 3–6" (7.5–15 cm) long and wide. *Star-shaped* or maplelike, with *5, sometimes 7, long-pointed, finely saw-toothed lobes* and 5 main veins from notched base; with *resinous odor when crushed;* leafstalks slender, nearly as long as blades. Shiny dark green above, turning reddish in autumn.
Bark: gray, deeply furrowed into narrow, scaly ridges.
Twigs: green to brown, stout, *often forming corky wings.*
Flowers: tiny; in *greenish, ball-like clusters;* male in several clusters along a

stalk, female in drooping cluster, on same tree; in spring.

Fruit: 1–1¼" (2.5–3 cm) in diameter; a long-stalked, drooping brown *ball,* composed of many individual fruits, each ending in *2 long, curved, prickly points* and each with 1–2 long-winged seeds; maturing in autumn and persistent into winter.

Habitat: Moist soils of valleys and lower slopes; in mixed woodlands. Often a pioneer after logging, clearing, and in old fields.

Range: Extreme SW. Connecticut south to central Florida, west to E. Texas, and north to S. Illinois; also a variety in E. Mexico; to 3000' (914 m) in southern Appalachians. Planted in the West.

An important timber tree, Sweetgum is second in production only to oaks among hardwoods. It is a leading furniture wood, used for cabinetwork, veneer, plywood, pulpwood, barrels, and boxes. In pioneer days, a gum was obtained from the trunks by peeling the bark and scraping off the resinlike solid. This gum was used medicinally as well as for chewing gum. Commercial storax, a fragrant resin used in perfumes and medicines, is from the related Oriental Sweetgum (*Liquidambar orientalis* Mill.) of western Asia.

SYCAMORE FAMILY
(Platanaceae)

Large trees with massive trunks; 1
genus (*Platanus*) with 9 or fewer species
in north temperate regions, 3 in North
America.
Leaves: deciduous, alternate, simple,
long-stalked, about as wide as long,
palmately 3- to 9-lobed, often toothed,
with star-shaped hairs and large,
paired, toothed stipules united at base.
Bark: smooth, light-colored, peeling in
large, thin flakes, becoming mottled.
Twigs: slender, zigzag, with ring scars,
conical buds covered by 1 hairless scale
and hidden inside enlarged base of
leafstalk.
Flowers: male and female on same tree
in spring with leaves; numerous, tiny,
greenish tinged with red, crowded in
1–6 balls or dense round heads on long
drooping stalk, with cup-shaped calyx,
no or few petals. Male with 3–7
stamens; female with 5–9 pistils.
Fruit: many 1-seeded, narrow, 4-angled
nutlets surrounded by long, stiff hairs,
crowded in ball or head.

---

220 **London Planetree**
Platanus ×*acerifolia* (Ait.) Willd.

Description: Large, introduced shade tree with
straight, stout trunk and broad, open
crown of spreading to slightly drooping
branches and coarse foliage.
Height: 70′ (21 m).
Diameter: 2′ (0.6 m).
Leaves: 5–10″ (13–25 cm) long and
wide. *Palmately 3- or 5-lobed;* shallow,
short-pointed lobes, with few large
teeth or none, *3 or 5 main veins* from
notched base, becoming hairless or
nearly so. Shiny green above, pale
beneath. Leafstalk long, stout, covering
side bud at enlarged base.
Bark: *smooth,* with patches of brown,

green, and gray, *peeling off in large flakes.*

Twigs: greenish, slender, zigzag, hairy, with ring scars at nodes.

Flowers: tiny; in *greenish, ball-like,* drooping *clusters;* male and female on separate twigs; in spring.

Fruit: 1″ (2.5 cm) in diameter; (usually) *2 bristly, brown balls* hanging on long stalk; composed of many narrow *nutlets* with hair tufts; maturing in autumn, separating in winter.

Habitat: Moist soils in humid temperate regions, hardy and tolerant of city conditions.

Range: Planted across the United States.

London Planetree is a hybrid between Sycamore of the eastern United States and Oriental Planetree (*Platanus orientalis* L.) of southeastern Europe and Asia Minor. It is also popular as a street tree in Europe, where it originated probably before 1700. The plants can be clipped into screens and arbors. The genus name *Platanus* is the classical Latin and Greek name of Oriental Planetree, from the Greek word for "broad" which describes the leaves. The specific name means "maple leaf," referring to the resemblance of London Planetree's leaves to those of maples.

---

224 California Sycamore
"Western Sycamore" "Aliso"
Platanus racemosa Nutt.

Description: Tree with enlarged base, stout trunk, often branched near base, and broad, irregular, open crown of thick, spreading branches.

Height: 40–80′ (12–24 m).

Diameter: 2–4′ (0.6–1.2 m), sometimes much larger.

Leaves: 6–9″ (15–23 cm) long and wide. Slightly *star-shaped; deeply divided* about halfway to base *into 5 (sometimes 3) narrow, long-pointed lobes,* wavy edges with *few large teeth, 5 (sometimes 3) main veins* from notched or blunt base. *Light green* above, paler and hairy beneath. Leafstalk *long,* stout, covering side bud at enlarged base.

Bark: whitish, smooth, thin on branches; *peeling in brownish flakes and mottled* on trunk, becoming dark gray or brown; rough, thick, deeply furrowed at base.

Twigs: slender, zigzag, light brown, with ring scars at nodes; hairy when young.

Flowers: tiny; male and female in 2–7 separate, ball-like clusters; in spring with leaves.

Fruit: ⅞" (22 m) in diameter; *2–7 balls* or heads hanging on long stalk, composed of many narrow nutlets with tuft of hairs at base; maturing in autumn, separating in winter.

Habitat: Wet soils of stream banks in valleys, foothills, and mountains.

Range: N. to S. California and N. Baja California; to 4000′ (1219 m).

Common in the valleys of California, it is also a shade and ornamental tree. Giant trees with massive, barrel-shaped trunks often lean and fork into picturesque shapes. The champion, at Santa Barbara, California, when measured in 1945, was 116′ (35.4 m) high, 27′ (8.2 m) in trunk circumference, and 158′ (48.2 m) in crown spread. The mottled bark and coarse, light green foliage are distinctive.

223 **Arizona Sycamore**
"Arizona Planetree" "Álamo"
Platanus wrightii Wats.

Description: Tree with stout trunk, often branched near base, and broad, irregular, open crown of thick, spreading branches. Height: 40–80′ (12–24 m).

Diameter: 2–4' (0.6–1.2 m).
Leaves: 6–9" (15–23 cm) long and wide. *Slightly star-shaped, deeply divided into 3 or 5 (sometimes 7) narrow, long-pointed lobes, usually without teeth,* generally 5 main veins from deeply notched base. *Light green* above, paler and hairy beneath.
Bark: whitish, smooth, thin on branches; *peeling in brownish flakes and mottled* on trunk; becoming dark gray, rough, thick, deeply and irregularly furrowed at base.
Twigs: slender, zigzag, light brown, with ring scars at nodes; hairy when young.
Flowers: tiny; male and female in (usually) 2–4 separate, ball-like clusters; in spring with leaves.

Fruit: ¾–1" (2–2.5 cm) in diameter; *2–4 balls* or heads hanging on long stalk; composed of many narrow nutlets with tuft of hairs at base; maturing in autumn, separating in winter.

Habitat: Wet soils along streams and canyons in foothills and mountains, deserts, desert grasslands, and with oaks.

Range: SW. New Mexico, Arizona, and NW. Mexico; at 2000–6000' (610–1829 m).

This common tree is one of the largest and most handsome deciduous trees in the Southwest and is valuable in preventing erosion along stream banks. Large trees with their spreading whitish branches and huge, mottled trunks are conspicuous along desert valleys and canyons. It is common in Sycamore Canyon near Williams in northern Arizona. Woodpeckers and other desert birds nest in the hollow trunks of old trees.

ROSE FAMILY
(Rosaceae)

About 2000 species of trees, shrubs, and herbs worldwide; approximately 77 native and 9 naturalized tree species and many species of shrubs and herbs in North America; including serviceberries (*Amelanchier*), hawthorns (*Crataegus*), apples (*Malus*), plums and cherries (*Prunus*), and mountain-ashes (*Sorbus*).

Leaves: alternate, generally simple, with paired stipules.

Flowers: small to large, bisexual, usually regular or sometimes slightly irregular, generally with cuplike base that bears 5 sepals and 5 petals and many (sometimes 20) separate stamens and with 1 to many simple pistils or 1 compound pistil.

Fruit: a drupe, pome, achene, or follicle.

Crataegus, commonly known as Hawthorn, Haw, Thorn-apple, Red Haw, or Hog-apple, is a large genus of many difficult-to-distinguish species. Many-branched shrubs and small spreading trees, they commonly have a dense, rounded or flattened crown and long thorns. 7 species (described here) are native in western United States and about 30 in eastern states, though several hundred have been named. The leaves are usually small, ovate to elliptical, saw-toothed and often also with shallow lobes. Hairy, especially when young, they often turn red or orange in autumn. On young twigs, leaves are larger and lobed. Leafstalks are short and slender. Bark is commonly gray or brown and scaly. Twigs are slender, often many-branched, and slightly zigzag, with long, slender spines or thorns. Flowers have a cuplike base with 5 narrow calyx lobes, 5 small, rounded white petals, and 5–20 stamens with pale yellow or

pink anthers, enclosing pistil with 1–5
styles. Flowers are usually in clusters at
end of twigs in spring. The commonly
red fruits are mostly small and round in
clusters, with 5 calyx lobes usually
persisting at tip, and are often sweet
and edible; 1–5 nutlets are each 1-
seeded. Fruit matures in late summer or
autumn and sometimes remains
attached in winter.

183, 364 Western Serviceberry
"Saskatoon" "Western Shadbush"
Amelanchier alnifolia (Nutt.) Nutt.

Description: Shrub or small tree, usually with
several trunks, and with star-shaped,
white flowers.
Height: 30′ (9 m).
Diameter: 8″ (20 cm).
Leaves: ¾–2″ (2–5 cm) long and
almost as broad. *Broadly elliptical to
nearly round,* rounded at both ends,
coarsely toothed above middle, usually with
7–9 straight veins on each side. Dark
green and becoming hairless above,
paler and hairy when young beneath.
Bark: gray or brown, thin, smooth or
slightly fissured.
Twigs: red-brown, slender, hairless.
Flowers: ¾–1¼″ (2–3 cm) wide; with
5 narrow, white petals; in small,
terminal clusters; in spring with leaves.

Fruit: ½″ (12 mm) in diameter; like a
small apple, purple or blackish, edible,
juicy and sweet; with several seeds; in
early summer.
Habitat: Moist soils in forests and openings.
Range: Central Alaska southeast to Manitoba,
W. Minnesota, and Colorado and west
to N. California; local east to SE.
Quebec; to 6000′ (1829 m).

The fruits of this and related species are
eaten fresh, prepared in puddings, pies,
and muffins, and dried like raisins and
currants. They are also an important

food for wildlife from songbirds to squirrels and bears. Deer and livestock browse the foliage.

186, 535   Birchleaf Cercocarpus
"Birchleaf Mountain-mahogany"
"Hardtack"
Cercocarpus betuloides Nutt.

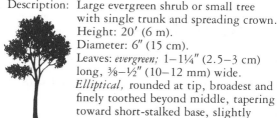

Description: Large evergreen shrub or small tree with single trunk and spreading crown.
Height: 20' (6 m).
Diameter: 6" (15 cm).
Leaves: *evergreen;* 1–1¼" (2.5–3 cm) long, ⅜–½" (10–12 mm) wide. *Elliptical,* rounded at tip, broadest and finely toothed beyond middle, tapering toward short-stalked base, slightly leathery, 5–8 *straight sunken veins* on each side. *Dark green* above, pale green or grayish and slightly hairy beneath.
Bark: dark brown, smooth, becoming scaly.
Twigs: reddish-brown, hairy when young.
Flowers: ⅜" (10 mm) long; *funnel-shaped,* slightly 5-lobed, *yellowish,* hairy, without petals, nearly stalkless; 1–3 at leaf base; in early spring.
Fruit: ⅜" (10 mm) long; narrowly cylindrical, with *twisted tail* 2–3¼" (5–8 cm) long, covered with *whitish hairs,* 1-seeded; maturing in late summer.

Habitat: Dry, rocky soils of mountain slopes; in chaparral and oak woodland.

Range: SW. Oregon south to N. Baja California and east to Arizona; at 3500–6500' (1067–1981 m).

Birchleaf Cercocarpus is a common shrub in chaparral vegetation, sprouting after fire. It is also an important browse plant for deer, cattle, and sheep. The common and scientific names both refer to the resemblance of the leaves to those of shrubby birches. It is called "Hardtack" perhaps from its ability to withstand cutting, fire, drought, and heavy browsing. *Cercocarpus,* from the Greek words for "tail" and "fruit," describes the hairy tails or plumes from the elongated flower style. These hairy fruits are carried long distances by the wind; animals harbor them in their fur, also aiding dispersal. After falling to the ground, the oddly shaped fruits twist into the soil.

185 Hairy Cercocarpus
"Hairy Mountain-mahogany"
Cercocarpus breviflorus Gray

Description: Evergreen shrub or small tree with open crown of widely spreading branches.
Height: 20′ (6 m).
Diameter: 6″ (15 cm).
Leaves: *evergreen;* ⅜–1″ (1–2.5 cm) long, ¼–½″ (6–12 mm) wide. Narrowly elliptical or obovate, *tapering to short-stalked base,* edges turned under; usually with *few rounded teeth* near tip, thick, with *few long side veins.* Gray-green above, pale and with fine hairs beneath.
Bark: gray, smooth; becoming reddish-brown, fissured, and scaly.
Twigs: long, slender, with reddish-brown hairs.
Flowers: ½″ (12 mm) long; *funnel-shaped,* slightly 5-lobed, *yellowish,* with white hairs, without petals, nearly stalkless; 1–3 at leaf base; in spring and summer.
Fruit: ¼″ (6 mm) long; narrowly cylindrical; with *twisted tail* 1–1½″ (2.5–4 cm) long, *covered with whitish hairs;* maturing in late summer and autumn.
Habitat: Dry mountain slopes; in chaparral and oak woodland.
Range: Trans-Pecos Texas west to Arizona; also N. Mexico; at 5000–8000′ (1524–2438 m).

Hairy Cercocarpus is an important browse plant for livestock. The species name describes the short flowers. The odd, silky tails characteristic of the fruit of cercocarpus are elongated styles from the flower. These plumes are straight when moist, but twist like a corkscrew upon drying, and aid in burying the seed in the soil.

82, 536 Curlleaf Cercocarpus
"Curlleaf Mountain-mahogany"
Cercocarpus ledifolius Nutt.

Description: Slightly resinous and aromatic evergreen shrub or small tree with compact, rounded crown of widely spreading, curved, and twisted branches and many stiff twigs.
Height: 15–30' (4.6–9 m).
Diameter: ½–1½' (0.15–0.5 m).
Leaves: *evergreen;* usually clustered; ½–1¼" (1.2–3 cm) long, less than ⅜" (10 mm) wide. *Narrowly lance-shaped* or elliptical, *thick and leathery, with edges rolled under,* slightly resinous and aromatic, almost stalkless. *Shiny dark green* with grooved midvein and *obscure side veins* above, pale and with fine hairs beneath.
Bark: reddish-brown, thick, deeply furrowed into scaly ridges.
Twigs: reddish-brown, hairy when young.
Flowers: ⅜" (10 mm) long; *funnel-shaped,* slightly 5-lobed, *yellowish,* hairy, without petals, stalkless; 1–3 at leaf bases; in early spring.
Fruit: ¼" (6 mm) long; narrowly cylindrical, hairy, with *twisted tail* 1½–3" (4–7.5 cm) long, covered with *whitish hairs;* maturing in summer.
Habitat: Dry, rocky mountain slopes; in grassland, with sagebrush, pinyons, and oaks, and in coniferous forests.
Range: Extreme SE. Washington east to S. Montana, south to N. Arizona, and

west to S. California; at 4000–10,500′ (1219–3200 m).

This species is a small tree characteristic of lower mountain slopes throughout the Great Basin. Deer browse the evergreen foliage year-round. The hard, heavy wood is an important source of fuel in local mining operations; it is also used for novelties, as it takes a high polish. The name "Mountain-mahogany" applied to this genus is misleading; these shrubby trees are not related to true mahogany (*Swietenia*), a valuable cabinetwood of tropical America. The dark reddish-brown, mahogany-colored heartwood may have led to this name. Navajo Indians made a red dye from the roots by grinding and then mixing them with juniper ashes and powdered alder bark.

Catalina Cercocarpus
"Bigleaf Mountain-mahogany"
Cercocarpus traskiae Eastw.

Description: Small evergreen tree with trunk often leaning and twisted and stout, widely spreading branches.
Height: 23′ (7 m).
Diameter: 8″ (20 cm).
Leaves: *evergreen;* 1½–2½″ (4–6 cm) long, 1–1½″ (2.5–4 cm) wide. *Broadly elliptical or rounded, wavy-toothed beyond middle, thick and leathery;* leafstalks short, stout. *Shiny dark green* with sunken veins above, covered with *white, woolly hairs* beneath.
Bark: light gray-brown, smooth or slightly fissured, with irregular, cream-colored spots.
Twigs: reddish-brown, stout, densely hairy when young.
Flowers: ⅝″ (15 mm) long; *funnel-shaped,* slightly 5-lobed, covered with *white hairs,* with many stamens and without petals; in nearly stalkless

clusters at leaf base; in early spring.
Fruit: ⅜″ (10 mm) long; narrowly
cylindrical, with *twisted tail* 1½–2½″
(4–6 cm) long, *covered with whitish
hairs.*

Habitat: Steep, narrow canyons; in chaparral.
Range: Only on Santa Catalina Island off
California; at 200–300′ (61–91 m).

Very rare and very local, this species is
on a proposed list of endangered plants.
The most beautiful species of its genus,
it is sometimes planted for ornament.

250, 323 Cliffrose
"Quinine-bush" "Stansbury Cliffrose"
Cowania mexicana D. Don

Description: Small-leaf, evergreen, resinous,
spreading shrub or small tree with
crooked trunk, irregular, open crown of
many short, stiff, erect branches, and
showy white flowers in spring and
summer.
Height: 20′ (6 m).
Diameter: 6″ (15 cm).
Leaves: *evergreen; crowded;* ¼–⅝″ (6–15
mm) long. *Wedge-shaped, divided into 3–
7 narrow lobes,* thick and leathery, edges
rolled under, with white *sticky* resin-
dots; bitter-tasting. *Dark green* and
often loosely hairy above, densely
covered with *white, woolly hairs
beneath.*
Bark: reddish-brown or gray; shreddy,
splitting into long, narrow strips.
Twigs: reddish-brown, hairy, gland-
dotted.
Flowers: ¾–1″ (2–2.5 cm) wide; with
*5 broad white or pale yellow, spreading
petals;* borne singly at end of side twigs;
fragrant; in spring and summer.
Fruit: ¼″ (6 mm) long; each with long,
whitish, feathery tail to 2″ (5 cm) long;
5–10 in cluster, developed from 1
flower; in autumn.
Habitat: Dry, rocky hills and plateaus, especially

limestone; in deserts and with oaks, pinyons, and junipers.

Range: SE. Colorado to N. Utah, E. California, and SW. New Mexico; also south to central Mexico; at 3500–8000' (1067–2438 m).

An attractive ornamental, Cliffrose is also planted for erosion control. It is an important browse plant for deer, cattle, and sheep, especially in winter. Indians used to make rope, sandals, and clothing from the shreddy bark and arrow shafts from the stems. It is called "Quinine-bush" because of the bitter-tasting foliage. This species is abundant on the south rim at Grand Canyon National Park, Arizona.

215, 472 **Fireberry Hawthorn**
"Roundleaf Hawthorn"
"Golden-fruit Hawthorn"
Crataegus chrysocarpa Ashe

Description: Many-branched shrub or small tree with broad, rounded crown.
Height: 20' (6 m).
Diameter: 8" (20 cm).
Leaves: 1½–2" (4–5 cm) long, 1–1½" (2.5–4 cm) wide. *Elliptical or nearly round,* finely and often doubly *saw-toothed,* lower teeth gland-tipped, *shallow-lobed* beyond middle, slightly thickened. *Shiny green with sunken veins* above, paler beneath.
Bark: dark brown, scaly.
Twigs: mostly hairless, with *many spines.*
Flowers: ⅝" (15 mm) wide; with 5 white petals, usually *5–10 pale yellow stamens,* and 3–4 styles; *many,* in broad, loose clusters; in spring.
Fruit: ⅜–½" (10–12 mm) in diameter; *rounded, dark red or rarely yellow;* thin, yellow, dry, sweet pulp; 3–4 nutlets; *many,* in drooping clusters; maturing in autumn.

Habitat: Moist valleys.
  Range: Alberta east to Newfoundland, south to
         Virginia, west to Missouri, and north
         to South Dakota; also south in western
         mountains to Colorado; to 2000′
         (610 m), higher in the West.

The scientific name of Fireberry
Hawthorn means "golden fruit,"
although the fruit is commonly dark
red and rarely yellow.

217, 473 Columbia Hawthorn
Crataegus columbiana Howell

Description: Many-branched shrub or small tree
             with rounded crown.
             Height: 16′ (5 m).
             Diameter: 4″ (10 cm).
             Leaves: 1–3½″ (2.5–9 cm) long, ¾–
             2½″ (2–6 cm) wide. *Ovate to nearly
             round,* short-pointed at tip, *finely saw-
             toothed,* with gland-tipped teeth except
             toward short-pointed base, *few shallow,
             short-pointed lobes* toward tip. Green
             with pressed hairs above, paler and
             hairless except on midvein and vein
             angles beneath.
             Bark: gray to reddish-brown, becoming
             furrowed into scaly ridges.
             Twigs: reddish-brown, hairless, with
             long, slender spines to 2½″ (6 cm)
             long.
             Flowers: ⅝″ (15 mm) wide; with 5
             white petals, *10 stamens,* and 2–4
             styles; many flowers in hairy clusters; in
             spring.
             Fruit: ⅜″ (10 mm) in diameter;
             purplish-red, often slightly hairy, with
             gland-toothed calyx at tip and 2–4
             nutlets; maturing in late summer.
  Habitat: Along streams and on moist hillsides in
           open areas.
    Range: From central British Columbia east to
           extreme SW. Saskatchewan, south to
           Montana and west to Oregon; at 2000–
           5000′ (610–1524 m).

This western relative of Fireberry Hawthorn (*C. chrysocarpa* Ashe) was named after the Columbia River, where it was discovered. One of the northernmost hawthorns, it can be planted for erosion control in dry areas where it is native.

219 Black Hawthorn
"Douglas Hawthorn"
"River Hawthorn"
Crataegus douglasii Lindl.

Description: Small tree with compact, rounded crown of stout, spreading branches; often a thicket-forming shrub.
Height: 30' (9 m).
Diameter: 1' (0.3 m).
Leaves: 1–3" (2.5–7.5 cm) long, ⅝–2" (1.5–5 cm) wide. *Obovate to ovate, broadest toward short-pointed tip, sharply saw-toothed and often slightly lobed.* Shiny dark green becoming nearly hairless above, paler beneath.
Bark: gray or brown, smooth or becoming scaly.
Twigs: *shiny red,* slender, hairless, often with straight or slightly curved spines to 1" (2.5 cm) long.
Flowers: ½" (12 mm) wide; with 5 white petals, *10–20 pink stamens,* and 3–5 styles; on long, slender stalks in broad, leafy clusters; in spring.
Fruit: ½" (12 mm) in diameter; turning *shiny black,* with thick, light yellow pulp and 3–5 nutlets; sweetish and mealy but somewhat insipid-tasting; several on long stalks in drooping clusters; maturing in late summer.

Habitat: Moist soils of mountain streams and valleys; with sagebrush and conifers.
Range: Local in S. Alaska and from British Columbia south to central California, east to New Mexico, and north to S. Saskatchewan; also local near Lake Superior; near sea level in north; to 6000' (1829 m) in south.

This species is a handsome ornamental with showy white flowers, glossy foliage, and odd, shiny black fruits. It is named for its discoverer, David Douglas (1798–1834), the Scottish botanical explorer. Cattle and sheep browse the foliage; pheasants, partridges, quail, and other birds consume the berries. The most widespread western member of its genus, Black Hawthorn is also the only species north to southeastern Alaska.

218 Cerro Hawthorn
Crataegus erythropoda Ashe

Description: Shrub or small tree with short trunk and widely spreading branches.
Height: 20' (6 m).
Diameter: 4" (10 cm).
Leaves: 1–2½" (2.5–6 cm) long, ⅝–1¾" (1.5–4.5 cm) wide. *Broadly ovate, coarsely saw-toothed,* with teeth gland-tipped, *often shallowly lobed.* Shiny green and becoming hairless or nearly so above; paler, prominently veined, often hairy beneath.
Bark: gray or reddish-brown, thin, scaly.
Twigs: brown, hairless, with many stout, straight or curved purplish spines to 2" (5 cm) long.
Flowers: nearly ¾" (2 cm) wide; with 5 white petals, about *10 pink or purplish stamens,* and 5 styles; 5–10 in compact clusters; in early spring.
Fruit: about ⅜" (10 mm) in diameter; *orange-red to brown or blackish,* with thin pulp and usually 5 large nutlets; maturing in early autumn.

Habitat: Along streams and in moist canyon soils; with pinyons, junipers, and Ponderosa Pine.
Range: S. Wyoming south to N. New Mexico and central Arizona; at 5000–8400' (1524–2560 m).

The scientific name, meaning "red-footed" or "red-stalked," refers to the leafstalks. *Cerro,* Spanish for "hill," is from another scientific name based on specimens collected at Cerro Summit, Colorado.

216, 471 Oneseed Hawthorn
"Single-seed Hawthorn"
"English Hawthorn"
Crataegus monogyna Jacq.

Description: Introduced shrub or small tree with dense, rounded crown of spreading branches.
Height; 25′ (7.6 m).
Diameter: 8″ (20 cm).
Leaves: 1–2″ (2.5–5 cm) long, ½–1″ (1.2–2.5 cm) wide. *Ovate* or *reverse ovate,* blunt at tip, broadest beyond middle, tapering to base; *deeply 3- to 7-lobed,* the lobes often slightly toothed, with side veins ending in both notches and lobes; hairless; leafstalks long, slender. Shiny green above, paler beneath; shedding late in autumn.
Bark: gray, scaly.
Twigs: brown or gray, slender, hairless or nearly so, with slender spines.
Flowers: ⅜–½″ (10–12 mm) wide; with 5 white petals, *20 red stamens,* and 1 style; many on slender stalks in broad, flat clusters; in spring.
Fruit: ⁵⁄₁₆–⅜″ (8–10 mm) in diameter; *elliptical* or rounded, bright *red,* with *1 nutlet;* maturing in early autumn.
Habitat: Scattered in moist soils, along roadsides and edges of forests, and in open areas and waste places.
Range: Native of Europe, N. Africa, and W. Asia. Planted, escaped from cultivation, and naturalized locally from S. British Columbia to California and east to Nova Scotia.
Long cultivated in Europe, this is the most common small tree over much of

Great Britain, because of its usefulness as a stock-proof hedge. When old, open fields were enclosed to make smaller farms in the 17th to the 19th centuries, wire was too expensive and stone often too scarce to use for fences. Instead, farmers planted countless miles of small hawthorns which grew into tough, stout hedges with sharp thorns that have successfully contained sheep and cattle ever since. Oneseed Hawthorn was brought to North America at an early date. This species is distinguishable by the single nutlet, or seed, instead of 2–5 as in other hawthorns. Cultivated varieties have double flowers, pink or red petals, many or no spines, variegated leaves, and a narrow, upright, drooping or dwarf habit. Oneseed Hawthorn is often confused with English Hawthorn (*Crataegus oxyacantha* L.), native to Europe and northern Africa, which is less common and not yet naturalized in the United States and has 3- or 5-lobed leaves, flowers with 2–3 styles, and fruit with 2–3 nutlets.

214, 360, 475   **Washington Hawthorn**
"Washington-thorn"
Crataegus phaenopyrum (L. f.) Medic.

Description:

Shrub or small tree with short trunk and regular, rounded crown of upright branches, abundant small flowers in the spring, many small, round red fruits, and brilliant autumn foliage; hairless throughout.

Height: 30' (9 m).
Diameter: 1' (0.3 m).
Leaves: 1½–2½" (4–6 cm) long, 1–1¾" (2.5–4.5 cm) wide. *Broadly ovate* to triangular or *3-lobed,* short-pointed at tip, nearly straight to slightly notched at base; *coarsely saw-toothed,* often with 5 shallow lobes; slightly hairy when young. Tinged with red,

becoming shiny dark green above, paler beneath; turning scarlet and orange in autumn.

Bark: light brown, smooth, thin, becoming scaly.

Twigs: shiny brown, with slender spines.

Flowers: more than ½″ (12 mm) wide; with 5 white petals, *20 pale yellow stamens,* and 3–5 styles; many flowers in compact, hairless clusters; in *late spring.*

Fruit: ¼″ (6 mm) in diameter; *shiny red* or scarlet, with ring scar from *shed calyx,* with thin dry pulp and 3–5 nutlets exposed at ends; maturing in autumn and *persisting* until spring.

Habitat: Moist valley soils.

Range: Virginia south to N. Florida, west to Arkansas, and north to S. Missouri; naturalized locally northeast to Massachusetts; to 2000′ (610 m). Planted for ornament in the West.

This is one of the showiest and most desirable hawthorns for planting. In the early 19th century, it was introduced into Pennsylvania from Washington, D.C., as a hedge plant and is thus called "Washington-thorn." The specific name refers to the pearlike foliage.

Willow Hawthorn
Crataegus saligna Greene

Description: Small tree with short trunk and crown of long, slender, spreading, and drooping branches; or a thicket-forming shrub with many stems.

Height: 20′ (6 m).

Diameter: 4″ (10 cm).

Leaves: mostly 1½–2″ (4–5 cm) long, ¾–1″ (2–2.5 cm) wide. *Elliptical, rounded at tip, finely saw-toothed, with rounded teeth* tipped with red glands; slightly thickened; hairy above when

young; short-stalked, with glands.
Becoming shiny dark green, paler
beneath; turning orange and red in
autumn. Leaves at ends of new twigs
larger, ovate, long-pointed; coarsely
saw-toothed, with few shallow lobes
and *veins extending to notches* as well as
lobes.

Bark: red or reddish-brown, becoming
gray and scaly.

Twigs: *shiny red,* slender, drooping,
with many straight spines to 1¼"
(3 cm).

Flowers: ⅝" (15 mm) wide; with 5
white petals, about *20 yellow stamens,*
and 5 styles; on short stalks in leafy
clusters; in spring.

Fruit: ¼" (6 mm) in diameter; shiny
red to *blue-black,* with calyx at tip;
thin, dry, sweet pulp and 5 nutlets; in
drooping clusters; maturing in fall.

Habitat: Moist soils along mountain streams.

Range: Only in W. Colorado; at 6000–8400'
(1829–2560 m).

This local species is handsome in
autumn, with its showy foliage and
abundant, fruits blue-black unlike
those of most other hawthorns. The
scientific and common names both refer
to the willowlike, slender, drooping
twigs and erect stems of shrubby
thickets and not to the shape of the
leaves.

184   **Fleshy Hawthorn**
       "Long-spine Hawthorn"
       "Succulent Hawthorn"
       Crataegus succulenta Schrad.

Description: Shrub or small tree with short trunk
and broad, irregular, dense crown of
stout branches.
Height: 20' (6 m).
Diameter: 6" (15 cm).
Leaves: 2–2½" (5–6 cm) long, 1–1½"
(2.5–4 cm) wide. *Elliptical,* gradually

narrowed from middle to base; doubly saw-toothed, *shallow-lobed beyond middle,* with 4–7 fine, *sunken veins* on each side; slightly thickened; hairy when young. Shiny dark green above, pale yellow-green with fine hairs on midvein beneath.

Bark: dark reddish-brown, scaly.

Twigs: stout, hairless, with many long, stout, slightly curved spines.

Flowers: ⅝″–¾″ (15–19 mm) wide; with 5 white petals, *10–20 white or pink stamens,* and 2–4 styles; in broad hairy clusters; in late spring.

Fruit: ½–⅝″ (12–15 mm) in diameter; *bright red,* with gland-toothed calyx at tip; *thick, juicy, sweet pulp* and 2–4 nutlets; in drooping clusters on long stalks; maturing in early autumn.

Habitat: Moist soils of valleys and open uplands.

Range: S. Manitoba east to Nova Scotia, south to W. North Carolina, and west to Kansas and Nebraska; to 3000′ (914 m); locally to Colorado at 5000–7000′ (1524–2134 m).

This handsome ornamental sometimes flowers when a low shrub. The common and Latin species names both refer to the succulent, soft fruit.

181, 470 Tracy Hawthorn
"Mountain Hawthorn"
Crataegus tracyi Ashe ex Eggl.

Description: Small bushy tree with rounded crown of spreading branches.
Height: 17′ (5 m).
Diameter: 8″ (20 cm).
Leaves: 1–1½″ (2.5–4 cm) long, ¾–1¼″ (2–3 cm) wide. *Reverse ovate,* ovate, or elliptical; *sharply and coarsely saw-toothed* to below middle, with gland-tipped teeth. Shiny dark green above; *slightly rough, hairy,* pale yellowish-green, and finely hairy beneath.

Bark: dark brown, smooth or becoming slightly scaly.

Twigs: gray, slender, hairy when young; with many straight spines to 2¼" (6 cm) long.

Flowers: more than ½" (12 mm) wide; with 5 white petals, *10–15 pink stamens,* and 2–3 styles; 7–10 in compact, slightly hairy clusters; in early spring.

Fruit: ⅜" (10 mm) in diameter; *elliptical, orange-red, with enlarged, gland-toothed calyx* at tip; thin pulp and 2–3 nutlets; maturing in early autumn.

Habitat: Rocky stream banks in warm regions.

Range: Central and Trans-Pecos Texas and N. Mexico; at 1500–5000' (457–1524 m).

This is a local southwestern relative of the widespread eastern species Cockspur Hawthorn (*C. crus-galli* L.). It is named after its discoverer, Samuel Mills Tracy (1847–1920), a U.S. horticulturist.

167, 350, 476 Toyon
"Christmas-berry"
"California-holly"
Heteromeles arbutifolia (Lindl.) M. J. Roem.

Description: One of the most beautiful native shrubs or small trees, evergreen, with short trunk, many branches, and rounded crown.

Height: 30' (9 m).

Diameter: 1' (0.3 m).

Leaves: *evergreen;* 2–4" (5–10 cm) long, ¾–1½" (2–4 cm) wide. Oblong lance-shaped, *sharply saw-toothed,* thick, short-stalked. *Shiny dark green* above, paler beneath.

Bark: light gray, smooth, aromatic.

Twigs: dark red, slender; hairy when young.

Flowers: ¼" (6 mm) wide; with 5 *white petals;* many in upright clusters 4–6" (10–15 cm) wide; in early summer.

Fruit: ¼–⅜″ (6–10 mm) long; *like small apples, red* (sometimes yellow), mealy and sour, usually 2-seeded; maturing in autumn and remaining attached in winter.

Habitat: Along streams and on dry slopes, often on sea cliffs; in chaparral and woodland zones.

Range: California in Coast Ranges and Sierra Nevada foothills and Channel Islands; also Baja California; to 4000′ (1219 m).

The only species in its genus, Toyon is very showy in winter with evergreen leaves and abundant red fruit and is popular for Christmas decorations. A pioneer plant on eroded soil, it sprouts vigorously after fire or cutting. The common name Toyon apparently is of American Indian origin.

88, 296, 351 Lyontree
"Lyonothamnus" "Catalina-ironwood"
Lyonothamnus floribundus Gray

Description: Evergreen shrub with several crooked stems or occasionally a small tree with straight trunk.
Height: 40′ (12 m).
Diameter: 1′ (0.3 m).
Leaves: *evergreen; opposite;* 4–7″ (10–18 cm) long, ½–1″ (1.2–2.5 cm) wide. Thick; *shiny dark green* above, yellow-green and densely hairy when young beneath. *Lance-shaped or oblong,* or pinnately compound with *3–7 narrowly lance-shaped leaflets deeply lobed* on edges.
Bark: *dark reddish-brown,* thin, papery and *shedding in narrow strips,* becoming gray.
Twigs: orange and hairy when young, becoming shiny red.
Flowers: ½″ (12 mm) wide; with 5 *rounded, white petals;* many in flattened clusters 4–8″ (10–20 cm) wide; in early summer.

Fruit: ³⁄₁₆″ (5 mm) long; *2 narrow, long-pointed capsules,* hairy, hard; each usually

4-seeded; maturing in late summer.

Habitat: Dry, rocky soils of canyons and chaparral.

Range: Only on Santa Rosa, Santa Cruz, Santa Catalina, and San Clemente islands of California; at 500–2000′ (152–610 m).

Lyontree occurs in two forms: the tree with mostly simple leaves is confined to Santa Catalina Island; the shrub (or sometimes tree) with pinnately compound leaves is found on the other islands. Both are grown as ornamentals in California. Indians made the tough wood, known locally as "Ironwood," into spear handles and shafts, and European settlers used it for making fishing poles and canes. It is placed alone in a distinct genus. The genus name, meaning "Lyon's shrub," honors the discoverer, William Scrugham Lyon (1852–1916), a U.S. horticulturist and forester.

149, 361   Oregon Crab Apple
"Pacific Crab Apple"
"Western Crab Apple"
Malus fusca (Raf.) Schneid.

Description: Small tree, often with several trunks and many branches, or a thicket-forming shrub; sometimes spiny.
Height: 30′ (9 m).
Diameter: 1′ (0.3 m).
Leaves: 1½–3½″ (4–9 cm) long, ¾–1½″ (2–4 cm) wide. *Ovate, elliptical, or lance-shaped;* sharply saw-toothed and *sometimes slightly 3-lobed* toward tip; with slender, hairy leafstalks. Shiny green and becoming hairless above, pale and usually slightly hairy beneath; turning orange and red in autumn.
Bark: gray, smooth to slightly scaly, thin.
Twigs: reddish, slender, sometimes spiny; hairy when young.

Flowers: ¾–1″ (2–2.5 cm) wide; with 5 *rounded white or pink petals;* several in long-stalked, upright clusters; in early summer.

Fruit: ½–¾″ (12–19 mm) long; *oblong,* applelike, *yellow or red,* sour, edible; with few seeds; maturing in late summer.

Habitat: Along streams and valleys; also on low slopes and in coniferous forests.

Range: S. Alaska south near Pacific Coast to NW. California; to 1000′ (305 m).

The only western species of crab apple has oblong fruit; the three eastern species have round fruit. The strong wood can be made into superior tool handles. The fruit is used for jellies and preserves and was once eaten by Indians; grouse and other birds consume the crab apples in quantity.

157, 368, 493 Apple
"Common Apple" "Wild Apple"
Malus sylvestris (L.) Mill.

Description: This familiar fruit tree, naturalized locally, has a short trunk, spreading, rounded crown, showy pink-tinged blossoms, and delicious red fruit.
Height: 30–40′ (9–12 m).
Diameter: 1–2′ (0.3–0.6 m).
Leaves: 2–3½″ (5–9 cm) long, 1¼–2¼″ (3–6 cm) wide. *Ovate* or elliptical, *wavy saw-toothed,* with hairy leafstalk. Green above, *densely covered with gray hairs beneath.*
Bark: gray, fissured and scaly.
Twigs: greenish, turning brown; densely covered with white hairs when young.
Flowers: 1¼″ (3 cm) wide; with 5 *rounded petals, white tinged with pink;* in early spring.
Fruit: 2–3½″ (5–9 cm) in diameter; the familiar *edible apple,* shiny red or yellow, sunken at ends; thick, sweet

pulp; *star-shaped core* contains up to 10
seeds; matures in late summer.

Habitat: Moist soils near houses, fences,
roadsides, and clearings.

Range: Native of Europe and W. Asia.
Naturalized locally across S. Canada, in
E. continental United States, and in
Pacific states.

The Apple has been cultivated since
ancient times. Numerous improved
varieties have been developed from this
species and from hybrids with related
species. Although well known, it is
sometimes not recognized when
growing wild. For nearly fifty years
Jonathan Chapman (1774–1845),
better known as Johnny Appleseed,
traveling mostly on foot, distributed
apple seeds to everybody he met. With
seeds from cider presses, he helped to
establish orchards from Pennsylvania to
Illinois. Wildlife consume quantities of
fallen fruit after harvest.

195, 367, 490 **American Plum**
"Red Plum" "River Plum"
Prunus americana Marsh.

Description: A thicket-forming shrub or small tree
with short trunk, many spreading
branches, broad crown, showy, large
white flowers, and red plums.
Height: 30' (9 m).
Diameter: 1' (0.3 m).
Leaves: 2½–4" (6–10 cm) long, 1¼–
1¾" (3–4.5 cm) wide. *Elliptical, long-
pointed* at tip, *sharply and often doubly
saw-toothed,* slightly thickened. *Dull
green* with *slightly sunken veins* above,
paler and often slightly hairy on veins
beneath.
Bark: dark brown, scaly.
Twigs: light brown, slender, hairless;
short twigs ending in spine.
Flowers: ¾–1" (2–2.5 cm) wide; with
5 rounded white petals; in clusters of

2–5 on slender, equal stalks; with slightly unpleasant odor; in early spring before leaves.
Fruit: ¾–1" (2–2.5 cm) in diameter; a *plum,* with thick red skin, juicy, sour, edible pulp, and large stone; maturing in summer.

Habitat: Moist soils of valleys and low upland slopes.

Range: SE. Saskatchewan east to New Hampshire, south to Florida, west to Oklahoma, and north to Montana; to 3000' (914 m) in the West and to 6000' (1829 m) in the Southwest.

The plums are eaten fresh and used in jellies and preserves and are also consumed by many kinds of birds. Numerous cultivated varieties with improved fruit have been developed. A handsome ornamental with large flowers and relatively big fruit, American Plum is also grown for erosion control, spreading by root sprouts.

194, 358, 465 Sweet Cherry
"Mazzard" "Mazzard Cherry"
Prunus avium (L.) L.

Description: Introduced fruit tree with tall trunk and cylindrical crown of stout gray branches, bearing sweet cherries.
Height: 50–70' (15–21 m).
Diameter: 1½' (0.5 m).
Leaves: 3–6" (7.5–15 cm) long, 1½–2½" (4–6 cm) wide. *Ovate or reverse ovate,* abruptly long-pointed; *coarsely and often doubly saw-toothed,* with blunt teeth; often drooping. Dull green and nearly hairless with slightly sunken veins above, paler beneath and with *soft hairs,* at least on veins. Slightly hairy leafstalks usually with 2 gland-dots.
Bark: reddish-brown, smooth, often peeling in horizontal strips, becoming thick and furrowed.

Twigs: stout, hairless.
Flowers: 1–1¼" (2.5–3 cm) wide; with 5 rounded white petals; 2–6 on long stalks; with leaves in early spring.
Fruit: ¾–1" (2–2.5 cm) in diameter; a *round or egg-shaped cherry* with *red* to purplish (sometimes yellow) skin, *sweet* edible pulp, and smooth, elliptical stone; maturing in summer.

Habitat: Along roadsides and borders of woods.
Range: Native of Europe and Asia. Naturalized locally in SE. Canada, NE. United States, and Pacific states.

Sweet Cherry is cultivated in many varieties. "Mazzard" is the common name of the wild form, used as a root stock in grafting improved varieties. The Latin species name, meaning "of birds," stresses the value of the fruit to wildlife.

145 Sour Cherry
"Pie Cherry" "Morello Cherry"
Prunus cerasus L.

Description: Small introduced fruit tree or thicket-forming shrub with short trunk, broad, rounded crown, and slender, spreading, and drooping branches, bearing sour cherries.
Height: 30' (9 m).
Diameter: 8" (20 cm).
Leaves: 2–3½" (5–9 cm) long, 1–2¼" (2.5–6 cm) wide. *Ovate or reverse ovate,* abruptly short-pointed; *finely, often doubly saw-toothed, with rounded teeth.* Light green and slightly shiny above, paler and hairless beneath.
Bark: gray, scaly, becoming rough.
Twigs: gray, stout, hairless.
Flowers: 1" (2.5 cm) wide; with 5 rounded white petals; *2–5 on equal stalks* in many clusters; in spring with leaves.

Fruit: ⅝–¾" (15–19 mm) in diameter; a *shiny red cherry,* with soft, juicy, *sour,*

edible pulp; round stone; maturing in summer.

Habitat: Along roadsides, fences, and borders of woods.

Range: Long cultivated in W. Asia and SE. Europe. Naturalized locally in SE. Canada and E. and NW. United States.

The widely cultivated, common "Pie Cherry" has several ornamental varieties; one is a double-flowered form. The scientific name is the classical Latin and Greek term for the cherry, long ago introduced into Europe from Crimea (ancient Cerasus).

147, 365, 491 **Garden Plum**
"Damson Plum" "Bullace Plum"
Prunus domestica L.

Description: Small introduced fruit tree with a spreading, rounded, open crown.
Height: 30' (9 m).
Diameter: 1' (0.3 m).
Leaves: 2–4" (5–10 cm) long, 1–2" (2.5–5 cm) wide. *Elliptical* or reverse ovate, *coarsely wavy saw-toothed, thickened.* Dull green (and hairy when young) above, *hairy with prominent network of veins beneath.*
Bark: gray, smooth or fissured, thick.
Twigs: reddish, stout, often hairy, a few short twigs ending in spines.
Flowers: ¾–1" (2–2.5 cm) wide; with 5 rounded white petals; 1–2 on hairy stalks in clusters; in early spring.
Fruit: 1–2" (2.5–5 cm) in diameter; a *large elliptical, bluish-black plum,* with thick, juicy, sweetish, edible pulp; and smooth *stone, free* from pulp; maturing in summer.

Habitat: Along roadsides and fence rows.

Range: Native of W. Asia and Europe. Naturalized locally in SE. Canada and NE. and NW. United States.

The name "Damson," meaning "of Damascus," alludes to the native range of this species. As the scientific name indicates, this is the common domesticated plum, long cultivated for its edible fruit. It was introduced by early British and French colonists. Numerous improved varieties have been developed for fruit; others, including a double-flowered form, are grown as ornamentals. The prune, a large, readily dried plum with firm flesh, is derived from this species.

148, 363 Bitter Cherry
"Quinine Cherry"
"Wild Cherry"
Prunus emarginata Dougl. ex Eaton

Description: Thicket-forming shrub or small tree with rounded crown, slender, upright branches, *bitter foliage,* and small, bitter cherries.
Height: 20′ (6 m).
Diameter: 8″ (20 cm).
Leaves: 1–2½″ (2.5–6 cm) long, ⅜–1¼″ (1–3 cm) wide. *Oblong to elliptical, rounded or blunt at tip,* short-pointed at *base, with 1–2 dotlike glands; finely saw-toothed, with blunt, gland-tipped teeth. Dark green* above, paler and sometimes hairy beneath.
Bark: dark brown, smooth, very bitter.
Twigs: *shiny red,* slender, hairy when young.
Flowers: ½″ (12 mm) wide; with 5 rounded, notched, white petals; 3–10 on slender stalks; in spring with leaves.

Fruit: ⁵⁄₁₆–⅜″ (8–10 mm) in diameter; a round *cherry,* with thick, *red to black* skin, thin, juicy, *bitter* pulp, and pointed stone; maturing in summer.
Habitat: Moist soils of valleys and on mountain slopes; in chaparral and coniferous forests.
Range: British Columbia, Washington, and W. Montana south to S. California and

SW. New Mexico; to 9000′ (2743 m) in south.

This is the most common western cherry. The scientific name describes the notched petals. As the common name indicates, the fruit is not edible; like the bark and leaves, it is intensely bitter. However, the fruit is consumed by many songbirds and mammals and the foliage is browsed by deer and livestock.

176, 355, 468 Hollyleaf Cherry
"Evergreen Cherry" "Islay"
Prunus ilicifolia (Hook. & Arn.) D. Dietr.

Description: Small evergreen tree with short trunk, dense crown of stout, spreading branches, spiny-toothed leaves, and red cherries; hairless throughout; often a shrub.
Height: 25′ (7.6 m).
Diameter: 1′ (0.3 m).
Leaves: *evergreen;* 1–2″ (2.5–5 cm) long, ¾–1¼″ (2–3 cm) wide. Ovate to rounded, short-pointed or rounded at ends, *coarsely spiny-toothed, thick* and leathery, crisp, with odor of almond when crushed. *Shiny green above,* paler beneath.
Bark: dark reddish-brown, fissured into small, square plates.
Twigs: gray or reddish-brown, slender.
Flowers: ¼″ (6 mm) wide; with 5 rounded white petals; in upright, narrow, unbranched clusters less than 2½″ (6 cm) long at leaf bases; in early spring.
Fruit: ½–⅝″ (12–15 mm) in diameter; a rounded *cherry, red,* sometimes purple or yellow, with thin, juicy, sweetish pulp and smooth stone marked by branching lines; maturing in late fall.
Habitat: Dry slopes and in moist soils along streams; in chaparral and foothill woodland.

Range: Pacific Coast from central California south to Baja California; to 5000′ (1524 m).

Hollyleaf Cherry has been planted as an ornamental and a hedge plant from the time of the Spanish settlement in California. Although sweetish and edible, the cherries are mostly stone and are consumed only by wildlife. Indians used to crack the dried fruit and prepare meal from the ground and leached seeds. The common and scientific names both refer to the hollylike leaves, which are used as Christmas decorations. *Islay* is the Spanish adaptation of the Indian name.

119, 160, **Catalina Cherry**
354, 467 *Prunus lyonii* (Eastw.) Sarg.

Description: Evergreen tree with broad crown of spreading branches, glossy leaves, and dark purple cherries.
Height: 40′ (12 m).
Diameter: 1½′ (0.5 m).
Leaves: *evergreen;* 2–4″ (5–10 cm) long, ½–2½″ (1.2–6 cm) wide. *Narrowly ovate,* short-pointed at tip, blunt or rounded at base; *mostly without teeth;* thick. *Shiny dark green* above, paler beneath.
Bark: dark reddish-brown, rough.
Twigs: yellow-green to reddish-brown, stout, hairless.
Flowers: ¼″ (6 mm) wide; with 5 rounded white petals; many in spreading, narrow clusters 2–4½″ (5–11 cm) long; at leaf bases; in early spring.
Fruit: ½–1″ (1.2–2.5 cm) in diameter; a rounded *cherry,* dark purple or blackish, with thick, sweetish, edible pulp and large stone; maturing in late autumn.
Habitat: Moist soils of canyons and on dry ridges; in chaparral.

Range: Channel Islands of California; also
reported in Baja California; to 3000'
(914 m).

Planted as an evergreen, ornamental,
and hedge plant in California, Catalina
Cherry grows rapidly and is drought-
resistant. Indians ate the fresh and
dried cherries, which can also be made
into jam; birds are also fond of the
fruit. It is closely related to Hollyleaf
Cherry, which is usually smaller and
shrubby and has smaller, rounded,
spiny-toothed leaves, fewer flowers, and
smaller fruit; some consider it an island
variety of that species.

161, 366 **Mahaleb Cherry**
"Mahaleb" "Perfumed Cherry"
Prunus mahaleb L.

Description: Small, introduced tree with a short,
often crooked trunk, open crown of
many spreading branches, and bitter
black cherries: the foliage, flowers,
seeds, bark, and wood are all *aromatic*.
Height: 20' (6 m).
Diameter: 8" (20 cm).
Leaves: 1¼–2¾" (3–7 cm) long, ¾–2"
(2–5 cm) wide. *Rounded to broadly ovate,*
and *finely saw-toothed* with blunt teeth.
Shiny dark green above, paler and hairy
on midvein beneath. Leafstalks slender,
with 1–2 glands.
Bark: dark gray or brown, rough, with
low, irregular ridges.
Twigs: stout, finely hairy when young.
Flowers: ⅝" (15 mm) wide; with 5
rounded white petals, very fragrant; 4–
10 along end of leafy axis, with small,
leaflike scales at base; in spring.
Fruit: ¼–⅜" (6–10 mm) in diameter;
a *small black cherry* (sometimes dark red)
with thin, *bitter, inedible pulp* and
rounded stone; maturing in early
summer.

Habitat: Along roadsides, fence rows, and

borders of woods.

Range: Native of Europe and W. Asia.
Naturalized locally in SE. Canada and
NE. and NW. United States.

Long cultivated as a stock for grafting
cherries and for hedges, Mahaleb Cherry
also has several ornamental varieties. In
Europe, the dark red, aromatic wood
was once used in cabinetwork and
woodenware; pipe stems and canes can
be made from young branches.
"Mahaleb" is the Arabic name.

138, 362, 469   Pin Cherry
"Fire Cherry" "Bird Cherry"
Prunus pensylvanica L. f.

Description: Small tree or shrub with horizontal
branches; narrow, rounded, open
crown; shiny red twigs; bitter, aromatic
bark and foliage; and *tiny red cherries.*
Height: 30' (9 m).
Diameter: 1' (0.3 m).
Leaves: 2½–4½" (6–11 cm) long, ¾–
1¼" (2–3 cm) wide. Broadly lance-
shaped, *long-pointed, finely and sharply
saw-toothed,* becoming hairless. Shiny
green above, paler beneath; turning
bright yellow in autumn. Slender
leafstalks often with 2 gland-dots near
tip.
Bark: reddish-gray, smooth, thin,
becoming gray and fissured into scaly
plates.
Flowers: ½" (12 mm) wide; with 5
rounded white petals; 3–5 on long,
equal stalks; in spring with leaves.
Fruit: ¼" (6 mm) in diameter; a *red
cherry,* with thin sour pulp and large
stone; in summer.
Habitat: Moist soils; often in pure stands on
burned areas and clearings and with
aspens, Paper Birch, and Eastern White
Pine.
Range: British Columbia and S. Mackenzie east
across Canada to Newfoundland, south

to N. Georgia, west to Colorado; to 6000′ (1829 m) in southern Appalachians.

This species is often called "Fire Cherry" because its seedlings come up after forest fires. The plants grow rapidly and can be used for fuel and pulpwood. It is also a "nurse" tree, providing cover and shade for the establishment of seedlings of the next generation of larger hardwoods. The cherries are made into jelly and are also consumed by wildlife.

137, 388, 492 Peach
Prunus persica Batsch

Description: A well-known, small fruit tree with a short trunk, spreading, rounded crown, showy pink blossoms, long, narrow leaves, and yellow to pink juicy fruit.
Height: 30′ (9 m).
Diameter: 1′ (0.3 m).
Leaves: 3½–6″ (9–15 cm) long, ¾–1″ (2–2.5 cm) wide. *Lance-shaped* or narrowly oblong, *finely saw-toothed, sides often curved up* from midvein; leafstalks short with glands near tip. *Shiny green above,* paler beneath. Crushed foliage has a strong odor and bitter taste.
Bark: dark reddish-brown, smooth, becoming rough, bitter.
Twigs: greenish turning reddish-brown, long, slender, hairless.
Flowers: 1–1¼″ (2.5–3 cm) wide; with *5 rounded pink petals, usually single* and *nearly stalkless;* in early spring before leaves.
Fruit: 2–3″ (5–7.5 cm) in diameter; a *nearly round peach;* grooved, with fine *velvety hairs* covering the yellow-to-pink skin; thick, sweet, edible pulp and large, elliptical, pitted stone; maturing in summer.
Habitat: Along roadsides and fence rows.
Range: Native of China; naturalized locally in

E. United States, S. Ontario, and California; mostly in the Southeast.

Peach has been grown as a fruit tree since ancient times. Numerous cultivated varieties include freestone peaches, with pulp separating from the stone, clingstones, with pulp adhering to the stone, and smaller, hairless fruits known as nectarines. Other varieties are ornamentals, some with double flowers and with white or red petals. Spanish colonists introduced the Peach into Florida, and American Indians then planted it widely.

146, 356, 464   Black Cherry
"Wild Cherry" "Rum Cherry"
Prunus serotina Ehrh.

Description: Aromatic tree with tall trunk, oblong crown, abundant, small white flowers, and small black cherries; crushed foliage and bark have distinctive, *cherrylike odor* and *bitter taste*.
Height: 80′ (24 m).
Diameter: 2′ (0.6 m).
Leaves: 2–5″ (5–13 cm) long, 1¼–2″ (3–5 cm) wide. *Elliptical, 1–2 dark red glands* at base; *finely saw-toothed, with curved or blunt teeth;* slightly thickened. Shiny dark green above, light green and often hairy along midvein beneath; turning yellow or reddish in autumn.
Bark: dark gray, smooth, with horizontal lines, becoming irregularly fissured and scaly, exposing reddish-brown inner bark, *bitter and aromatic.*
Twigs: reddish-brown, slender, hairless
Flowers: ⅜″ (10 mm) wide; 5 rounded, white petals; *many, along spreading or drooping axis* of 4–6″ (10–15 cm) at end of leafy twig; in late spring.
Fruit: ⅜″ (10 mm) in diameter; a *cherry, dark red turning blackish;* with slightly *bitter, juicy, edible* pulp and elliptical stone; maturing in summer.

Habitat: On many sites except very wet or very dry soils; sometimes in pure stands.

Range: S. Quebec to Nova Scotia, south to central Florida, west to E. Texas, and north to Minnesota; varieties from central Texas west to Arizona and south to Mexico; to 5000' (1524 m) in southern Appalachians and at 4500–7500' (1372–2286 m) in the Southwest.

This widespread species is the largest and most important native cherry. The valuable wood is used particularly for furniture, paneling, professional and scientific instruments, handles, and toys. Wild cherry syrup, a cough medicine, is obtained from the bark, and jelly and wine are prepared from the fruit. One of the first New World trees introduced into English gardens, it was recorded as early as 1629. As many as 5 geographical varieties have been distinguished.

182, 359 Klamath Plum
"Sierra Plum" "Pacific Plum"
Prunus subcordata Benth.

Description: Thicket-forming shrub or small tree with short trunk, stiff, crooked, nearly horizontal branches, and reddish-purple plums.
Height: 20' (6 m).
Diameter: 8" (20 cm).
Leaves: 1–2½" (2.5–6 cm) long, ½–1¾" (1.2–4.5 cm) wide. *Elliptical to rounded,* tip blunt or rounded, base blunt to notched; *sharply saw-toothed;* becoming nearly hairless; slightly thickened; prominent side veins. *Dark green* above, pale beneath; turning red to yellow in autumn.
Bark: gray-brown, furrowed into long, scaly plates.
Twigs: bright red, short, often ending in *spines.*

Flowers: ⅝″ (15 mm) wide; with 5
rounded white petals, turning pink; 2–
4 *on slender, equal stalks;* in early spring
before leaves.

Fruit: ⅝–1″ (1.5–2.5 cm) in diameter;
a rounded *plum, reddish-purple,* or
sometimes yellow, with sour, edible
pulp and slightly flattened, smooth
stone; maturing in late summer.

Habitat: Dry, rocky slopes in moist valleys.

Range: W. and S. Oregon south to central
California in Coast Ranges and Sierra
Nevada; to 6000′ (1829 m).

This is the only wild plum in the
Pacific states and is easily identified
when in fruit. The fruit, often borne in
quantities, is eaten fresh or dried and in
preserves and jellies. Sheep and deer
browse the foliage.

162, 357, 466 **Common Chokecherry**
"Eastern Chokecherry"
"Western Chokecherry"
Prunus virginiana L.

Description: Shrub or small tree, often forming
dense thickets, with dark red or
blackish chokecherries.

Height: 20′ (6 m).

Diameter: 6″ (15 cm).

Leaves: 1½–3¼″ (4–8 cm) long, ⅝–
1½″ (1.5–4 cm) wide. *Elliptical,* finely
and *sharply saw-toothed,* slightly
thickened. Shiny dark green above,
light green and sometimes slightly
hairy beneath; turning yellow in
autumn. Leafstalks slender, usually
with 2 gland-dots.

Bark: brown or gray, smooth or
becoming scaly.

Twigs: brown, slender, with
disagreeable odor and bitter taste.

Flowers: ½″ (12 mm) wide; with 5
rounded white petals; in unbranched
clusters to 4″ (10 cm) long; in late
spring.

Fruit: ¼–⅜" (6–10 mm) in diameter; a
chokecherry, shiny dark red or blackish
skin; juicy, *astringent or bitter* pulp;
large stone; maturing in summer.

Habitat: Moist soils, especially along streams in
mountains and clearings and along
forest borders and roadsides.

Range: N. British Columbia east to
Newfoundland, south to W. North
Carolina, and west to S. California; to
8000' (2438 m) in the Southwest.

As the common name suggests,
chokecherries are astringent or puckery,
especially when immature or raw. They
can be made into preserves and jelly,
however the fruit stones are poisonous.
This species is sometimes divided into
3 geographic varieties based on minor
differences of leaves and fruits.

158, 369, 494   **Pear**
"Common Pear"
Pyrus communis L.

Description: The well-known, naturalized fruit tree

with broad crown of shiny green
foliage, white flowers in early spring,
and edible pears in autumn.
Height: 40' (12 m).
Diameter: 1' (0.3 m).
Leaves: 1½–3" (4–7.5 cm) long, 1–2"
(2.5–5 cm) wide. *Broadly ovate to
elliptical,* finely *wavy saw-toothed,*
becoming nearly *hairless;* often crowded
on spur twigs; leafstalks slender. *Shiny
green* above, paler beneath.
Bark: gray, smooth, becoming scaly.
Twigs: both long and short, *hairless,*
the many stout spurs sometimes ending
in spines.
Flowers: 1¼" (3 cm) wide; with 5
rounded white petals; in long-stalked
clusters; in early spring with leaves.
Fruit: 2½–4" (6–10 cm) long; a *pear,*
with green to brown skin; thick, juicy,
sweet, edible pulp, and star-shaped,

gritty core; maturing in late summer.

Habitat: Moist soils near houses, fences, roadsides, clearings, and borders of forests.

Range: Native of Europe and W. Asia. Naturalized locally from Maine to Missouri, Florida, and Texas and in the Northwest.

The Pear has been cultivated since ancient times. Numerous varieties have been developed from this species and from hybrids.

289, 477 European Mountain-ash
"Rowan-tree"
Sorbus aucuparia L.

Description: Introduced tree with open, spreading, rounded crown, showy white flowers, and many bright red berries.
Height: 20–40' (6–12 m).
Diameter: 1' (0.3 m).
Leaves: pinnately compound; 4–8" (10–20 cm) long. 9–17 leaflets 1–2" (2.5–5 cm) long, ½–¾" (12–19 mm) wide; *oblong* or lance-shaped, *short-pointed, saw-toothed* except near base, stalkless. Dull green above, with *white hairs beneath;* turning reddish in autumn.
Bark: dark gray, smooth, with horizontal lines, aromatic.
Twigs: stout, *densely covered with white hairs when young.*
Flowers: ⅜" (10 mm) wide; with 5 rounded white petals; in clusters 3–6" (7.5–15 cm) wide of *75–100 flowers* on stalks with white hairs; in late spring.

Fruit: 5⁄16" (8 mm) in diameter; *like small apples;* with *bright red* skin and bitter pulp; numerous, in clusters; maturing in late spring.

Habitat: Along roads, forming thickets.

Range: Native of Eurasia. Naturalized from SE. Alaska and across S. Canada to Newfoundland and from Maine to Minnesota and California.

Introduced in colonial times, this handsome ornamental has showy red fruit that remains on the tree into early winter, providing food for birds which then spread the seeds. The Latin species name, meaning "to catch birds," refers to the early use by fowlers of the sticky fruit to make bird lime, which they smeared on branches to trap birds. The only tree naturalized in Alaska, it was introduced in the southeastern part of the state and has escaped around towns. The word "rowan" is derived from an old Scandinavian word meaning "red," and refers to the brightly colored berries so characteristic of this genus.

290 Greene Mountain-ash
"Western Mountain-ash"
Sorbus scopulina Greene

Description: A shrub forming dense clumps, or rarely a small tree, with many small white flowers and small, applelike fruit.
Height: 20' (6 m).
Diameter: 4" (10 cm).
Leaves: pinnately compound; 4–9" (10–23 cm) long. *11–15 stalkless leaflets* 1¼–2½" (3–6 cm) long; *lance-shaped,* pointed at tip, sharply *saw-toothed almost to unequal, rounded base; becoming hairless. Shiny dark green* above, slightly paler beneath.
Bark: gray or reddish, smooth.
Twigs: light brown, slender, with whitish hairs when young.
Flowers: ⅜" (10 mm) wide; with 5 *white rounded petals,* fragrant; many in upright, rounded *clusters* 1¼–3" (3–7.5 cm) wide; in early summer.

Fruit: less than ⅜" (10 mm) in diameter; *like a tiny apple, shiny red,* bitter, with several seeds; *fewer than 25 in a cluster;* maturing in summer and persisting into winter.

Habitat: Moist soils, in openings and clearings; in coniferous forests.

Range: Alaska east to SW. Mackenzie and south in mountains to central California and S. New Mexico, locally beyond; down to sea level in north; at 4000–9000′ (1219–2743 m) in south.

This shrubby species takes the form of a small tree in southeastern Alaska. The common name honors Edward Lee Greene (1843–1915), the United States botanist who prepared the description. Birds and mammals eat the fruit.

Sitka Mountain-ash
"Western Mountain-ash"
Sorbus sitchensis Roem.

Description: Shrub or small tree with rounded crown and with many fragrant, small white flowers in early summer, followed by small, applelike fruit.
Height: 20′ (6 m).
Diameter: 6″ (15 cm).
Leaves: pinnately compound; 4–8″ (10–20 cm) long. *Usually 9–11* (sometimes 7–13) stalkless *leaflets* 1¼–2½″ (3–6 cm) long; *elliptical or oblong,* rounded or blunt at ends, *sharply saw-toothed above middle. Dull blue-green* above, *pale and hairless* or nearly so beneath.
Bark: gray, smooth.
Twigs: with rust-colored hairs when young.
Flowers: ¼″ (6 mm) wide; with 5 *rounded, white petals;* many in upright, rounded *clusters* 2–4″ (5–10 cm) wide; in early summer.
Fruit: ⅜–½″ (10–12 mm) in diameter; *like a small apple, red, turning orange or purple,* bitter, with few seeds; several in cluster; maturing in late summer.

Habitat: Moist soils in coniferous forests.
Range: SW. Alaska southeast along coast to Washington and in mountains to central California and N. Idaho; to

timberline northward and at 5000–
10,000' (1524–3048 m) in south.

This species is named for Sitka, Alaska,
where it was discovered. Grosbeaks,
cedar waxwings, grouse, and other
birds consume the fruit and spread the
seeds.

---

140 **Torrey Vauquelinia**
"Arizona-rosewood"
Vauquelinia californica (Torr.) Sarg.

Description: Evergreen, many-branched shrub or
small tree with stiff, twisted branches.
Height: 20' (6 m).
Diameter: 8" (20 cm).
Leaves: *evergreen;* 1½–4" (4–10 cm)
long, ¼–⅝" (6–15 mm) wide.
Narrowly lance-shaped, long-pointed at
tip, short-pointed at base; *saw-toothed;
thick and leathery;* short-stalked. Bright
yellow-green with *sunken midvein* and
many fine side veins above, covered with
fine white hairs beneath.
Bark: dark reddish-brown, thin, broken
into small square scales or shaggy.
Twigs: reddish-brown, slender, densely
hairy.
Flowers: ⅜" (10 mm) wide; with 5
spreading, rounded white petals, turning
red; many crowded in terminal,
upright, hairy clusters 2–3" (5–7.5
cm) wide; in late spring.

Fruit: ¼" (6 mm) long; a *hard, hairy
capsule,* splitting into 5 parts, each with
2 winged seeds; maturing in summer,
remaining attached during winter.
Habitat: Canyons and mountains in upper desert
and oak woodland.
Range: Mountains of S. Arizona; also Baja
California; at 2500–5000' (762–
1524 m).

The beautiful, dark brown wood
streaked with red is hard and very
heavy; however, its scarcity, small size,

and slow growth have limited
commercial use. This genus is
dedicated to Louis Nicholas Vauquelin
(1763–1827), a French chemist.
Despite its Latin species name, which was
given before boundaries were definite,
it is not native to California.

LEGUME FAMILY
(Leguminosae)

The third largest family of seed plants, characterized by pealike flowers and pods; with approximately 12,000 species of herbs, shrubs, and trees worldwide. About 44 native and 6 naturalized tree species and numerous species of shrubs and herbs in North America.

Leaves: alternate; compound (mostly pinnately compound; also bipinnate and with 3 leaflets), rarely simple; with paired stipules sometimes becoming spines.

Flowers: small to large and showy; often in racemes, spikes, and heads; bisexual, mostly irregular in shape of bean flower or butterfly; calyx usually tubular with 5 lobes, corolla of 5 unequal petals, 10 or more stamens distinct or united at base, and 1 pistil.

Fruit: generally a pod (legume), which opens on 2 lines and contains 1 or more elliptical, bean-shaped seeds.

---

265 Green Wattle
"Green-wattle Acacia"
Acacia decurrens Willd.

Description: Evergreen tree with thin, spreading, rounded crown; planted in subtropical regions and escaping from cultivation.
Height: 40' (12 m).
Diameter: 8" (20 cm).
Leaves: *evergreen; finely divided; bipinnately compound;* 3–6" (7.5–15 cm) long; with slender axis and 4–15 pairs of side axes. *30–80 leaflets* ³⁄₁₆–³⁄₈" (5–10 mm) long; *crowded featherlike* on each side axis; not paired, stalkless, *very narrow; gray-green* to dark green.
Bark: brown to gray, smooth to finely fissured.
Twigs: brownish, angled, with tiny hairs.

Flowers: *tiny,* crowded together and stalkless in *small, light yellow balls* ¼" (6 mm) in diameter; fragrant; few balls on short stalks at leaf bases; in early spring.

Fruit: 2–4" (5–10 cm) long, ¼" (6 mm) wide; clustered, *flattened pods,* gray-brown to blackish, finely hairy, slightly narrowed between seeds; ripening in summer and splitting open; several beanlike, dull black seeds.

Habitat: Moist areas but adapted also to dry soils; in subtropical regions.

Range: Native of Australia. Extensively planted in California and escaping.

One of the hardiest acacias, this handsome ornamental grows rapidly and sprouts from roots. Although planted as a street tree, it is short-lived. Two varieties also in California are Black Wattle (var. *mollis* Lindl.), with shiny dark green leaves and pale yellow flowers, and Silver Wattle (var. *dealbata* (Link) F. Muell.), with silvery-gray leaves and deep yellow flowers.

264, 331 Huisache
"Sweet Acacia" "Cassie"
Acacia farnesiana (L.) Willd.

Description: Spiny, many-branched shrub or small tree with a widely spreading, flattened crown and fragrant yellow balls of tiny flowers.

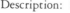

Height: 16' (5 m).
Diameter: 4" (10 cm).
Leaves: alternate or clustered; *bipinnately compound;* 2–4" (5–10 cm) long; usually with 3–5 pairs of side axes. *10–20 pairs of leaflets* ⅛–¼" (3–6 mm) long; oblong, mostly hairless, stalkless; gray-green.
Bark: grayish-brown, thin, smooth or scaly.
Twigs: slightly zigzag, slender, covered with fine hairs when young; with

straight, slender, paired white spines at nodes.

Flowers: ³⁄₁₆″ (5 mm) long; *yellow or orange, very fragrant,* including many tiny stamens clustered in stalked *balls* ½″ (12 mm) in diameter; mainly in late winter and early spring.

Fruit: 1½–3″ (4–7.5 cm) long, ³⁄₈–½″ (10–12 mm) in diameter; a *cylindrical* pod, short-pointed at ends, dark brown or black, hard; maturing in summer, remaining attached, often opening late; many elliptical, flattened, shiny brown seeds.

Habitat: Sandy and clay soils, especially in open areas and along borders of woodlands and roadsides.

Range: S. Texas and local in S. Arizona. Cultivated and naturalized from Florida west to Texas and S. California; also in Mexico; to 5000′ (1524 m).

In southern Europe this species is extensively planted for the "cassie" flowers, which are a perfume ingredient. After drying in the shade, the flowers can be used in sachets to keep clothes smelling fragrant. The tender foliage and pods are browsed by livestock; it is also a honey plant. Mucilage is produced from the gum of the trunk and tannin and dye from the pods and bark.

255, 329 **Gregg Catclaw**
"Devilsclaw" "Catclaw Acacia"
Acacia greggii Gray

Description: Spiny, many-branched, thicket-forming shrub; occasionally a small tree with a broad crown.

Height: 20′ (6 m).
Diameter: 6″ (15 cm).
Leaves: clustered; *bipinnately compound;* 1–3″ (2.5–7.5 cm) long; the slender axis usually with 2–3 pairs of side axes. *3–7 pairs of leaflets* ⅛–¼″ (3–6 mm)

long; oblong, rounded at ends; thick; usually hairy; almost stalkless; dull green.

Bark: gray, thin, becoming deeply furrowed.

Twigs: brown, slender, angled, covered with fine hairs; with many *scattered, stout spines* ¼" (6 mm) long, *hooked* or curved backward.

Flowers: ¼" (6 mm) long; light yellow, fragrant; stalkless; including many tiny stamens in long, narrow clusters 1–2" (2.5–5 cm) long; in early spring and irregularly in summer.

Fruit: 2½–5" (6–13 cm) long, ½–¾" (12–19 mm) wide; thin, flat, ribbonlike, *oblong pod;* brown, *curved, much twisted,* often narrowed between seeds; maturing in summer and shedding in winter, remaining closed; several beanlike, nearly round, flat brown seeds.

Habitat: Along streams, in canyons, and on dry, rocky slopes of plains and foothills.

Range: Central, S., and Trans-Pecos Texas and NE. Mexico; to 2000' (610 m), sometimes higher.

This is one of the most despised southwestern shrubs. As indicated by the common names (including the Spanish, *uña de gato*), the sharp, stout, hooked spines, like a cat's claws, tear clothing and flesh. The hard, heavy wood with reddish-brown heartwood and yellow sapwood is used for souvenirs and locally for tool handles and fuel. Catclaw honey (also known as Uvalde honey, from the Texas county of that name) is made from the flowers of this and related species. Indians once made meal called "pinole" from the seeds.

262, 524 Roemer Catclaw
"Roemer Acacia"
Acacia roemeriana Scheele

Description:
Spiny, many-branched, irregularly spreading shrub with slender, weak stems and whitish-yellow balls of tiny flowers; sometimes a small tree with rounded crown.
Height: 3–6' (0.9–1.8 m), rarely 15' (4.6 m).
Diameter: 1' (0.3 m).
Leaves: *bipinnately compound;* 1½–4" (4–10 cm) long; with 1–3 pairs of side axes, each with *4–12 pairs of stalkless leaflets* ³⁄₁₆–³⁄₈" (5–10 mm) long; *elliptical to oblong,* rounded at ends, hairless, gray-green.
Bark: gray to brown, smooth but becoming scaly.
Twigs: gray or brown, slender, hairless; usually with *single or paired, short, curved spines.*
Flowers: ³⁄₁₆" (5 mm) long including many tiny stamens; *whitish-yellow,* fragrant, clustered in stalked *balls* ³⁄₈" (10 mm) in diameter at leaf bases; from early spring to early summer.
Fruit: 2–4" (5–10 cm) long, ¾–1¼" (2–3 cm) wide; *flat, oblong pods,* slightly curved, *reddish-brown,* leathery and flexible; several small, flattened seeds; maturing in summer and splitting open.

Habitat:
Dry, rocky and gravelly slopes and bluffs, mainly limestone outcrops; forming thickets, or "brush," with junipers and evergreen oaks.

Range:
Central, SW., and Trans-Pecos Texas, SE. New Mexico and NE. Mexico; at 500–4500' (152–1372 m).

Common to abundant as a thicket-forming shrub on uplands, Roemer Catclaw becomes a small tree in moist areas. It was named for Karl Ferdinand Roemer (1818–91) of Germany, who collected plants in Texas in 1846–47.

266, 386 Silktree
"Mimosa-tree" "Powderpuff-tree"
Albizia julibrissin Durazzini

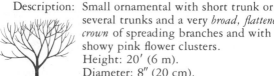

Description: Small ornamental with short trunk or several trunks and a very *broad, flattened crown* of spreading branches and with showy pink flower clusters.
Height: 20' (6 m).
Diameter: 8" (20 cm).
Leaves: *bipinnately compound;* 6–15" (15–38 cm) long; *fernlike,* with *5–12 pairs* of side axes covered with fine hairs. Each axis has *15–30 pairs* of oblong, pale green leaflets ⅜–⅝" (10–15 mm) long.
Bark: blackish or gray, nearly smooth.
Twigs: brown or gray, often angled.
Flowers: more than 1" (2.5 cm) long; with long, *threadlike, pink stamens* whitish toward base; crowded in long-stalked, *ball-like clusters* 1½–2" (4–5 cm) wide; grouped at ends of twigs; throughout summer.

Fruit: 5–8" (13–20 cm) long; flat, pointed, *oblong pod,* yellow-brown; maturing in summer, remaining closed; several beanlike, flattened, shiny brown seeds.

Habitat: Open areas including wasteland and dry, gravelly soils.

Range: Native from Iran to China. Naturalized from Maryland to S. Florida, west to E. Texas, north to Indiana; to 2000' (610 m). Planted in the West.

The hardiest tree of its genus, Silktree has an unusually long flowering period. It is often called "Mimosa-tree" because the flowers are similar to those of the related herbaceous sensitive-plants (genus *Mimosa*). Silktree leaflets fold up at night; those of sensitive-plants fold up when touched.

273 Carob
"St. Johns-bread" "Algarroba"
Ceratonia siliqua L.

Description: Introduced, small to medium-sized
evergreen tree with spreading, broad,
rounded crown.
Height: 40′ (12 m).
Diameter: 1′ (0.3 m).
Leaves: *evergreen;* pinnately compound;
4–8″ (10–20 cm) long. *4–10 paired
leaflets* 1–2″ (2.5–5 cm) long, 1–1¼″
(2.5–3 cm) wide; *broadly elliptical,*
rounded at ends, edges often slightly
wavy; thick and *leathery;* short-stalked.
Shiny dark green above, paler beneath.
Bark: dark reddish-brown, rough.
Twigs: gray, slender, with fine hairs
when young.
Flowers: ⁵⁄₁₆″ (8 mm) long; without
petals; *yellowish* from *red buds;* in narrow
clusters 1–2″ (2.5–5 cm) long at leaf
bases; male and female on same or
separate trees; in spring.
Fruit: 4–10″ (10–25 cm) long, ¾″ (19
mm) wide; dark brown, slightly
flattened, leathery pods; edible, *sugary
pulp;* several flattened seeds; maturing
and shedding in autumn, without
splitting open.

Habitat: Various soils including alkali and clay
in hot, dry subtropical regions.

Range: Native probably in eastern
Mediterranean region of Asia Minor and
Syria. Planted in Florida, S. Texas, S.
Arizona, and California.

Carob is a handsome shade tree along
streets in hot, dry regions, thriving
with some water from irrigation;
however, it does not withstand severe
frost. It has long been cultivated in the
Mediterranean Basin as a forage crop;
the edible pods are rich in protein as
well as sugar and are ground as feed for
livestock. Many cultivated varieties
differ in pod size and shape and in
yield. The sweetish pulp tastes like
chocolate and is used to flavor ice cream

and in the manufacture of syrups and drinks. Powdered pulp is marketed as a substitute for cocoa in baked goods. In the parable of the Prodigal Son, Carob pods were the "husks" fed to the swine. The name "St. Johns-bread" is from the mistaken belief that the seeds and sugary pulp were the locusts and wild honey which St. John the Baptist found in the wilderness. The seeds were the original carat weights for gold and precious stones.

254, 326　Blue Paloverde
Cercidium floridum Benth. ex Gray

Description: Spiny, small tree, *leafless most of the year,* with a short blue-green trunk and widely spreading, very open crown.
Height: 30' (9 m).
Diameter: 1½' (0.5 m).
Leaves: few and scattered; *bipinnately compound;* 1" (2.5 cm) long, with short axis forking into 2 side axes. *2 or 3 pairs of leaflets on each side axis,* ¼" (6 mm) long; oblong; pale *blue-green;* appearing in spring but soon shedding.
Bark: *trunk and branches blue-green and smooth;* base of large trunks becoming brown and scaly.
Twigs: *blue-green, smooth,* slightly zigzag, hairless, with *straight, slender spine* less than ¼" (6 mm) long *at each node.*
Flowers: ¾" (19 mm) wide; with 5 *bright yellow petals,* the largest with a few *red spots;* 4–5 flowers in a cluster less than 2" (5 cm) long; covering the tree in spring, sometimes again in late summer.
Fruit: 1½–3¼" (4–8 cm) long; narrowly *oblong, flat, thin* pods; short-pointed at ends, yellowish-brown; maturing and falling in summer; 2–8 beanlike seeds.
Habitat: Along washes and valleys and

sometimes on lower slopes of deserts and desert grasslands.

Range: Central and S. Arizona, SE. California, and NW. Mexico; 4000' (1219 m).

Although leaves are absent most of the year, photosynthesis, the manufacture of food, is performed by the blue-green branches and twigs. Indians cooked and ate the immature beanlike pods and ground the mature seeds for meal. Twigs and pods of paloverdes serve as browse for wildlife and emergency food for livestock; the seeds are consumed by rodents and birds; and the flowers are a source of honey. This species is useful for erosion control along drainages. The Spanish common name, *palo verde*, means "green tree" or "green pole."

253, 325, 521 Yellow Paloverde
"Foothill Paloverde"
"Littleleaf Paloverde"
Cercidium microphyllum (Torr.) Rose & Johnst.

Description: Small, spiny tree that is *leafless most of the year*, with yellow-green trunk and wide, many-branched, open crown.
Height: 25' (7.6 m).
Diameter: 1' (0.3 m).
Leaves: few; *bipinnately compound* but appear as if a pinnately compound pair; ¾–1" (2–2.5 cm) long; consisting of very short axis with 2 forks, each with *3–7 pairs of minute leaflets:* elliptical, slightly hairy; yellow-green; appearing in spring but soon shedding.
Bark: *yellow-green, smooth.*
Twigs: short, stiff, *ending in long, straight spines* about 2" (5 cm) long.
Flowers: about ½" (12 mm) wide; with *5 pale yellow petals,* largest petals white or cream; in clusters to 1" (2.5 cm) long; covering the tree in spring.
Fruit: 2–3" (5–7.5 cm) long; *cylindrical pods,* ending in *long, narrow points;*

constricted between seeds; remaining
attached.

Habitat: Associated with Saguaro on desert
plains and rocky slopes of foothills and
mountains.

Range: Arizona, SE. California, and NW.
Mexico; at 500–4000' (152–1219 m).

The Spanish common name, *palo verde,*
meaning "green tree" or "green pole,"
describes the smooth, green branches
and twigs. During most of the year, in
the absence of leaves, the branches and
twigs manufacture food. This
adaptation to a desert habitat exposes
less surface to the sun, aiding moisture
retention. Yellow Paloverde, which
gets its name from the yellowish hue of
the branches and foliage, occupies drier
foothills. Blue Paloverde, which has
blue-green branches and foliage, occurs
chiefly along drainages, blooms earlier,
and has larger, deeper yellow flowers.
Indians ground and ate the beanlike
seeds of both species.

123, 389 **Eastern Redbud**
"Judas-tree"
Cercis canadensis L.

Description:  Planted tree with short trunk, rounded
crown of spreading branches, and pink
flowers that cover the twigs in spring.
Height: 40' (12 m).
Diameter: 8" (20 cm).
Leaves: 2½–4½" (6–11 cm) long and
broad. *Heart-shaped, with broad, short
point;* without teeth, with 5–9 main
veins; long-stalked. Dull green above,
paler and sometimes hairy beneath;
turning yellow in autumn.
Bark: dark gray or brown, smooth,
becoming furrowed into scaly plates.
Twigs: brown, slender, angled.
Flowers: ½" (12 mm) long; pea-shaped,
with 5 *slightly unequal purplish-pink
petals,* rarely white; 4–8 flowers in a

cluster on slender stalks; in early spring before leaves.
Fruit: 2½–3¼" (6–8 cm) long; *flat, narrowly oblong pods;* pointed at ends, pink, turning blackish, splitting open on 1 edge; falling in late autumn or winter. Several beanlike, flat, elliptical, dark brown seeds.

Habitat: Moist soils of valleys and slopes and in hardwood forests.

Range: New Jersey south to central Florida, west to S. Texas, and north to SE. Nebraska; also N. Mexico; to 2200' (671 m). Planted as an ornamental in western United States.

Very showy in early spring, when the leafless twigs are covered with masses of pink flowers, Eastern Redbud is often planted as an ornamental. The flowers can be eaten as a salad or fried. According to a myth, Judas Iscariot hanged himself on the related Judas-tree (*Cercis siliquastrum* L.) of western Asia and southern Europe, after which the white flowers turned red with shame or blood.

122, 390, 516 **California Redbud**
"Western Redbud" "Judas-tree"
Cercis occidentalis Torr. ex Gray

Description: Large flowering shrub or small tree with rounded crown of many spreading branches.
Height: 16' (5 m).
Diameter: 4" (10 cm).
Leaves: 1½–3½" (4–9 cm) long and broad. *Nearly round,* long-stalked, with *7–9 veins from notched base,* thickened, mostly hairless. Dark green above, paler beneath.
Bark: gray, smooth, and becoming fissured.
Twigs: reddish-brown when young, becoming dark gray; hairless.
Flowers: ½" (12 mm) long; pealike,

with 5 *slightly unequal purplish-pink petals,* rarely white; 2–5 in cluster on slender stalks; scattered along twigs; in early spring before leaves.

Fruit: 2–3½″ (5–9 cm) long; narrowly *oblong, flat, thin pods;* brown or purplish; maturing in late summer and splitting open on 1 edge; many hanging in clusters along twigs; several beanlike seeds.

Habitat: Canyons and slopes of foothills and mountains.

Range: N. California east to S. Utah and south to S. Arizona; at 500–6000′ (152–1829 m).

California Redbud is common in Grand Canyon National Park, Arizona, and is a handsome ornamental with showy flowers that cover the twigs in early spring. Indians used to make bows from the wood. Deer browse the foliage.

80, 398   Smokethorn
"Smoketree"
"Indigobush"
Dalea spinosa Gray

Description: Spiny, many-branched shrub or small tree with short, crooked trunk, compact or irregular crown of smoky-gray branches, and *small leaves; leafless most of the year.*

Height: 20′ (6 m).

Diameter: 1′ (0.3 m).

Leaves: ⅜–1″ (1–2.5 cm) long, ⅛–½″ (3–12 mm) wide. *Reverse lance-shaped,* rounded at tip, long-pointed at base, edges wavy; *stalkless* or nearly so; *gray,* densely hairy and gland-dotted. Few in early spring, shedding after a few weeks and before flowering.

Bark: dark gray-brown, furrowed and scaly.

Twigs: *smoky-gray,* with dense, pressed hairs and brown gland-dots; zigzag,

slender, ending in slender, sharp *spines*.
Flowers: ½" (12 mm) long; *pea-shaped*,
with 5 unequal, *dark purple or violet
petals*, fragrant, gland-dotted; few, in
clusters to 1¼" (3 cm) long, along
twigs; in late spring or early summer
when leafless.

Fruit: ⅜" (10 mm) long; a *small, egg-
shaped pod* ending in curved point, hairy
and gland-dotted; containing usually 1
brown, beanlike seed; maturing in late
summer and not opening.

Habitat: Sandy and gravelly washes in desert,
mainly with Creosotebush; in
subtropical regions.

Range: W. Arizona, extreme S. Nevada, SE.
California, and NW. Mexico; from
below sea level to 1500' (457 m).

The common names describe the hazy
appearance of the leafless plants when
seen from a distance. The smoky-gray
twigs produce most of the food, by
photosynthesis, since the plants have
leaves for only a few weeks each year.
Showy when covered with flowers,
Smokethorn is sometimes grown as an
ornamental in frost-free regions.

---

258 Kidneywood
Eysenhardtia polystachya (Ortega) Sarg.

Description: Many-branched shrub or small tree
with short trunk, many slender
branches, and *gland-dotted, aromatic,
resinous twigs, foliage, and flowers*.
Height: 20' (6 m).
Diameter: 6" (15 cm).
Leaves: pinnately compound; 2–5" (5–
13 cm) long. *Leaflets generally in 10–20
pairs* (or sometimes last unpaired);
almost stalkless; ⅜–¾" (10–19 mm)
long, ⅛–¼" (3–6 mm) wide; *narrowly
oblong*, rounded or slightly notched at
tip; thick. Pale *gray-green* and hairless
or nearly so above, finely hairy with
brown gland-dots beneath.

Bark: light gray, thin, scaly and peeling.

Twigs: gray, slender, hairy.

Flowers: ⅜″ (10 mm) long; with 5 *narrow, nearly equal white petals* and gland-dotted calyx; many crowded in upright, lateral clusters to 4″ (10 cm) long and almost stalkless; in late spring and summer.

Fruit: ½″ (12 mm) long; a *narrow, flat, slightly curved, hairless, brown pod; hanging downward;* abundant; maturing in late summer and not opening; usually 1 beanlike, brown seed.

Habitat: Dry hillsides and rocky canyons, upper desert grassland and lower oak woodland.

Range: Mountains of extreme SW. New Mexico and SE. Arizona and south to S. Mexico; at 4500–6500′ (1372–1981 m).

The name Kidneywood refers to an earlier use as a remedy for kidney and bladder diseases. Water in which the reddish-brown heartwood has been soaked has a peculiar fluorescence, changing from golden-yellow to orange and, against a black background, appearing bluish. A yellow-brown dye can be obtained from the wood. Livestock and wildlife, especially deer, readily browse the resinous foliage.

269, 518   **Honeylocust**
"Sweet-locust" "Thorny-locust"
Gleditsia triacanthos L.

Description: Large spiny tree with open, flattened crown of spreading branches.

Height: 80′ (24 m).

Diameter: 2½′ (0.8 m).

Leaves: *pinnately and bipinnately compound;* 4–8″ (10–20 cm) long; the axis often with 3–6 pairs of side axes or forks; in late spring. *Many oblong leaflets* ⅜–1¼″ (1–3 cm) long; paired and

stalkless, with *finely wavy edges.* Shiny dark green above, dull yellow-green and nearly hairless beneath; turning yellow in autumn.

Bark: gray-brown or black, fissured in long, narrow, scaly ridges; with *stout brown spines, usually branched,* sometimes 8″ (20 cm) long, with 3 to many points.

Twigs: shiny brown, stout, zigzag, with long spines.

Flowers: ⅜″ (10 mm) wide; *bell-shaped,* with 5 petals, *greenish-yellow,* covered with fine hairs; in short narrow clusters at leaf bases in late spring; usually male and female on separate twigs or trees.

Fruit: 6–16″ (15–41 cm) long, 1¼″ (3 cm) wide; a *flat pod,* dark brown, hairy, *slightly curved and twisted,* thick-walled; shedding unopened in late autumn; many beanlike, flattened, dark brown seeds in *sweetish, edible pulp.*

Habitat: Moist soils of river flood plains in mixed forests; sometimes on dry, upland limestone hills; also in waste places.

Range: Extreme S. Ontario to central Pennsylvania, south to NW. Florida, west to SE. Texas, and north to SE. South Dakota; naturalized eastward; to 2000′ (610 m). Planted in western states.

Livestock and wildlife consume the honeylike, sweet pulp of the pods. Honeylocust is easily recognized by the large, branched spines on the trunk; thornless forms, however, are common in cultivation and are sometimes found wild. The spines have been used as pins. This hardy species is popular for shade and hedges and for attracting wildlife.

270, 332, 514   Littleleaf Leucaena
"Littleleaf Leadtree" "Wahoo-tree"
Leucaena retusa Benth.

Description: Evergreen, slender shrub or small tree
with many showy balls of golden-
yellow flowers throughout spring and
summer.
Height: 20' (6 m).
Diameter: 6" (15 cm).
Leaves: *evergreen; bipinnately compound;*
3–4" (7.5–10 cm) long, the slender
axis with 2–4 pairs of side axes, each
with *4–9 pairs of leaflets* and 1 round,
white gland between each pair. Leaflets
nearly stalkless, ⅜–1" (1–2.5 cm)
long, ¼–½" (6–12 mm) wide; *oblong or
elliptical,* rounded with tiny point at
tip, rounded and unequal at base;
without teeth; hairy when young; *blue-
green.*
Bark: gray to brown, smooth,
becoming scaly.
Twigs: brown, slender, finely hairy
when young.
Flowers: ⅜" (10 mm) long; very
narrow, each with 10 long stamens,
bright yellow; many clustered in *long-
stalked balls* ¾–1" (2–2.5 cm) in
diameter; throughout spring and
summer.
Fruit: 6–10" (15–25 cm) long, about
½" (12 mm) wide; few *long, narrow pods*
at end of long stalk; pointed at ends,
flat, dark reddish-brown; maturing in
late summer, splitting open; several
flattened, beanlike seeds.
Habitat: Dry, rocky soils, mostly limestone;
often with junipers and oaks.
Range: SW. and Trans-Pecos Texas, SE. New
Mexico, and N. Mexico; at 1500–
5500' (457–1676 m).

This attractive, small tree is sometimes
cultivated in warm regions for the
showy yellow flowers, conspicuous
pods, and evergreen foliage. It is a good
browse plant for cattle.

261 Tesota
"Arizona Ironwood" "Desert Ironwood"
Olneya tesota Gray

Description: Spiny evergreen tree with short trunk
and widely spreading, rounded, dense
crown often broader than high and with
numerous purplish, pea-shaped flowers
in late spring.
Height: 30' (9 m).
Diameter: 2' (0.6 m).
Leaves: *evergreen* or nearly so; densely
clustered; pinnately compound; 1–2¼"
(2.5–6 cm) long. *2–10 pairs of leaflets*
¼–¾" (6–19 mm) long; *oblong*,
generally rounded at tip and short-
pointed at base, without teeth, thick,
short-stalked; *blue-green*, with fine
pressed hairs.
Bark: gray, smooth, thin, becoming
much fissured, scaly, and *shreddy*.
Twigs: greenish, slender; covered with
gray hairs when young; with short,
slender, straight *spines paired at nodes*.
Flowers: ½" (12 mm) long; *pea-shaped*,
with 5 *unequal purple petals*; few, in
short clusters along twigs; fragrant,
abundant and showy; in late spring
with new leaves.
Fruit: 2–2½" (5–6 cm) long; a
cylindrical pod; short-pointed, *slightly
narrowed between seeds*, light brown;
covered with sticky hairs, thick-walled;
with 1–5 beanlike, shiny brown seeds;
maturing in late summer, splitting in 2
parts.
Habitat: Sandy and gravelly washes in rocky
foothills in deserts.
Range: S. and SW. Arizona, SE. California,
and NW. Mexico; to 2500' (762 m).

Tesota, from the American Indian
name, is the single species of its genus
named for Stephen Thayer Olney
(1812–78), a businessman and botanist
of Rhode Island. A characteristic and
common desert tree, it is regarded as an
indicator in selecting favorable sites for
citrus orchards, since it grows only in

subtropical areas with warm, mild winters. It is known locally as "Ironwood" and in Spanish as *palo de hierro*. The hard, dark brown wood with thin, yellow sapwood is easily polished but dulls tools used to work it. It is made into novelties such as bowls and small boxes and is excellent fuel. It is one of the heaviest native woods; only Leadwood (*Krugiodendron ferrum* (Vahl) Urban), a small tropical tree of southern Florida, is heavier. The beanlike seeds can be roasted and eaten. Desert animals also consume the seeds, and livestock browse the foliage. A parasitic mistletoe on the branches, with reddish, juicy berries, attracts birds, such as the phainopepla; the birds in turn spread the sticky seeds of the parasite to other trees, mostly in the legume family.

267, 327, 520 Jerusalem-thorn
"Horsebean" "Mexican Paloverde"
Parkinsonia aculeata L.

Description: Spiny tree with very open, spreading crown of drooping twigs and narrow evergreen "streamers"; *appearing leafless* most of the year.
Height: 40' (12 m).
Diameter: 1' (0.3 m).
Leaves: *bipinnately compound* but appearing pinnately compound, with short, spine-tipped axis and 1–3 pairs of wiry, flattened, *narrow, evergreen, drooping* axes or *"streamers"* 8–20" (20–51 cm) long. *25–30 pairs* of leaflets ¼" (6 mm) or less in length; narrowly *oblong, light green;* remaining on tree only a short time before falling.
Bark: *trunk and branches yellow-green, smooth;* base of large trunks becoming scaly.
Twigs: *yellow-green, smooth,* slightly zigzag, slender; finely hairy when young; with *paired, short spines at nodes*

bordering a third, *larger brownish spine* of leaf axis.

Flowers: ¾" (19 mm) wide; with 5 *rounded golden-yellow petals,* the *largest red-spotted* and turning red in withering; in loose, upright clusters to 8" (20 cm) long, showy; in spring and summer or nearly continuously in tropical climates.

Fruit: 2–4" (5–10 cm) long; a *narrowly cylindrical,* dark brown *pod,* long-pointed at ends, narrowed between seeds, hanging down; 1–8 beanlike seeds; maturing in summer and autumn and remaining closed.

Habitat: Moist valley soils; rare in foothills and mountain canyons of desert and desert grassland.

Range: S. to Trans-Pecos Texas and local in S. Arizona; to 4500' (1372 m). Planted and becoming naturalized across southern border of United States, sometimes as a weed. Widely distributed in tropical America.

Jerusalem-thorn is a popular, fast-growing tree widely used as an ornamental and hedge plant in warm regions. The foliage and pods have been used as emergency forage for livestock, as well as by wildlife. Bees produce fragrant honey from the flowers. The word "Jerusalem" does not refer to the Israeli city but is a corruption of the Spanish and Portuguese word *girasol,* meaning "turning toward the sun."

259, 517 Honey Mesquite
Prosopis glandulosa Torr.

Description: Spiny, large thicket-forming shrub or small tree with short trunk, open, spreading crown of crooked branches, and narrow, beanlike pods.
Height: 20' (6 m).
Diameter: 1' (0.3 m).
Leaves: *bipinnately compound;* 3–8" (7.5–

20 cm) long; the short axis bearing *1 pair of side axes or forks,* each fork with *7–17 pairs* of stalkless *leaflets,* ⅜–1¼" (1–3 cm) long, ⅛" (3 mm) wide; narrowly oblong, hairless or nearly so, yellow-green.

Bark: dark brown, rough, thick, becoming shreddy.

Twigs: slightly zigzag, with stout, yellowish, mostly *paired spines* ¼–1" (0.6–2.5 cm) long at enlarged nodes, which afterwards bear short spurs.

Flowers: ¼" (6 mm) long; nearly stalkless, light yellow, fragrant; crowded in *narrow clusters* 2–3" (5–7.5 cm) long; in spring and summer.

Fruit: 3½–8" (9–20 cm) long, less than ⅜" (10 mm) wide; *narrow pod* ending in long narrow point, slightly flattened, wavy-margined between seeds; sweetish pulp; maturing in summer, remaining closed; several beanlike seeds within 4-sided case.

Habitat: Sandy plains and sandhills and along valleys and washes; in short grass, desert grasslands, and deserts.

Range: E. Texas and SW. Oklahoma west to extreme SW. Utah and S. California; also N. Mexico; naturalized north to Kansas and SE. Colorado; to 4500' (1372 m).

The seeds are disseminated by livestock that graze on the sweet pods, and the shrubs have invaded grasslands. Cattlemen regard mesquites as range weeds and eradicate them. In sandy soils, dunes often form around shrubby mesquites, burying them except for a rounded mass of branching tips. The deep taproots, often larger than the trunks, are grubbed up for firewood. Southwestern Indians prepared meal and cakes from the pods. As the common name indicates, this species is also a honey plant. The word "mesquite" is a Spanish adaptation of the Aztec name *mizquitl.*

260, 330, 525 Screwbean Mesquite
"Screwbean" "Tornillo"
Prosopis pubescens Benth.

Description: Spiny shrub or small tree with long,
slender branches and odd, screwlike
pods.
Height: 20' (6 m).
Diameter: 8" (20 cm).
Leaves: clustered; *bipinnately compound;*
2–3" (5–7.5 cm) long, with stalk of
½" (12 mm) and *1 pair* (sometimes 2)
of side axes. 5–8 *pairs of leaflets ¼–⅜"
(6–10 mm) long,* ⅛" (3 mm) wide;
oblong, short-pointed, finely covered
with gray hairs, *dull green,* stalkless.
Bark: light brown, smooth, thick,
separating in long, fibrous strips and
becoming *shaggy.*
Twigs: slender; covered with gray hairs
when young; with slender, *whitish,
paired spines* about ⅜" (10 mm) long
united with base of leafstalk at nodes.
Flowers: ³⁄₁₆" (5 mm) long; *light yellow;*
many crowded in *narrow clusters* about
2" (5 cm) long; in spring and summer.
Fruit: 1–2" (2.5–5 cm) long; a pod
tightly coiled into a narrow spiral like a
large screw, pale *yellow or light brown,*
hard, with sweetish pulp and many
tiny, beanlike seeds; several to many
crowded on a stalk, often abundant;
maturing in summer, not splitting
open, and shedding in autumn.

Habitat: Along streams and valleys in deserts;
often forming thickets.

Range: Trans-Pecos Texas west to extreme SW.
Utah and SE. California; also adjacent
N. Mexico; to 5500' (1676 m).

Screwbean Mesquite is easily recognized
by the unusual pods, which are the
basis of both the English and Spanish
common names. The sweetish,
nutritious pods can be eaten and are
browsed by livestock and wildlife.
Indians made meal, cakes, and syrup
from the pods and prepared a treatment
for wounds from the root bark. The

durable hard wood is used for
fenceposts, tool handles, and fuel.

263, 523 **Velvet Mesquite**
Prosopis velutina Woot.

Description: Spiny tree with short, forking trunk,
open, spreading crown of crooked
branches, and finely hairy or *velvety
foliage, twigs, and pods.*
Height: 20–40' (6–12 m).
Diameter: 1–2' (0.3–0.6 m).
Leaves: generally clustered; *bipinnately
compound;* 5–6" (13–15 cm) long; *finely
hairy,* the slender axis with *1 or 2 pairs*
of side axes. Leaflets crowded, *15–20
pairs,* ¼–½" (6–12 mm) long; narrowly
oblong, dull green, stalkless.
Bark: dark brown, rough and thick,
separating into long, narrow strips.
Twigs: light brown, slightly zigzag,
covered with fine velvety hairs; with
stout, yellowish, generally *paired spines*
¼–1" (0.6–2.5 cm) long at enlarged
nodes.
Flowers: ¼" (6 mm) long; *light yellow,*
nearly stalkless, fragrant; crowded in
hairy, long, *narrow clusters* 2–3" (5–7.5
cm) long of many flowers; in spring and
summer.
Fruit: 4–8" (10–20 cm) long, less than
⅜" (10 mm) wide; *a narrow pod; short-
pointed,* slightly flattened, wavy-
margined between seeds; finely hairy;
sweetish pulp; maturing in summer,
not splitting open; several beanlike
seeds within 4-sided case.
Habitat: Along washes and valleys and on slopes
and mesas in desert, desert grassland,
and occasionally with oaks.
Range: Extreme SW. New Mexico west to
central Arizona and NW. Mexico; at
500–5500' (152–1676 m).

The medium-sized tree mesquite of
central and southern Arizona, Velvet
Mesquite reaches larger size than

related species. The wood is used for fenceposts and novelties and is one of the best in the desert for fuel; even the large, deep taproots are grubbed up for that use. Southwestern Indians prepared meal and cakes from the sweet pods and livestock browse them, disseminating the seeds. Bees produce a fragrant honey from mesquites.

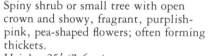

271, 385, 515 New Mexico Locust
"New Mexican Locust"
"Southwestern Locust"
Robinia neomexicana Gray

Description:

Spiny shrub or small tree with open crown and showy, fragrant, purplish-pink, pea-shaped flowers; often forming thickets.
Height: 25′ (7.6 m).
Diameter: 8″ (20 cm).
Leaves: pinnately compound; 4–10″ (10–25 cm) long; *13–21 leaflets,* paired except at end, ½–1½″ (1.2–4 cm) long, ¼–1″ (0.6–2.5 cm) wide; *elliptical,* rounded at ends, with *tiny bristle tip;* not toothed; finely hairy when young; pale *blue-green;* nearly stalkless.
Bark: light gray, thick, furrowed into scaly ridges.
Twigs: brown; with rust-colored gland hairs when young; with stout, brown, *paired spines* ¼–½″ (6–12 mm) long at nodes.
Flowers: ¾″ (19 mm) long; *pea-shaped,* with 5 unequal *purplish-pink petals;* on stalks with sticky hairs; many in large, unbranched, drooping clusters at base of leaves; in late spring and early summer.
Fruit: 2½–4½″ (6–11 cm) long; a *narrowly oblong, flat,* thin brown *pod* with *bristly* and often glandular *hairs;* splitting open, maturing in early autumn; 3–8 beanlike, flattened, dark brown seeds.

Habitat: Canyons and moist slopes; with Gambel Oak, Ponderosa Pine, and pinyons.

Range: Mountains from SE. Nevada east to S. and central Colorado, south and east to Trans-Pecos Texas, and west to SE. Arizona; also N. Mexico; at 4000–8500' (1219–2591 m).

Spectacular flower displays of New Mexico Locust can be seen at the north rim of Grand Canyon National Park in early summer. It is sometimes planted as an ornamental for the handsome flowers and is also valuable for erosion control, sprouting from roots and stumps and rapidly forming thickets. Livestock and wildlife browse the foliage and cattle relish the flowers. Indians also ate the pods and flowers.

268, 373, 519 **Black Locust**
"Yellow Locust" "Locust"
Robinia pseudoacacia L.

Description: Medium-sized, spiny tree with a forking, often crooked, and angled trunk and irregular, open crown of upright branches; introduced.
Height: 40–80' (12–24 m).
Diameter: 1–2' (0.3–0.6 m).
Leaves: pinnately compound; 6–12" (15–30 cm) long; *7–19 leaflets* 1–1¾" (2.5–4.5 cm) long, ½–¾" (12–19 mm) wide; paired except at end, *elliptical,* with *tiny bristle tip,* without teeth, hairy when young; drooping and folding at night. Dark blue-green above, pale and usually hairless beneath.
Bark: light gray, thick, deeply furrowed into long, rough, forking ridges.
Twigs: dark brown, with stout *paired spines* ¼–½" (6–12 mm) long at nodes.
Flowers: ¾" (19 mm) long; *pea-shaped,* with 5 unequal *white petals,* the largest yellow near base; very fragrant; in

showy, drooping clusters 4–8" (10–
20 cm) long at base of leaves; in late
spring.

Fruit: 2–4" (5–10 cm) long; a dark
brown, *narrowly oblong, flat pod;*
maturing in autumn, remaining
attached into winter, splitting open; 3–
14 dark brown, flattened, beanlike
seeds.

Habitat: Moist to dry sandy and rocky soils,
especially in old fields and other open
areas, and in woodlands.

Range: Central Pennsylvania and S. Ohio south
to NE. Alabama and from S. Missouri
to E. Oklahoma; naturalized from
Maine to California and in S. Canada;
from 500' (152 m) to above 5000'
(1524) in southern Appalachians.

Black Locust is widely planted for
ornament and shelterbelts. It is also
used for erosion control, particularly on
strip-mined areas. Although it grows
rapidly and spreads by sprouts like a
weed, it is short-lived. Virginia Indians
made bows from the wood and
apparently planted the trees eastward.
British colonists at Jamestown
discovered this species in 1607 and
named it for its resemblance to the
Carob or Old World Locust (*Ceratonia
siliqua* L.). Posts of this durable timber
served as cornerposts for the colonists'
first homes.

272, 397, 522 Mescalbean
"Texas Mountain-laurel" "Frijolillo"
Sophora secundiflora (Ortega) Lag. ex DC.

Description: Evergreen shrub or sometimes a small,
many-branched tree with narrow
crown, bearing bright red, beanlike,
poisonous seeds and showy, purple
flowers.

Height: usually less than 5' (1.5 m),
sometimes 20' (6 m).

Diameter: 6" (15 cm).

Leaves: *evergreen;* pinnately compound; 3–6" (7.5–15 cm) long. Usually *5–11 leaflets* ¾–2" (2–5 cm) long, ⅜–1" (1–2.5 cm) wide; paired (except at end); *elliptical,* rounded or slightly notched at tip; without teeth; *thick and leathery. Shiny dark green* and becoming hairless above, paler and hairless or nearly so beneath.

Bark: dark gray, fissured into narrow, flattened, scaly ridges, becoming rough and shaggy.

Twigs: densely covered with white hairs when young, becoming light brown.

Flowers: ¾–1" (2–2.5 cm) long; pea-shaped, with 5 *unequal bluish-purple or violet petals;* very fragrant; crowded on 1 side of stalk in clusters 2–4½" (5–11 cm) long of many flowers each at twig ends; in early spring with new leaves, sometimes also in autumn.

Fruit: 1–5" (2.5–13 cm) long, about ⅝" (15 mm) in diameter; a *cylindrical pod,* densely covered with brown hairs, pointed at ends, *slightly narrowed between seeds,* hard and thick-walled; maturing in late summer and not opening; usually 3–4 (sometimes 1–8) small *bright red,* rounded, beanlike *seeds, extremely poisonous.*

Habitat: Moist soils of streams, canyons, and hillsides, mainly on limestone; forming thickets.

Range: Central to SW. and Trans-Pecos Texas and SE. New Mexico; also south to central Mexico; to 6500' (1981) m).

This species is often cultivated in warm regions for the shiny, evergreen foliage and large showy flowers. However, further planting of this dangerous plant is not recommended. One seed is sufficiently toxic to kill an adult and children have become ill from eating the flowers. Indians made necklaces as well as a narcotic powder from the seeds. The foliage and seeds are also poisonous to livestock.

CALTROP FAMILY
(Zygophyllaceae)

About 200 species of shrubs, herbs, and
a few trees in mostly tropical or
subtropical regions; 2 native tree
species and several species of shrubs and
herbs in North America.
Leaves: opposite; pinnately compound
with even number of leaflets which are
asymmetrical with unequal sides, not
toothed, often leathery; with paired
stipules.
Twigs: with rings at enlarged nodes.
Flowers: solitary or few, bisexual,
regular; generally with 5 sepals and
5 yellow or blue petals, usually with
disk, 5–10 (15) stamens often with
scales at base, and 1 pistil with 4- or 5-
celled ovary mostly angled or winged.
Fruit: usually an angled or winged
capsule with few seeds.

256, 399, 484 Texas Lignumvitae
"Texas Porliera" "Soapbush"
Guaiacum angustifolium Engelm.

Description: Evergreen shrub or small tree with a

compact head of short, stout, crooked
branches and blue or purple flowers.
Height: 20′ (6 m).
Diameter: 8″ (20 cm).
Leaves: *evergreen; opposite* or crowded;
pinnately compound; 1–3″ (2.5–7.5 cm)
long; with paired, *spiny-pointed* scales or
stipules at base and with hairy axis.
8–16 paired leaflets ¼–⅝″ (6–15 mm)
long, ¹⁄₁₆–⅛″ (1.5–3 mm) wide;
narrowly oblong, unequal at base,
stalkless, thickened, with network of
veins; folding at night and often also at
midday. *Shiny dark green* on both
surfaces.
Bark: gray or black; becoming rough,
fissured, and scaly.
Twigs: gray, short, stout, stiff with
enlarged ring nodes.

Flowers: ½–¾" (12–19 mm) wide; with 5 *rounded, blue or purple petals;* fragrant; on long stalks at leaf bases; in spring and summer.

Fruit: ½" (12 mm) in diameter; a flat, *heart-shaped capsule* with 2 (sometimes 3–4) lobes and cells; brown, slightly winged, splitting open; usually 2 shiny yellow-brown, *beanlike seeds* with thick, reddish cover.

Habitat: Thickets in valleys and canyons.

Range: S. to central and Trans-Pecos Texas; NE. Mexico; to 3000' (914 m).

Texas Lignumvitae is the northernmost of about 5 tropical species of lignumvitae; another, Roughbark Lignumvitae (*G. sanctum* L.), occurs rarely on the Florida Keys. The bark has been used as a soap substitute. The name "lignumvitae" (Latin for "tree of life") refers to the former medicinal use of the wood extract of a related tree.

RUE (CITRUS) FAMILY
(Rutaceae)

Shrubs and trees, rarely herbs, with leaves, fruit, and bark aromatic with pungent citrus odor. About 1000 species in tropical and warm temperate regions; 12 native and 4 naturalized tree and several native shrub species in North America.

Leaves: simple, pinnately or palmately compound; with gland-dots, generally hairless, without stipules.

Flowers: usually regular; mostly white or greenish; small to large and showy; commonly bisexual, sometimes male and female on separate plants; generally with 5 or 4 sepals and 5 or 4 petals, 8–10 (or as few as 3, sometimes more than 10) stamens, and generally 1 pistil with ovary usually 5- or 4-celled.

Fruit: usually a capsule or berry, sometimes a drupe, follicle, or winged key (samara).

---

163 Sour Orange
"Seville Orange"
Citrus aurantium L.

Description: Small ornamental evergreen tree with rounded crown of glossy foliage, aromatic (and spicy in taste), with fragrant, white blossoms and sour, orange-colored fruit.

Height: 30' (9 m).

Diameter: 1' (0.3 m).

Leaves: *evergreen;* 2½–5½" (6–14 cm) long, 1½–4" (4–10 cm) wide. *Ovate,* long-pointed, with many tiny, *rounded teeth,* slightly leathery, with tiny *gland-dots;* leafstalks broadly *winged,* jointed with blade. *Shiny green* above, pale light green beneath.

Bark: brown, smooth.

Twigs: *green; angled* when young; with sharp *spines* to 1" (2.5 cm) long, single at leaf bases.

Flowers: 1" (2.5 cm) wide; with 5 narrow, white, curved petals and many stamens; very fragrant; few, in short-stalked clusters at base of leaf; in spring.

Fruit: 2½–4½" (6–11 cm) in diameter; a more or less *rough orange,* with thick peel and partly hollow, pulpy core; *sour,* bitter, and *inedible;* many whitish seeds; ripening in early autumn.

Habitat: Moist soils in subtropical regions.

Range: Native of SE. Asia. Naturalized in Florida and Georgia; also planted as an ornamental in S. Texas, S. Arizona, California, Hawaii, and Puerto Rico.

Sour Orange is only one of the citrus fruits that are cultivated in the United States. Others are Orange (*Citrus sinensis* Osbeck), Grapefruit (*C. paradisi* Macf.), Lemon (*C. limon* (L.) Burm.), Lime (*C. aurantifolia* (Christmann in L.) Swingle), and Mandarin Orange or Tangerine (*C. reticulata* Blanco), as well as many varieties and hybrids. Sour Orange is preferred as an ornamental not only for its hardiness but because the inedible, showy fruit remains on the tree indefinitely. This species has been employed as the stock for budding the other species. Orange marmalade is made from peels and the juice has been used in home remedies as an antiseptic and to check bleeding.

301, 343, 532   Common Hoptree
"Wafer-ash"
Ptelea trifoliata L.

Description: Aromatic shrub or small tree with a rounded crown; bark, crushed foliage, and twigs have a slightly lemonlike, *unpleasant odor.*

Height: 20' (6 m).

Diameter: 6" (15 cm).

Leaves: palmately compound, 4–7" (10–18 cm) long, with 3 *leaflets* at end of long leafstalk. Leaflets 2–4″ (5–10 cm) long, ¾–2″ (2–5 cm) wide; *ovate* or elliptical; long-pointed at tip; finely wavy-toothed or not toothed; with *tiny gland-dots;* hairy when young. Shiny dark green above, paler and sometimes hairy beneath; turning yellow in autumn.

Bark: brownish-gray, thin, smooth or slightly scaly, bitter.

Twigs: brown, slender, covered with fine hairs, slightly warty.

Flowers: ⅜″ (10 mm) wide; 4–5 narrow *greenish-white* petals; in terminal, branched clusters including male, female, and bisexual flowers; in spring.

Fruit: ⅞″ (22 mm) in diameter; numerous, *disk-shaped,* waferlike keys, with rounded wing, yellow-brown; in drooping clusters; maturing in summer, remaining closed and attached in winter; 2–3 reddish-brown, long-pointed seeds.

Habitat: Dry, rocky uplands; also in valleys and canyons.

Range: Extreme S. Ontario east to W. New York and New Jersey, south to Florida, west to Texas and north to S. Wisconsin; also local west to Arizona and S. Utah and in Mexico; to 8500′ (2591 m) in West.

This widespread species includes many varieties with leaflets of differing size and shape. The common name refers to an early use of the bitter fruit as a substitute for hops in brewing beer. The bitter bark of the root, like other aromatic barks, has been used for home remedies. This is the northernmost New World representative of the Rue (Citrus) family.

QUASSIA FAMILY
(Simaroubaceae)

Shrubs and trees often with very bitter bark, wood, and other parts. About 150 species in tropical and subtropical regions; 5 native and 1 naturalized tree and 2 native shrub species in North America.

Leaves: alternate; pinnately compound, with leaflets generally not toothed; without stipules.

Flowers: mostly small; often in large branched clusters; generally male and female and usually on separate plants; regular, with 3- to 8-lobed calyx; 3–8 petals (sometimes none), with same number or twice as many stamens as petals; usually 2–5 simple pistils with 1-celled ovary on disk.

Fruit: generally a capsule, drupe, or winged key (samara).

288, 344, 529 Ailanthus
"Tree-of-Heaven"
Ailanthus altissima (Mill.) Swingle

Description: A hardy, introduced tree with a spreading, rounded, open crown of stout branches and coarse foliage; male flowers and crushed leaves have *disagreeable odor.*

Height: 50–80′ (15–24 m).

Diameter: 1–2′ (0.3–0.6 m).

Leaves: pinnately compound; 12–24″ (30–61 cm) long; *13–25 leaflets* (sometimes more) 3–5″ (7.5–13 cm) long, 1–2″ (2.5–5 cm) wide; paired (except at end), broadly *lance-shaped,* with 2–5 *teeth near broad, 1-sided base* and gland-dot beneath each tooth; covered with fine hairs when young. Green above, paler beneath.

Bark: light brown, smooth, becoming rough and fissured.

Twigs: light brown, *very stout,* with fine hairs when young; with *brown pith.*

Flowers: ¼" (6 mm) long; with 5 *yellowish-green petals;* in terminal, branched clusters 6–10" (15–25 cm) long; male and female usually on separate trees; in late spring and early summer.

Fruit: 1½" (4 cm) long; showy, reddish-green or reddish-brown key, narrow, flat, winged, 1-seeded; 1–6 from a flower; maturing in late summer and autumn.

Habitat: Widespread in waste places, spreading rapidly by suckers.

Range: Native of China but widely naturalized across temperate North America; from near sea level to high mountains.

Ailanthus is widely planted as an ornamental, for shade and in shelterbelts for its rapid growth and coarse foliage reminiscent of tropical trees. However, it is no longer recommended for good sites where other trees will grow. Male flowers have an objectionable odor, and some people are allergic to their pollen, which may produce symptoms of hayfever. The roots, which are classed as poisonous, get into drains, springs, and wells. The weak branches are easily broken by storms. Tolerant of crowded dusty cities and smoky factory districts, plants often even grow out of cracks in concrete.

BURSERA FAMILY
(Burseraceae)

Trees and shrubs, often with smooth bark and with aromatic resin. About 500 species in tropical regions, especially America and northeastern Africa; 3 native tree species in North America.

Leaves: generally alternate and odd pinnately compound, sometimes with winged axis, without stipules.

Flowers: tiny, generally in clusters, bisexual or male and female, regular, with calyx or 3–5 sepals or lobes, 3–5 petals, disk, 3–5 (sometimes 8–10) separate stamens, and 1 pistil with 2- to 5-celled ovary.

Fruit: a berry, drupe or capsule with 1–5 seeds.

257, 462 **Elephant-tree**
"Elephant Bursera"
"Small-leaf Elephant-tree"
Bursera microphylla Gray

Description: *Aromatic* shrub or tree with *short, very thick, sharply tapered trunk* with stout, crooked, tapering branches and a widely spreading but sparse, open crown.

Height: 16' (5 m).

Diameter: 1' (0.3 m).

Leaves: pinnately compound; 1–1¼" (2.5–3 cm) long; with winged axis, *aromatic.* 15–30 leaflets about ¼" (6 mm) long; narrowly oblong, short-pointed at base, stalkless, not toothed; dull light green on both surfaces.

Bark: *papery,* peeling in thin flakes, *white on outside;* next thin layers green, inner layers red and corky.

Twigs: *reddish-brown.*

Flowers: less than ¼" (6 mm) wide; with 5 whitish petals, short-stalked; *1– 3 at leaf base;* male and female on same tree; in early summer.

Fruit: ¼" (6 mm) long; *ellipitical, red,* aromatic, *3-angled,* splitting into 3 parts; drooping on slender, curved stalk; with 1 nutlet; maturing in autumn.

Habitat: Dry, rocky slopes of desert mountains.

Range: SW. Arizona and extreme S. California; also NW. Mexico; to 2500′ (762 m).

As the common name suggests, the stout trunk and branches recall the legs and trunk of an elephant. The northernmost representative of a small tropical family, it is very susceptible to frost; young plants are killed back by cold weather. The innermost bark exudes a reddish sap used as dye and tannin.

MAHOGANY FAMILY
(Meliaceae)

Trees and shrubs with bitter, astringent
bark and often aromatic wood. About
1000 species in tropical regions; 1
native and 1 naturalized tree species in
North America.

Leaves: alternate; generally pinnately
compound; leaflets mostly paired,
usually not toothed and often oblique;
without stipules. Naked buds with
tiny, young leaves often in form of
hand.

Flowers: small, generally in branched
clusters, bisexual; regular, with 4- to 5-
lobed calyx and corolla of 4–5 petals or
lobes, 8–10 stamens united in a tube
and around a disk, and 1 pistil with 2-
to 5-celled ovary.

Fruit: generally a capsule or berry with
seeds often winged.

295, 394 Chinaberry
"Chinatree" "Pride-of-India"
Melia azedarach L.

Description: A naturalized shade tree, deciduous or

nearly evergreen in the far South, with
dense, spreading crown and clusters of
round, yellow, poisonous fruit; *foliage has
bitter taste and strong odor* when crushed.
Height: 40′ (12 m).
Diameter: 1′ (0.3 m).
Leaves: *bipinnately compound;* 8–18″
(20–46 cm) long; with slender green
axis and few paired forks. *Numerous
leaflets* 1–2″ (2.5–5 cm) long, ⅜–¾″
(10–19 mm) wide; paired except at
end; *lance-shaped* or ovate, *saw-toothed,*
wavy or lobed; hairless or nearly so.
Dark green above, paler beneath.
Bark: dark brown or reddish-brown;
furrowed, with broad ridges.
Twigs: green, stout, hairless or nearly
so.
Flowers: ¾″ (19 mm) wide; with 5 *pale*

purple petals and narrow, violet tube;
fragrant; on slender stalks in *showy,*
branched clusters 4–10″ (10–25 cm)
long; in spring.

Fruit: ⅝″ (15 mm) in diameter; a
yellow, poisonous berry, becoming slightly
wrinkled; thin, juicy pulp; maturing in
autumn, remaining attached in winter;
hard stone contains 3–5 seeds.

Habitat: Dry soils near dwellings, in open areas
and clearings; sometimes within forests.

Range: Native of S. Asia. Naturalized from SE.
Virginia to Florida, west to Texas, and
north to SE. Oklahoma; also in
California; usually below 1000′
(305 m).

One of the hardiest members of the
tropical Mahogany family, Chinaberry
grows rapidly but is short-lived. One
cultivated variety known as "Umbrella-
tree," with a very dense, compact,
flattened crown like an umbrella, is a
popular ornamental. The fruit stones
can be made into beads, but the
abundant fruit produces litter and is
toxic or narcotic.

CASHEW FAMILY
(Anacardiaceae)

Trees, shrubs, and few woody vines with resinous sap often in bark and other parts; in a few species the resin or volatile oil is caustic and poisonous to the skin. About 600 species in tropical and north temperate regions; 15 native and 3 naturalized tree, 6 native shrub, and 1 woody vine species in North America.

Leaves: alternate, pinnately compound with 3 leaflets or simple; without stipules.

Flowers: tiny or small; commonly white, many in large branched clusters; bisexual or male or female; mostly regular, with 3–5 sepals united at base and 3–5 petals (or no petals); generally 10 stamens, and 1 pistil with superior 1-celled (to 5-celled) ovary, 1 style, and 3 stigmas.

Fruit: a 1-seeded, resinous drupe.

287, 345, 486 Smooth Sumac
"Scarlet Sumac" "Common Sumac"
Rhus glabra L.

Description:

The most common sumac; a large shrub or sometimes a small tree with open, flattened crown of a few stout, spreading branches and with whitish sap.

Height: 20' (6 m).

Diameter: 4" (10 cm).

Leaves: pinnately compound; 12" (30 cm) long; with slender axis. *11–31 leaflets* 2–4" (5–10 cm) long; *lance-shaped,* saw-toothed, *hairless,* almost stalkless. Shiny green above, *whitish beneath;* turning reddish in autumn.

Bark: brown, smooth or becoming scaly.

Twigs: gray, with whitish bloom; few; very stout, *hairless.*

Flowers: less than ⅛" (3 mm) wide;

with 5 whitish petals; crowded in large, upright clusters to 8″ (20 cm) long, with *hairless branches;* male and female usually on separate plants; in early summer.

Fruit: more than ⅛″ (3 mm) in diameter; rounded, 1-seeded, numerous, crowded in upright clusters; dark red, covered with *short, sticky, red hairs;* maturing in late summer, remaining attached in winter.

Habitat: Open uplands including edges of forests, grasslands, clearings, roadsides, and waste places, especially in sandy soils.

Range: Nearly throughout the United States; to 4900′ (1372 m) in the East; to 7000′ (2134 m) in the West.

Smooth sumac is the only shrub or tree species native to all 48 contiguous states. One cultivated variety has dissected or bipinnate leaves. The raw, young sprouts were eaten by the Indians as salad. The sour fruit, which is mostly seed, can be chewed to quench thirst or prepared as a drink similar to lemonade. Deer browse the twigs and fruit throughout the year.

117, 479 **Lemonade Sumac**
"Lemonade-berry"
"Mahogany Sumac"
Rhus integrifolia (Nutt.) Brewer & Wats.

Description: Evergreen, aromatic, rounded, thicket-forming shrub; rarely a small tree with a short, stout trunk and many branches.
Height: 20′ (6 m).
Diameter: 8″ (20 cm).
Leaves: *evergreen;* 1–2½″ (2.5–6 cm) long, ¾–1½″ (2–4 cm) wide. *Elliptical, mostly without teeth* or with short, sharp teeth, sometimes with 1–2 lobes at base or rarely with 3 leaflets; *thick* and stiff, nearly hairless. *Shiny*

dark green above, paler with raised veins beneath.

Bark: reddish-brown, shedding in large plates or scales.

Twigs: reddish, stout, finely hairy.

Flowers: ¼″ (6 mm) wide; with *5 pink to white petals;* many crowded in much-branched clusters 1–3″ (2.5–7.5 cm) long; upright at end of twig; mainly in late winter and early spring. Male and female usually on different plants.

Fruit: ½″ (12 mm) long; elliptical, berrylike, flattened, *sour, dark red, densely hairy,* covered with *whitish secretion,* resinous and sticky, 1-seeded; maturing in summer.

Habitat: Dry, sandy and rocky soils of beaches, ocean bluffs, and canyons; in coastal sage scrub and chaparral.

Range: Coastal S. California including islands; also Baja California; to 2500′ (762 m).

As the common name implies, a sour, refreshing drink can be made from the acid fruit. The scientific name means "leaf entire" or "without teeth"; however, a variation has spiny-toothed leaves. The fruit is consumed by roadrunners and other birds.

---

286 Prairie Sumac
"Prairie Flameleaf Sumac"
"Texan Sumac"
Rhus lanceolata (Gray) Britton

Description: Large shrub or small tree with short trunk and open, rounded crown of foliage, flame-colored in autumn.

Height: 25′ (7.6 m).

Diameter: 6″ (15 cm).

Leaves: pinnately compound; to 9″ (23 cm) long; with *flat, narrowly winged axis.* Usually *13–19 leaflets* 1–2½″ (2.5–6 cm) long, less than ½″ (12 mm) wide; paired except at end, *narrowly lance-shaped, slightly curved,* long-pointed at tip, blunt and unequal at base, usually without teeth. Shiny dark green above, *paler, covered with fine hairs,* and with prominent veins beneath; turning reddish-purple in autumn.

Bark, gray or brown; smooth or becoming scaly.

Twigs: green or reddish and hairy when young, becoming gray and hairless; stout, ending in whitish, hairy bud.

Flowers: $\frac{3}{16}''$ (5 mm) wide; with 5 greenish-white petals; many crowded in upright clusters to 6″ (15 cm) long, at ends of twigs; male and female usually on separate plants; in summer.

Fruit: about $\frac{3}{16}''$ (5 mm) in diameter; slightly flattened, dark red, covered with short, sticky red hairs, 1-seeded; numerous, crowded in clusters; maturing in early autumn and falling in early winter.

Habitat: Dry, rocky slopes and hills, especially limestone; often forming thickets.

Range: Texas and S. New Mexico, local in S. Oklahoma and in NE. Mexico; to 2500′ (762 m); locally to 4000′ (1219 m).

Birds, especially bobwhites, grouse, and pheasants, consume quantities of the fruit in winter, and deer browse the foliage. The leaves contain tannin and have been used in tanning leather.

115, 349 Laurel Sumac
Rhus laurina Nutt.

Description: Aromatic evergreen shrub or small tree with dense, rounded crown of glossy foliage and odor of bitter almonds.
Height: 16′ (5 m).
Diameter: 6″ (15 cm).
Leaves: *evergreen;* 2–4″ (5–10 cm) long, ¾–2″ (2–5 cm) wide. *Lance-shaped* or ovate; short and sharp-pointed at tip, rounded at base, not toothed, *thick* and leathery; edges *folded upward. Shiny*

green with reddish veins above, pale green and whitish beneath.
Bark: brown or reddish, smooth.
Twigs: reddish, turning brown; slender, hairless.
Flowers: less than $\frac{1}{16}''$ (1.5 mm) wide; with *5 tiny white petals;* numerous, crowded in many-branched clusters 2–6″ (5–15 cm) long, at end of twig; mainly in late spring and early summer.
Fruit: less than $\frac{1}{8}''$ (3 mm) in diameter; *rounded, whitish, hairless,* waxy, with 1 flattish seed; maturing in late summer.

Habitat: Dry slopes, in chaparral and coastal sage scrub.

Range: S. California including Santa Catalina and San Clemente islands, south to Baja California Sur; to 3000′ (914 m).

Planted as an ornamental in California, it grows rapidly and is drought-resistant but is damaged by cold winters. The common and scientific names refer to the resemblance of the foliage to that of Laurel (*Laurus nobilis* L.), an unrelated small tree of the Mediterranean region. Songbirds eat the fruit.

116, 379 Sugar Sumac
"Sugarbush" "Chaparral Sumac"
Rhus ovata Wats.

Description: Evergreen shrub or small tree with rounded crown.
Height: 15′ (4.6 m).
Diameter: 5″ (13 cm).
Leaves: *evergreen;* 1½–3¼″ (4–8 cm) long, 1–2″ (2.5–5 cm) wide. *Ovate,* short-pointed at tip, rounded at base, *without teeth, thick and leathery;* curved or *folded up at midvein;* shiny light green on both surfaces.
Bark: gray-brown, rough, shaggy and very scaly.
Twigs: reddish, stout, hairless.

Flowers: ¼" (6 mm) wide; with 5 *rounded, whitish petals,* from pink or reddish buds; many crowded in clusters 2" (5 cm) long, at end of twig; in early spring.

Fruit: more than ¼" (6 mm) in diameter; slightly flattened, red, covered with *short, sticky, red hairs,* 1-seeded; many in clusters; maturing in summer.

Habitat: Dry slopes in chaparral zone.

Range: Mountains of central Arizona and S. California including Santa Cruz and Santa Catalina islands south to N. Baja California; from near sea level to 2500' (762 m); in Arizona to 5000' (1524 m).

Sometimes planted for erosion control and landscaping in mountainous areas, this common species is also an attractive ornamental. The edible fruit with thin pulp is sweet and was used as a sweetener by Indians; however, the large seeds are not eaten. Birds also consume the fruit.

284, 483 **Peppertree**
"California Peppertree"
"Peru Peppertree"
Schinus molle L.

Description: Cultivated and naturalized evergreen tree with short, often gnarled trunk and widely spreading, rounded crown of *drooping branches* and fine foliage; aromatic and resinous, with *milky sap.*
Height: 40' (12 m) high.
Diameter: 1' (0.3 m).
Leaves: pinnately compound; 6–12" (15–30 cm) long; *drooping,* with milky sap. *19–41 leaflets, generally paired,* 1–2" (2.5–5 cm) long; narrowly lance-shaped, slightly curved at tip, sometimes slightly toothed, hairless or nearly so, stalkless; *yellow-green on both surfaces.*

Bark: light brown, scaly.
Twigs: brownish, slender, hairless or finely hairly.
Flowers: ⅛" (3 mm) wide; with 5 *yellowish-white petals;* many in branched clusters 5–6" (13–15 cm) long; male and female on separate trees.
Fruit: ³⁄₁₆–¼" (5–6 mm) in diameter; *berrylike, reddish* or pinkish, shiny, *resembling beads,* resinous, juicy, 1-seeded; many, hanging down on short stalks; remaining attached in winter.

Habitat: Hardy in a wide range of soils including alkaline; in subtropical regions.

Range: Native of South America. Planted in S. Texas, Arizona, and California, becoming naturalized in extreme S. Texas and California.

This exotic ornamental from Peru and nearby countries is often called "California Peppertree" because of its popularity for planting along streets and highways in that state. The abundant red fruit has a peppery taste and is showy in autumn. It is consumed by various songbirds, including robins, mockingbirds, and cedar waxwings. Although drought-resistant, Peppertree has shallow roots that crack pavements and damage sewers. The bark, gum from the trunk, leaves, and fruit have long served in home medicines from Mexico to South America.

HOLLY FAMILY
(Aquifoliaceae)

Shrubs and trees, small to medium-sized, rarely large. 300–350 species, nearly all in the holly genus (*Ilex*) in tropical and temperate regions, especially tropical America; 14 native tree and 2 native shrub species in North America.

Leaves: alternate, simple, generally leathery and evergreen, sometimes with tiny stipules.

Flowers: small, few clustered along twigs, whitish or greenish, regular, generally male and female on separate plants or bisexual; calyx with 4 (sometimes 5) tiny sepals or teeth, 4 (5) rounded whitish petals sometimes united at base, 4 (5) alternate stamens inserted at base of corolla, without disk, and 1 pistil with superior ovary of 3 (3–5) cells of 1–2 ovules each, usually without style, and 3–5 stalkless stigmas.

Fruit: a round drupe or berry, red, black or yellow, with stalkless stigmas, bitter pulp, and 3–5 nutlets.

180, 489 English Holly
"European Holly"
Ilex aquifolium L.

Description: Cultivated evergreen tree with dense, conical crown of short, spreading branches and shiny red berries.
Height: 50' (15 m).
Diameter: 1½' (0.5 m).
Leaves: *evergreen;* 1¼–2¾" (3–7 cm) long, ¾–1½" (2–4 cm) wide. *Elliptical,* spiny-pointed, wavy-edged, with *large spiny teeth,* blunt at base, *stiff* and leathery. *Shiny dark green* above, paler beneath.
Bark: gray, smooth or nearly so.
Twigs: greenish or purplish; angled, hairless or with short hairs.

Flowers: ¼" (6 mm) wide; with 4
rounded, white petals; several on short
stalks at base of previous year's leaves;
male and female on separate trees; in
late spring.

Fruit: ¼–⅜" (6–10 mm) in diameter;
berrylike, shiny red, clustered at leaf
bases, short-stalked, with 4 nutlets;
maturing in autumn, remaining
attached in winter.

Habitat: Moist soils in humid, temperate
regions.

Range: Native of S. Europe, N. Africa, and
W. Asia. Planted across the United
States, mainly in Atlantic,
southeastern, and Pacific states.

Numerous horticultural varieties differ
in leaf size, shape, spines, color, and
tree habit. Several varieties grown in
orchards for Christmas decorations have
larger berries and larger leaves than
native hollies. To assure fruit, a male
plant is needed to pollinate the female.
Cultivated since ancient times, it is
propagated by cuttings and seeds. The
wood is used for veneers and inlays.

BITTERSWEET FAMILY
(Celastraceae)

Shrubs, woody vines, and mostly small trees. Widespread, about 700 species; 7 native tree species and several shrub species in North America.

Leaves: alternate or opposite, sometimes whorled, simple, with tiny stipules or none.

Flowers: tiny; usually in clusters with stalks mostly jointed; greenish; bisexual or male and female; regular, with 4–5 sepals united at base and persisting at base of fruit and 4–5 petals, 4–5 alternate stamens inserted on or below the large disk, and 1 pistil with superior ovary of 2–5 cells each with 2 ovules, short style, and stigma often with 2–5 lobes.

Fruit: a capsule, berry, or drupe; the seed generally with colored covering.

---

79 **Canotia**
"Crucifixion-thorn"
Canotia holacantha Torr.

Description: Spiny, spreading shrub or small tree with short trunk, *many upright,* flexible, *yellow-green branches,* and *twigs in broomlike masses; leafless* most of year.
Height: 18' (5.5 m).
Diameter: 8" (20 cm).
Leaves: *very small, scalelike, greenish, very short-lived.*
Bark: yellow-green, smooth; becoming gray, rough, slightly fissured and shreddy at base.
Twigs: ⅛" (3 mm) in diameter; inconspicuously grooved, with small black rings at forks, often *ending in spines* or dead tips.
Flowers: ⁵⁄₁₆" (8 mm) wide; with 5 *rounded, greenish-white petals;* in small clusters near ends of twigs; in spring and early summer.
Fruit: ¾" (19 mm) or more in length;

egg-shaped, long-pointed capsules, upright,
reddish-brown, hard; 5-celled and
splitting open along 10 lines; maturing in
autumn, remaining attached until
spring.

Habitat: Dry, rocky slopes and hillsides in desert
and chaparral.

Range: Arizona and extreme S. Utah; local in
N. Mexico; at 2000–5000′ (610–
1524 m).

Canotia is distinguishable from
paloverdes, which it replaces to the
north, by the more crowded, upright
branches and twigs in broomlike
masses. It is the most common of the 3
spiny, many-branched shrubs called
crucifixion-thorns and the only one that
commonly reaches tree size. The green
branches and twigs, like those of
paloverdes, manufacture food, requiring
less water than leaves would. *Canotia* is
the Mexican name, while *holacantha* is
from Greek words meaning "wholly"
and "thorn," referring to the spiny,
leafless branches.

MAPLE FAMILY
(Aceraceae)

Trees and shrubs; about 125 species, nearly all in the maple genus (*Acer*), in north temperate regions south into tropical mountains; 13 native tree species in North America.

Leaves: deciduous; opposite; long-stalked; mostly simple, broad, and palmately lobed and veined; toothed; sometimes pinnately compound; without stipules; the sap sometimes sweetish or milky.

Flowers: commonly male and female on separate trees or bisexual, in often-branched clusters; small; with 5 or 4 colored sepals separate or sometimes united at base, corolla of 5 or 4 overlapping petals or corolla absent, 4–10 stamens, usually 8, from edge of large disk, and 1 pistil with 2-celled 2-lobed ovary with 2 ovules in each cell, and 2-forked style.

Fruit: paired, flat, long-winged, 1-seeded keys (samaras).

231 Vine Maple
Acer circinatum Pursh

Description: Shrub or small tree with short trunk or several branches turning and twisting from base, *often vinelike* and leaning or sprawling.

Height: 25′ (7.6 m).

Diameter: 8″ (20 cm).

Leaves: opposite; 2½–4½″ (6–11 cm) long and wide. Rounded, with *7–11 long-pointed lobes, sharply doubly toothed,* with 7–11 main veins from notched base; long leafstalks with enlarged bases joined. Bright green above, paler with tufts of hairs in vein angles beneath; turning orange and red in autumn.

Bark: gray or brown; smooth or finely fissured.

Twigs: green to reddish brown, with whitish bloom; slender.

Flowers: ½" (12 mm) wide; *spreading purple sepals and whitish petals;* in broad, branched clusters at end of short twigs; with new leaves in spring; usually male and female on same plant.

Fruit: 1½" (4 cm) long; *paired,* long-winged *keys spreading* almost horizontally; *reddish* when young; 1-seeded; maturing in autumn.

Habitat: Moist soils, especially along shaded stream banks; in understory of coniferous forests.

Range: Pacific Coast from SW. British Columbia south to N. California; to 5000' (1524 m).

This handsome ornamental is dramatically colored in most seasons with bright green foliage turning orange and red in autumn, purple and white flowers in spring, and young red fruit in summer. The seeds of this and other maples are consumed by songbirds, game birds, and large and small mammals. The scientific name, meaning "rounded" or "circular," refers to the leaf shape.

230, 300 Rocky Mountain Maple
"Dwarf Maple"
"Mountain Maple"
Acer glabrum Torr.

Description: Shrub or small tree with short trunk and slender, upright branches, hairless throughout.

Height: 30' (9 m).

Diameter: 1' (0.3 m).

Leaves: opposite; 1½–4½" (4–11 cm) long and wide, sometimes smaller. *3 short-pointed lobes* (sometimes 5) or divided into 3 lance-shaped leaflets; *doubly saw-toothed,* 3 or 5 main veins from base; *long, reddish leafstalks. Shiny dark green* above, paler or whitish beneath; turning red or yellow in autumn.

Bark: gray or brown; smooth, thin.

Twigs: reddish-brown, slender.

Flowers: ¼" (6 mm) wide; *greenish-yellow;* 4 narrow sepals and 4 petals on drooping stalks; in branched clusters; male and female usually on separate plants; with new leaves in spring.

Fruit: ¾–1" (2–2.5 cm) long; *paired, forking, long-winged keys; reddish,* turning light brown; 1-seeded; maturing in late summer or autumn.

Habitat: Moist soils, especially along canyons and mountain slopes in coniferous forests.

Range: SE. Alaska, British Columbia, and SW. Alberta, south mostly in mountains to S. New Mexico and S. California; to 5000–9000' (1524–2743 m) in south.

The northernmost maple in the New World, it extends through southeastern Alaska. Deer, elk, cattle, and sheep browse the foliage. The Latin species name, meaning "hairless," refers to the leaves.

226 Canyon Maple
"Bigtooth Maple" "Sugar Maple"
Acer grandidentatum Nutt.

Description: Small to medium-sized tree with short trunk and spreading, rounded, dense crown; often a shrub.

Height: 40' (12 m).

Diameter: 8" (20 cm).

Leaves: opposite; 2–3¼" (5–8 cm) long and wide. *3 broad, blunt lobes and 2 small* basal lobes; *few blunt teeth,* 3 or 5 main veins from notched or straight base; *slightly thickened;* long leafstalks. Shiny dark green above, pale and finely hairy beneath; turning red or yellow in autumn.

Bark: gray or dark brown; thin; smooth or scaly.

Twigs: reddish, slender, hairless.
Flowers: ³⁄₁₆″ (5 mm) long; with *bell-shaped,* 5-lobed, *yellow calyx;* in
drooping clusters on long, slender,
hairy stalks; male and female in same or
different clusters; with new leaves in
early spring.

Fruit: 1–1¼″ (2.5–3 cm) long; *paired,*
forking, long-winged *keys;* reddish or
green, mostly hairless, 1-seeded;
maturing in autumn.

Habitat: Moist soils of canyons in mountains and
plateaus; in woodlands.

Range: SE. Idaho south to Arizona and east to
S. New Mexico and Trans-Pecos Texas;
local in Edwards Plateau of S. central
Texas, SW. Oklahoma, and N. Mexico;
at 4000–7000′ (1219–2134 m); locally
to 1500′ (457 m).

The western relative of Sugar Maple,
Canyon Maple has sweetish sap used
locally to prepare maple sugar. The
wood provides good fuel. The showy
autumn foliage makes it suitable as an
ornamental. The scientific name,
meaning "large-toothed," refers to the
leaves.

---

225 Bigleaf Maple
"Broadleaf Maple" "Oregon Maple"
Acer macrophyllum Pursh

Description: Small to large tree with broad, rounded
crown of spreading or drooping
branches and the *largest leaves of all
maples.*

Height: 30–70′ (9–21 m).
Diameter: 1–2½′ (0.3–0.8 m).
Leaves: opposite; 6–10″ (15–25 cm)
long and wide. Rounded, with 5 *deep,
long-pointed lobes* (sometimes 3); edges
with *few small, blunt lobes* and teeth; 5
main veins; slightly thickened. Shiny
dark green above, paler and hairy
beneath; turning orange or yellow in
autumn. Leafstalks to 10″ (25 cm);

stout, with *milky sap* when broken.
Bark: brown, furrowed into small 4-sided plates.
Twigs: green, stout, hairless.
Flowers: ¼" (6 mm) long; many on slender stalks; *yellow;* fragrant; male and female together in narrow, drooping clusters to 6" (15 cm) long at end of leafy twig; in spring.

Fruit: 1–1½" (2.5–4 cm) long; *paired, long-winged keys;* brown, with *stiff yellowish hairs;* 1-seeded; maturing in autumn.

Habitat: Stream banks and in moist canyon soils; sometimes in pure stands.

Range: SW. British Columbia to S. California; to 1000' (305 m) in north; at 3000–5500' (914–1676 m) in south.

The common and scientific names describe the very large leaves. A handsome shade tree and particularly showy in autumn, it is popular on the Pacific Coast. The only western maple with wood of commercial importance, it is used for veneer, furniture, handles, woodenware, and novelties. Indians made canoe paddles from the wood, and maple sugar can be obtained from the sap.

299, 528 Boxelder
"Ashleaf Maple" "Manitoba Maple"
Acer negundo L.

Description: Small to medium-sized tree with short trunk and broad, rounded crown of light green foliage.
Height: 30–60' (9–18 m).
Diameter: 2½' (0.8 m).
Leaves: opposite; *pinnately compound;* 6" (15 cm) long; with slender axis. *3–7 leaflets* sometimes slightly lobed, 2–4" (5–10 cm) long, 1–1½" (2.5–4 cm) wide; paired and short-stalked (except at end); *ovate or elliptical, long-pointed* at tip, short-pointed at base; *coarsely saw-*

toothed, sometimes lobed. *Light green* and mostly hairless above, paler and varying in hairiness beneath; turning yellow (or sometimes red) in autumn.
Bark: light gray-brown; with many narrow ridges and fissures, becoming deeply furrowed.
Twigs: *green,* often whitish or purplish; slender, ringed at nodes, mostly hairless.
Flowers: ³⁄₁₆″ (5 mm) long; with very small, *yellow-green* calyx of 5 lobes or sepals; several clustered on slender, drooping stalks; male and female on separate trees; before leaves in early spring.
Fruit: 1–1½″ (2.5–4 cm) long; *paired, slightly forking keys* with flat, *narrow* body and *long, curved wing, pale yellow,* 1-seeded; maturing in summer and remaining attached in winter.

Habitat: Wet or moist soils along stream banks and in valleys, with various hardwoods; also naturalized in waste places and roadsides.

Range: S. Alberta east to extreme S. Ontario and New York, south to central Florida, and west to S. Texas also scattered from New Mexico to California and naturalized in New England; to 8000′ (2438 m) in the Southwest.

Boxelder is classed with maples, having similar, paired key fruits, but is easily distinguishable by the pinnately compound leaves. Hardy and fast-growing, it is planted for shade and shelterbelts but is short-lived and easily broken in storms. Common and widely distributed, it is spreading in the East as a weed tree. Plains Indians made sugar from the sap. The common name indicates the resemblance of the foliage to that of elders (*Sambucus*) and the whitish wood to that of Box (*Buxus sempervirens* L.).

228, 335 Norway Maple
Acer platanoides L.

Description:

Introduced shade tree with rounded crown of dense foliage and with milky sap in leafstalks.

Height: 60' (18 m).

Diameter: 2' (0.6 m).

Leaves: opposite; 4–7" (10–18 cm) long and wide. *Palmately 5-lobed;* the shallow lobes and edges with *scattered long teeth;* 5 or 7 main veins from notched base. Dull green with sunken veins above, paler and hairless (except in vein angles) beneath; turning bright yellow in autumn. Long slender leafstalk, with *milky sap* at end when broken off.

Bark: gray or brown; becoming rough and furrowed into narrow ridges.

Twigs: brown, hairless.

Flowers: ⁵⁄₁₆" (8 mm) wide; with 5 *greenish-yellow* petals; in upright or spreading clusters; usually male and female on separate trees; in early spring before leaves.

Fruit: 1½–2" (4–5 cm) long; *paired keys* with long wing and *flattened body, spreading widely,* light brown, hanging on long stalk; maturing in summer.

Habitat: A street tree, escaping along roadsides; in humid temperate regions.

Range: Native across Europe from Norway to Caucasus and N. Turkey. Widely planted across the United States.

Norway Maple is fast-growing and tolerant of city smoke and dust. Varieties have columnar and low, rounded habits and reddish and variegated foliage. The species name, meaning "like *Platanus,*" indicates the similarity of the leaves to those of Sycamore and Planetree, to which it is not related.

221 Planetree Maple
"Sycamore Maple"
Acer pseudoplatanus L.

Description: Large introduced shade tree with widely
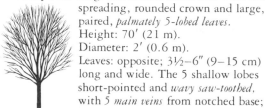
spreading, rounded crown and large,
paired, *palmately 5-lobed leaves.*
Height: 70′ (21 m).
Diameter: 2′ (0.6 m).
Leaves: opposite; 3½–6″ (9–15 cm)
long and wide. The 5 shallow lobes
short-pointed and *wavy saw-toothed,*
with *5 main veins* from notched base;
long, slender leafstalk. Dull dark green
with sunken veins above, pale with
raised, sometimes hairy veins beneath;
turning brown in autumn.
Bark: gray, smooth or with broad, flaky
scales.
Twigs: gray, hairless.
Flowers: ³⁄₁₆″ (5 mm) wide; with 5
greenish-yellow petals; male and bisexual;
in narrow, branched, *drooping* clusters
5″ (13 cm) long; in early spring.
Fruit: 1¼–2″ (3–5 cm) long; *paired keys*
with *elliptical body* and long wing, light
brown; maturing in summer.

Habitat: Hardy in exposed places and adapted to
seashore gardens, tolerant of salt spray;
sometimes escaping along roadsides.

Range: Native of Europe and W. Asia. Planted
across the United States.

Although fast-growing, this species is
not as hardy northward as Norway
Maple. It is an important timber and
shade tree in Europe, where it is called
Sycamore. The species name, meaning
"false *Platanus,*" refers to the
resemblance of the foliage to that of the
sycamore or planetree genus.

227, 317, 318, **Red Maple**
526 "Scarlet Maple" "Swamp Maple"
Acer rubrum L.

Description: Large planted tree with narrow or
rounded, compact crown and *red flowers,
fruit, leafstalks,* and *autumn foliage.*
Height: 60–90' (18–27 m).
Diameter: 2½' (0.8 m).
Leaves: opposite; 2½–4" (6–10 cm)
long and nearly as wide. Broadly ovate,
with *3 shallow, short-pointed lobes*
(sometimes with 2 smaller lobes near
base); irregularly and wavy *saw-toothed,*
with 5 main veins from base; long red
or green leafstalk. Dull green above,
whitish and hairy *beneath;* turning red,
orange, and yellow in autumn.
Bark: gray, thin, smooth, becoming
fissured into long, thin, scaly ridges.
Twigs: reddish, slender, hairless.
Flowers: ⅛" (3 mm) long; *reddish;*
crowded in nearly stalkless clusters
along twigs; male and female in
separate clusters; in *late winter* or very
early spring before leaves.
Fruit: ¾–1" (2–2.5 cm) long including
long wing; *paired, forking keys, red
turning reddish-brown,* 1-seeded;
maturing in spring.

Habitat: Wet or moist soils of stream banks,
valleys, swamps, and uplands and
sometimes on dry ridges; in mixed
hardwood forests.

Range: Extreme SE. Manitoba east to E.
Newfoundland, south to S. Florida, and
west to E. Texas; to 6000'
(1829 m). Planted in the West.

While most spectacular in autumn, the
red flowers, fruit, and twigs make this
species handsome much of the year.
Pioneers made ink and cinnamon-
brown and black dyes from a bark
extract.

229, 316, 527 Silver Maple
"Soft Maple" "White Maple"
Acer saccharinum L.

Description: Large planted tree with short, stout
trunk; few large forks; spreading, open,
irregular crown of long, curving
branches; and graceful cut-leaves.
Height: 50–80′ (15–24 m).
Diameter: 3′ (0.9 m).
Leaves: opposite; 4–6″ (10–15 cm)
long and nearly as wide. Broadly ovate,
deeply 5-lobed and *long-pointed* (middle
lobe often 3-lobed); *doubly saw-toothed,*
with 5 main veins from base; becoming
hairless; slender, drooping, reddish
leafstalk. Dull green above, *silvery-white
beneath;* turning pale yellow in autumn.
Bark: gray, becoming furrowed into
long, scaly, shaggy ridges.
Twigs: light green to brown; long,
spreading, and often slightly drooping;
hairless; with slightly unpleasant odor
when crushed.
Flowers: ¼″ (6 mm) long; *reddish buds
turning greenish-yellow;* crowded in nearly
stalkless clusters; male and female in
separate clusters; in *late winter* or very
early spring before leaves.
Fruit: 1½–2½″ (4–6 cm) long
including long, broad wing; *paired,
widely forking* keys, *light brown,* 1-
seeded; maturing in spring.

Habitat: Wet soils of stream banks, flood plains,
and swamps with other hardwoods.

Range: S. Ontario east to New Brunswick,
south to NW. Florida, west to E.
Oklahoma, north to N. Minnesota; to
2000′ (610 m), higher in mountains.
Cultivated in the West.

Its rapid growth makes Silver Maple a
popular shade tree; however, its form is
not generally pleasing, its brittle
branches are easily broken in
windstorms, and the abundant fruit
produce litter. Sugar can be obtained
from the sweetish sap, but yield is low.

BUCKEYE (HORSECHESTNUT) FAMILY
(Hippocastanaceae)

Trees and shrubs; 15 species in north temperate regions and mountains of tropical America. 6 native tree species, all in the buckeye genus (*Aesculus*), in North America.

Leaves: deciduous (evergreen in tropics); opposite; long-stalked; palmately compound, with 5–9 (sometimes 3) elliptical or lance-shaped leaflets, mostly saw-toothed; without stipules.

Flowers: many in large upright branched clusters; showy, often large; irregular; slightly bell-shaped; both bisexual and male; with tubular, 5-lobed calyx and corolla of 4–5 generally unequal rounded white, pink, red, or yellow petals, 6–8 long curved stamens, and 1 pistil composed of 3-celled ovary with 2 ovules in each cell and long, curved style.

Fruit: a large, rounded, brown capsule, often spiny, with hard, thick wall splitting into 3 parts. 1–3 large, rounded or slightly angled, shiny brown seeds with gray scar at base, poisonous or inedible.

302, 382 **California Buckeye**
Aesculus californica (Spach) Nutt.

Description: Thicket-forming shrub or small tree with a short trunk often enlarged at base, a broad, rounded crown of crooked branches, and many showy flowers.

Height: 25′ (7.6 m).

Diameter: 1′ (0.3 m).

Leaves: opposite; *palmately compound;* long-stalked. Generally 5 *leaflets* (sometimes 4–7) 3–6″ (7.5–15 cm) long, 1–2″ (2.5–5 cm) wide; *narrowly elliptical,* finely saw-toothed, short-stalked. Dark green above, paler with

whitish hairs on veins beneath; turning dull brown and shedding in late summer.
Bark: light gray, smooth, thin.
Twigs: reddish-brown, stout, ending in *resinous bud.*
Flowers: 1–1¼" (2.5–3 cm) long; with 4–5 *nearly equal, white* or sometimes *pale pink petals* and 5–7 much *longer stamens;* fragrant; in upright, narrow clusters 4–8" (10–20 cm) long; in late spring and early summer.

Fruit: 2–3" (5–7.5 cm) long; *pear-shaped capsules,* pale brown, *smooth,* splitting usually on 3 lines; maturing in late summer; usually 1 large, rounded, shiny brown, *poisonous* seed.

Habitat: Moist soils of canyons and on hillsides in chaparral and oak woodland.

Range: California in Coast Ranges and Sierra Nevada foothills; to 4000' (1219 m).

The only native buckeye in the West, this species is sometimes grown as an ornamental. California Indians made flour from the poisonous seeds after leaching out the toxic element with boiling water. The ground, untreated seeds were thrown into pools of water to stupefy fish, which then rose to the surface and were easily caught. Chipmunks and squirrels consume the seeds, but bees are poisoned by the nectar and pollen.

303, 380, 500 **Horsechestnut**
Aesculus hippocastanum L.

Description:

Introduced shade and ornamental tree with spreading, elliptical to rounded crown of stout branches and coarse foliage.
Height: 70' (21 m).
Diameter: 2' (0.6 m).
Leaves: opposite; *palmately compound;* with leafstalks 3–7" (7.5–18 cm) long. 7 *leaflets* (sometimes 5), spreading

fingerlike, 4–10″ (10–25 cm) long, 1–3½″ (2.5–9 cm) wide; *obovate* or elliptical, broadest toward abrupt point, tapering to stalkless base, *sawtoothed.* Dull dark green above, paler beneath.

Bark: gray or brown, thin, smooth, becoming fissured and scaly.

Twigs: light brown, stout, hairless, ending in large, blackish, sticky bud.

Flowers: 1″ (2.5 cm) long; narrowly *bell-shaped,* with 4–5 spreading, *narrow white petals, red- and yellow-spotted* at base; many flowers in *upright, branched clusters* 10″ (25 cm) long; in late spring.

Fruit: 2–2½″ (5–6 cm) in diameter; a brown, *spiny* or warty *capsule,* splitting into 2–3 parts; 1–2 large, rounded, shiny brown, *poisonous seeds;* maturing in late summer.

Habitat: A shade and street tree in rich, moist soils.

Range: Native of SE. Europe. Widely planted across the United States and escaped in the Northeast.

Horsechestnut is showy when bearing masses of whitish flowers for a few weeks in spring. It is easily propagated from seed and tolerant of city conditions, although the stout branches are broken by winds. Turks reportedly used the seeds to concoct a remedy given to horses suffering from cough, hence the common and scientific names.

SOAPBERRY FAMILY
(Sapindaceae)

About 1500 species of trees, shrubs, and woody vines with tendrils; rarely herbs; in tropical and subtropical regions; 7 native tree species, 3 of woody vines, and 3 of herbaceous vines in North America.

Leaves: alternate; generally pinnately compound, sometimes with 3 leaflets; without stipules (except in vines); the leaflets commonly alternate.

Flowers: numerous; tiny; in branched clusters; usually male and female or bisexual; regular or irregular, with 5 sepals, usually 5 petals often with scale or gland at base within, generally 8 or 10 stamens inserted in a disk, and 1 pistil with superior ovary usually 3-celled with 1–2 ovules in each cell and style.

Fruit: often large, 3-celled capsule, berry, drupe, or winged key; the seed often with covering. Fruit and seeds of a few species are edible; those of some are poisonous.

283, 480 Western Soapberry
"Wild Chinatree" "Jaboncillo"
Sapindus drummondii Hook. & Arn.

Description:

Poisonous tree with rounded crown of upright branches; or large, spreading shrub.

Height: 20–40' (6–12 m).

Diameter: 1' (0.3 m).

Leaves: pinnately compound; 5–8" (13–20 cm) long; with long, slender axis. *11–19 leaflets* 1½–3" (4–7.5 cm) long, ⅜–¾" (10–19 mm) wide; paired except at end, *lance-shaped, curved and slightly one-sided, long-pointed* at tip, blunt and unequal at base; *without teeth;* slightly thickened; short-stalked. *Dull yellow-green* above, slightly hairy with prominent veins beneath.

Bark: light gray, becoming rough and furrowed.

Twigs: *yellow-green;* covered with fine hairs.

Flowers: ⅛" (3 mm) wide; usually with 5 rounded, *yellowish-white* petals; almost stalkless; generally male and female; in upright clusters 6–9" (15–23 cm) long; in late spring or summer.

Fruit: ⅜–½" (10–12 mm) in diameter; *berrylike,* sometimes paired, *yellow* or orange, *nearly transparent,* leathery; maturing in autumn, turning black and remaining attached in winter; *poisonous.* Single, round, dark brown seed.

Habitat: Moist soils along streams and on limestone uplands, in and bordering hardwood forests; westward in plains and mountains, grassland, upper desert, and oak woodland zones.

Range: SW. Missouri south to Louisiana, west to S. Arizona, and northeast to SE. Colorado; also N. Mexico; to 6000' (1829 m).

The poisonous fruit, containing the alkaloid saponin, has been used as a soap substitute for washing clothes. The foliage is unpalatable and may be poisonous to livestock, and the fruit causes a rash on some people's skin. Necklaces and buttons are made from the round, dark brown seeds; baskets are made from the wood, which splits easily.

285, 499 **Mexican-buckeye**
"Texas-buckeye" "Spanish-buckeye"
Ungnadia speciosa Endl.

Description: Shrub or small tree with irregular crown of upright branches and showy pink flowers.

Height: 25' (7.6 m). ·
Diameter: 8" (20 cm).
Leaves: pinnately compound; 5–12"

(13–30 cm) long; with slender axis.
5–9 leaflets 3–5″ (7.5–13 cm) long, 1–
1½″ (2.5–4 cm) wide; paired except at
end, *ovate, long-pointed* at tip, rounded
and often unequal at base, *wavy saw-
toothed;* hairy when young; *leathery;*
stalkless or nearly so. *Shiny dark green*
above, light green and nearly hairless
beneath.

Bark: light gray, thin, fissured.

Twigs: light brown, slender, slightly
zigzag, covered with fine hairs.

Flowers: 1″ (2.5 cm) wide; with *4–5
unequal, pink or purplish-pink petals;*
male, female, and bisexual; slender-
stalked; in clusters crowded along
twigs; before or with leaves in spring.

Fruit: 1½–2″ (4–5 cm) wide; a long-
stalked, drooping *capsule;* broadly pear-
shaped, *3-lobed, reddish-brown,* rough
and leathery, 3-celled; *3 round, poisonous
seeds;* maturing and splitting open in
autumn.

Habitat: Moist soils of rocky canyons, slopes,
and ridges.

Range: Central to Trans-Pecos Texas and S.
New Mexico; also in NE. Mexico; to
5000′ (1524 m).

From a distance the plants in full flower
resemble redbuds or peaches. The
sweetish but poisonous seeds are
sometimes used by children as marbles.
Livestock seldom browse the toxic
foliage, but bees produce fragrant
honey from the flowers. Although not a
true buckeye, it is so called because of
the similar large capsules and seeds.
This distinct plant, alone in its genus,
commemorates Baron Ferdinand von
Ungnad, Austrian ambassador at
Constantinople, who introduced
Horsechestnut into western Europe in
1576.

BUCKTHORN FAMILY
(Rhamnaceae)

Shrubs, woody vines, and small to large trees, rarely herbs, often spiny. About 700 species worldwide; 15 native and 3 naturalized tree, about 50 shrub, and 1 woody vine species in North America.
Leaves: mostly alternate, also opposite; simple; often with 3 or more veins from base; usually with tiny stipules.
Flowers: small; greenish or yellowish; mostly in clusters along twigs; usually bisexual; regular, with concave cuplike base with 5 or 4 sepals touching by edges (not overlapping) in bud; with 5 or 4 small petals (sometimes not present) concave and very narrow with narrow base, 5 or 4 opposite stamens enclosed by petals, and 1 pistil with superior 2- to 4-celled ovary within the disk, style, and 1–5 stigmas.
Fruit: a berry, drupe, or capsule, often opening in 3 parts.

---

198 **Feltleaf Ceanothus**
"Catalina Ceanothus" "Island-myrtle"
Ceanothus arboreus Greene

Description:

Evergreen large shrub or small tree with short, straight trunk, rounded crown of many stout, spreading branches, and abundant, pale blue flowers.
Height: 25′ (7.6 m).
Diameter: 1′ (0.3 m).
Leaves: *evergreen;* 1–3″ (2.5–7.5 cm) long, ¾–1½″ (2–4 cm) wide. Broadly ovate to elliptical, with *3 prominent, sunken main veins, finely wavy saw-toothed,* slightly thickened. *Dull green* and finely hairy above, paler and *densely covered with white hairs* beneath.
Bark: gray, smooth, thin; becoming dark brown and fissured into small, square, scaly plates.
Twigs: light brown, slender, slightly

angled, covered with *soft hairs.*
Flowers: ⅛" (3 mm) wide; with 5 *pale
blue petals;* fragrant; in branched clusters
2–6" (5–15 cm) long, crowded on
slender, hairy stalks at ends of leafy
twigs; in early spring.

Fruit: ¼" (6 mm) in diameter; rough,
black capsules; 3-lobed, splitting into 3
1-seeded nutlets; maturing in summer.

Habitat: Chaparral on dry slopes and in canyons.

Range: Restricted to Santa Rosa, Santa Cruz,
and Santa Catalina islands off
California; near sea level.

This distinctive species grows wild only
on three California islands. It is planted
along the coast as an ornamental and a
screen and along roadsides. One of the
largest species of its genus, it bears a
suitable scientific name meaning
"treelike."

---

165 Greenbark Ceanothus
"California-lilac" "Redheart"
Ceanothus spinosus Nutt.

Description: Evergreen, spiny shrub or small tree
with irregular, spreading, open crown
and many showy flowers resembling
lilacs.

Height: 20' (6 m).

Diameter: 6" (15 cm).

Leaves: *evergreen;* ½–1¼" (1.2–3 cm)
long, ⅜–¾" (15–19 mm) wide.
*Elliptical to oblong, generally not toothed,
1 midvein, thick and stiff; shiny green* on
both surfaces. On young shoots and
seedlings, leaves are larger, toothed,
and 3-veined.

Bark: *olive green, smooth;* becoming dark
red-brown, rough, and scaly.

Twigs: bright green, slender, widely
forking, some ending in *short, sharp
spines.*

Flowers: ⅛" (3 mm) wide; with 5 *pale
blue to whitish petals;* fragrant; crowded
on slender stalks in branched clusters

2–6″ (5–15 cm) long, at ends of new leafy twigs; for many weeks in early spring.

Fruit: ¼″ (6 mm) in diameter; sticky, *black capsules;* slightly 3-lobed, splitting into 3 1-seeded nutlets.

Habitat: Dry hillsides and mountain canyons usually near coast; in chaparral and sage shrub.

Range: Pacific Coast Ranges of SW. California and NW. Baja California; to 3000′ (914 m).

This species is grown as an attractive ornamental and screen along roadsides and is useful for erosion control. However, old plants become irregular, sprawling, and untidy with the dried remains of the capsules. It is called "Redheart" because of the dark red wood, which makes good fuel. The shrubs are drought-resistant and sprout from stumps but are killed back in cold winters. The foliage of this and related species is browsed by deer and elk.

153, 396 Blueblossom
"Blue-myrtle" "Bluebrush"
Ceanothus thyrsiflorus Eschsch.

Description: Large evergreen shrub or small tree with short trunk, many spreading branches, and showy blooms resembling lilacs.
Height: 20′ (6 m).
Diameter: 8″ (20 cm).
Leaves: *evergreen;* ¾–2″ (2–5 cm) long, ½–¾″ (12–19 mm) wide. *Oblong or elliptical, rounded to short-pointed* at both ends; *finely wavy saw-toothed; 3 main veins;* slightly thickened. Shiny green above, paler and slightly hairy on raised veins beneath.
Bark: red-brown, fissured into narrow scales.
Twigs: pale yellow-green, *angled;*

slightly hairy when young.
Flowers: ³⁄₁₆″ (5 mm) wide; with 5 *light to deep blue petals* (rarely almost white); fragrant; crowded on slender stalks in branched clusters 1–3″ (2.5–7.5 cm) at base of upper leaves; in spring.

Fruit: ³⁄₁₆″ (5 mm) in diameter; smooth, sticky, *black capsules;* slightly 3-lobed, splitting into 3 1-seeded nutlets; maturing in summer.

Habitat: Mountain slopes and in canyons in chaparral, Redwood, and mixed evergreen forests.

Range: SW. Oregon south to S. California in outer Pacific Coast Range; to 2000′ (610 m).

This is the hardiest and largest ceanothus. Each spring the highways of the West Coast display masses of Blueblossom flowers. Plants can be grown in screens, in hedges, and against walls. Elk and deer browse the foliage. The shrubs form dense thickets after fires and logging. The scientific name, meaning "thyrse-flower," refers to the compact, branched flower cluster; *thyrsus* is the name of the staff, adorned with leaves and berries, that belonged to Bacchus, the Greek god of wine.

84 Bitter Condalia
Condalia globosa I. M. Johnst.

Description: Spiny, much-branched shrub or small tree with short trunk and irregular, thin crown of tangled, spreading branches.
Height: 20′ (6 m).
Diameter: 8″ (20 cm).
Leaves: alternate or usually in *clusters of 2–7* on short twigs; ⅛–½″ (3–12 mm) long and half as wide. *Spoon-shaped,* without teeth, thin but stiff, almost stalkless, finely covered with rough hairs or hairless. Pale yellow-green,

with raised veins beneath.

Bark: brownish-gray, thin, much fissured and shreddy.

Twigs: light gray or brown, *forking nearly at right angles,* very slender, stiff; some short and *ending in slender spine.*

Flowers: less than ⅛" (3 mm) wide; *cup-shaped,* with 5 yellow-green, pointed sepals and no petals, short-stalked, fragrant; few clustered together on short twig; in early spring or autumn, following irregular rains.

Fruit: ³⁄₁₆" (5 mm) in diameter; *black or dark blue,* with thin, juicy pulp, often very *bitter;* 1-seeded; ripening in spring.

Habitat: Dry, sandy plains, rocky slopes, and along washes in creosotebush desert.

Range: SW. Arizona and SE. California; also in NW. Mexico; at 500–2500' (152–762 m).

The Latin species name *globosa* refers to the rounded fruit. Bitter Condalia is plentiful at Organ Pipe Cactus National Monument in southwestern Arizona. It was originally described in 1924 as a shrub from Baja California and was later found in Arizona as a tree new to the United States.

193 Cascara Buckthorn
"Cascara Sagrada" "Chittam"
Rhamnus purshiana DC.

Description: Large shrub or small tree with short trunk and crown of many stout, upright branches.

Height: 30' (9 m).

Diameter: ½–1' (0.15–0.3 m).

Leaves: often clustered near ends of twigs; 2–6" (5–15 cm) long, 1–2½" (2.5–6 cm) wide. *Broadly elliptical, finely wavy-toothed,* with many nearly straight side veins. *Dull green* and nearly hairless above, paler and slightly hairy beneath; turning pale yellow in late autumn.

Bark: gray or brown, thin, fissured into short, thin scales; inner bark yellow, turning brown upon exposure; *very bitter*.

Twigs: gray, slender; hairy when young; ending in bud of tiny brown leaves covered with rust-colored hairs.

Flowers: ³⁄₁₆" (5 mm) wide; *bell-shaped*, with 5 pointed, *greenish-yellow* sepals; in clusters at leaf bases; in spring and early summer.

Fruit: ⅜" (10 mm) in diameter; *berrylike; red,* turning to *purplish-black;* with thin, juicy, sweetish pulp and 2–3 seeds; maturing in late summer or autumn.

Habitat: Moist soils in open areas, along roadsides, and in understory of coniferous and mixed evergreen forests.

Range: S. British Columbia south to N. California; also Rocky Mountain region south to N. Idaho and W. Montana; to 5000' (1524 m).

The bark is the source of the laxative drug, Cascara Sagrada, meaning "sacred bark" in Spanish. It is harvested commercially in Washington and Oregon by stripping bark from wild trees. When a tree is cut down, several sprouts grow from the stump. The berries are consumed by songbirds and bears, raccoons, and other mammals; hence this species is sometimes called "Bearberry."

BASSWOOD (LINDEN) FAMILY
(Tiliaceae)

Trees, or in tropical regions, also
shrubs and herbs, deciduous except in
tropics, commonly with fibrous bark;
more than 400 species; represented by
the basswood or linden genus (*Tilia*)
with 3 native species and by a few
tropical herbs in North America.
Leaves: alternate in 2 rows, simple,
often oblique or unequal and with 3 or
more main veins from base, often
toothed, commonly with star-shaped
hairs and with paired stipules.
Flowers: in branched clusters (hanging
from a strap-shaped stalk in basswood,
Tilia), bisexual, regular, with calyx of
5 sepals usually separate, 5 petals
(sometimes none), many stamens
usually united at base in groups of 5–
10, and 1 pistil with 2- to 10-celled
ovary, 1 style, and as many stigmas as
cells.
Fruit: nutlike (rounded and 1- to 3-
seeded in basswood), capsule or
drupelike.

---

197 **European Linden**
"Common Linden"
Tilia ×*europaea* L.

Description: Large cultivated tree with dense,

pyramid-shaped crown.
Height: 70′ (21 m).
Diameter: 2′ (0.6 m).
Leaves: in 2 rows; 2–4″ (5–10 cm) long
and wide. *Heart-shaped or rounded,*
abruptly long-pointed at tip, unequal
and slightly notched at base; *sharply
saw-toothed,* palmately veined, long-
stalked. Dull dark green above, pale
green and *nearly hairless* with tufts of
hairs in vein angles beneath.
Bark: gray, smooth, becoming fissured
into scaly ridges.
Twigs: brown, slender, hairless.

Flowers: ½″ (12 mm) wide; with 5 *pale yellow petals;* fragrant; in long-stalked clusters hanging from middle of leafy, greenish bract; in early summer.

Fruit: ¼″ (6 mm) in diameter; nutlike, *round,* short-pointed, gray, covered with fine hairs, hard; maturing in late summer.

Habitat: Moist soils in humid temperate regions.

Range: Cultivated across the United States, especially in the Northeast and Pacific Northwest.

A hybrid of two European species, Littleleaf Linden (*Tilia cordata* Mill.) and Bigleaf Linden (*Tilia platyphyllos* Scop.), European Linden is one of several introduced shade trees known as lindens and related to the native basswoods. The flowers are a source of honey.

STERCULIA FAMILY
(Sterculiaceae)

Shrubs, vines, and trees; about 700 species in tropical and subtropical regions; 2 native and 1 naturalized tree and a few native shrub species in North America.

Leaves: alternate; simple; often palmate-veined and palmately lobed, sometimes palmately compound; with star-shaped hairs; the leafstalk often with enlargement at tip; with stipules.

Flowers: usually in branched clusters along or at end of twigs or sometimes along trunks; generally bisexual and regular with parts in groups of 5; calyx of 3–5 lobes; 5 petals (sometimes none present); and 5 stamens united in a tube or separate, sometimes with 5 sterile stamens, and 1 pistil composed of superior ovary generally with 5 (sometimes 1–4) cells with 2 or more ovules and 1–5 styles often lobed; sometimes the stamens and pistils on a long stalk.

Fruit: a capsule or berry or 5 follicles.

232 Chinese Parasoltree
"Bottletree" "Japanese Varnish-tree"
Firmiana simplex (L.) W. F. Wight

Description: Ornamental and naturalized tree with crown rounded like an umbrella and with large, lobed leaves.

Height: 30′ (9 m).

Diameter: 6″ (15 cm).

Leaves: 6–12″ (15–30 cm) long and wide. *Heart-shaped* or rounded; with *3 or 5 long-pointed lobes;* 5 main veins from base; without teeth; very long-stalked. Dull green above, often hairy beneath.

Bark: *gray-green, smooth.*

Twigs: gray-green, stout.

Flowers: ½″ (12 mm) long; numerous, with 5 *yellow sepals, turning red* (petals absent); male and female in separate, upright clusters 8–16" (20–41 cm) long, with scurfy, hairy, branched stalks; in late spring and early summer. Fruit: 2–4" (5–10 cm) long; 5 produced from 1 flower; podlike, long-pointed, greenish turning light brown, stalked; splitting open *like a leaf, exposing on edges* several *pealike seeds;* maturing in late summer.

Habitat: Escaping along roadsides and in mixed hardwood forests.

Range: Native of China. Cultivated and naturalized locally across S. United States from North Carolina south to Florida and west to California.

The opening fruit releases a brownish-black liquid. It is reported that a tea can be prepared from the roasted seeds. This fast-growing member of a tropical family is popular as an ornamental for its large leaves.

237, 324 California Fremontia
"Flannelbush"
"California Slippery-elm"
Fremontodendron californicum (Torr.) Cov.

Description: Evergreen, many-branched, thicket-forming shrub or small tree with short trunk, open crown, and large, showy, bright yellow flowers.
Height: 20' (6 m).
Diameter: 8" (20 cm).
Leaves: *evergreen;* ½–1½" (1.2–4 cm) long and wide. *Rounded or broadly ovate,* usually with *3 blunt lobes* and 3 main veins; thick; long-stalked; mostly on short side-twigs. Dull dark green and sparsely hairy above, *covered with rust-colored and scurfy hairs* and showing raised veins beneath.
Bark: brownish-gray, fissured and scaly; inner bark *mucilaginous.*
Twigs: stout, stiff; densely covered with rust-colored hairs when young; becoming reddish-brown.

Flowers: 1¼–2″ (3–5 cm) wide; with 5 *broad, bright yellow calyx lobes* and without petals; borne singly on short side-twigs or opposite leaves; in spring and early summer.

Fruit: 1–1¼″ (2.5–3 cm) long; pointed, densely hairy, *egg-shaped capsule;* 4- or 5-celled; ripening in late summer and splitting open on 4–5 lines; many elliptical, dark reddish-brown seeds.

Habitat: Dry, rocky mountain slopes and canyons; with chaparral, pinyons, junipers, and Ponderosa Pine.

Range: California, central Arizona, and N. Baja California; at 3000–6500′ (914–1981 m).

The beautiful masses of large yellow flowers and odd, small leaves covered with scurfy hairs make this plant an attractive ornamental. When flowering, the showy plants are conspicuous from a distance. The name "Flannelbush" refers to the densely hairy foliage. Under a hand lens, the many-branched hairs on various parts look like tiny stars. This species is also called "California Slippery-elm" because of the mucilaginous bark, which is sometimes used as a poultice. It was discovered by Gen. John Charles Frémont (1813–90), the politician, soldier, and explorer of the western United States.

TEA FAMILY
(Theaceae)

About 500 species of trees and shrubs,
mostly in tropical and subtropical but
also in warm northern temperate
regions; 4 native species in southeastern
United States.

Leaves: alternate, evergreen, simple,
usually leathery, sometimes with lines
parallel to midvein, without stipules.

Flowers: often large, showy and
aromatic; generally solitary or a few
clustered along twigs; bisexual; regular;
often with 2 scales at base, with calyx
of 5–7 sepals usually separate,
overlapping, and persisting at base of
fruit; corolla of 5 commonly white or
pink petals, separate or united at base,
overlapping; many stamens often united
to corolla in 5 opposite groups; and 1
pistil composed of 2- to 5-celled ovary
generally superior with 2 or more
ovules in each cell and 2–5 persistent
styles often united at base.

Fruit: usually a hard capsule with
central, persistent column, a berry or
a drupe.

164, 393 **Common Camellia**
"Japanese Camellia"
Camellia japonica L.

Description: Evergreen cultivated shrub or small tree
with *large, showy flowers* and narrow or
rounded crown, hairless throughout.
Height: 30′ (9 m).
Diameter: 6″ (15 cm).
Leaves: evergreen; 2–4″ (5–10 cm)
long, 1–1¾″ (2.5–4.5 cm) wide.
Elliptical, long-pointed at tip, short-
pointed at base, *finely saw-toothed, thick
and leathery;* short-stalked. *Shiny dark
green* above, paler beneath.
Bark: gray, smooth.
Twigs: gray; slender.
Flowers: 2–5″ (5–13 cm) wide; with

5–7 *spreading, rounded, waxy petals;* red
to pink, white, or variegated; with
many stamens, or with stamens
replaced by many narrow petals and
"double" flowers; borne *singly,* upright
and almost stalkless at base of leaves;
from autumn to spring.
Fruit: 1¼″ (3 cm) in diameter; a hard,
brown, *rounded capsule,* thick-walled, 2-
to 3-celled and splitting on 2–3 lines;
few large rounded seeds; in autumn.

Habitat: Grown as an ornamental in humid,
warm temperate regions.

Range: Native of Japan, Korea, and Taiwan.
Planted in the Southeast and Pacific
states.

A popular ornamental shrub with
beautiful large flowers and glossy
evergreen foliage, it is propagated by
cuttings and grafting. More than 2000
varieties exist, differing mainly in color
and size of flowers. The generic name
honors George Joseph Kamel (Latinized
as Camellius), a Jesuit priest of the
17th century who lived in the
Philippines.

TAMARISK FAMILY
(Tamaricaceae)

Generally small trees and shrubs of dry
and salty areas. About 100 species of
trees and shrubs in temperate and
subtropical Eurasia; 3 naturalized tree
species in North America.
Leaves: tiny, scalelike, alternate,
pressed against twig, without stipules.
Twigs: very slender, often drooping.
Flowers: tiny, usually crowded and
short-stalked in unbranched clusters,
bisexual, regular; composed of 4–5
sepals, 4–5 petals, as many or twice as
many stamens as petals, and 1 pistil
with 1-celled ovary.
Fruit: a capsule with many hairy seeds.

2 **Athel Tamarisk**
"Athel" "Evergreen Tamarisk"
Tamarix aphylla (L.) Karst.

Description: Evergreen, introduced tree of warm,
dry areas with dense, rounded or
irregular, spreading, gray-green crown
of many stout branches and long,
drooping twigs that appear leafless.
Height: 60′ (18 m).
Diameter: 2½′ (0.8 m).
Leaves: *tiny evergreen scales* ¹⁄₁₆″ (1.5
mm) long; borne singly, circling twig
and forming joint, ending in point;
hairless, *gray-green;* those on each twig
shedding together.
Bark: light gray-brown or reddish-
brown; becoming thick and deeply
furrowed into long narrow ridges;
purplish-brown and smooth on
branches.
Twigs: *wiry, very slender,* less than ¹⁄₁₆″
(1.5 mm) in diameter; *jointed;* shedding
or becoming greenish-brown.
Flowers: *very small,* less than ⅛″ (3
mm) long; with 5 pointed, *whitish-pink
petals;* numerous, crowded together and
almost stalkless in narrow clusters 1¼–

2½" (3–6 cm) long; several clusters drooping at end of twigs; in summer. Fruit: ³⁄₁₆" (5 mm) long; *pointed capsules* splitting into 3 *narrow parts;* maturing in late summer; many tiny seeds ending in tuft of *whitish hairs.*

Habitat: Warm, dry areas.

Range: Native from N. and E. Africa to SW. Asia. Introduced, cultivated, and sometimes escaping (but not naturalized) from S. Texas west to S. Arizona and California.

Athel Tamarisk is a handsome tree used for dense shade, in shelterbelts or windbreaks, and in hedges. The fast-growing trees are drought-resistant and tolerant of alkaline and salty soils; however, the heavy branches of weak, brittle wood break easily and become hazardous unless removed. The wood is used for fuel and produces a fragrant odor when burned. The light brown wood takes a high polish.

62, 383 **Tamarisk**
"Saltcedar" "Five-stamen Tamarisk"
Tamarix chinensis Lour.

Description: Naturalized shrub or small tree with slender, upright or spreading branches and narrow or rounded crown; resembling a juniper, though not evergreen.
Height: 16' (5 m).
Diameter: 4" (10 cm).
Leaves: about ¹⁄₁₆" (1.5 mm) long; *scalelike, crowded,* narrow and pointed; dull *blue-green.*
Bark: reddish-brown, smooth, becoming furrowed and ridged.
Twigs: green, becoming purplish; long, slender, hairless; usually shedding with leaves.
Flowers: less than ⅛" (3 mm) long and wide; with 5 *pink petals;* numerous, crowded together in narrow clusters

¾–2″ (2–5 cm) long at ends of twigs; in spring and summer.
Fruit: ⅛″ (3 mm) long; *narrow, pointed,* reddish-brown *capsules;* splitting into 3–5 parts; many *tiny, hairy seeds;* maturing in summer.

Habitat: Wet, open areas along streams, irrigation ditches, and reservoirs, including sand banks and alkali and salty soils.

Range: Native of Asia and SE. Europe. Extensively naturalized from SW. Nebraska west to Nevada and south to S. California and S. Texas; local beyond; also in N. Mexico; to 5000′ (1524 m).

Introduced at the beginning of this century as an ornamental and for erosion control, Tamarisk has become an undesirable weed in many places. It is considered a phreatophyte (literally, a "well-plant," having deep roots and high water use). Eradication has been difficult, since the plants spread by seeds and cuttings and grow rapidly. However, the large thickets provide cover for doves and other wildlife. *Tamarix,* the classical Latin name, may allude to Tamaris, a river in Spain; the genus is related neither to junipers ("cedars") nor to Tamarack.

CACTUS FAMILY
(Cactaceae)

Succulent, spiny plants of dry areas
including herbs, shrubs, a few small
trees, epiphytes, and vines; herbaceous
or woody, with enlarged, cylindrical or
flattened stems often jointed, with
many clustered spines and hairs
spreading from a center. About 1800
species in tropical and warm temperate
America, especially dry regions; 3
native and 2 naturalized tree and many
native shrub species in North America.
Leaves: reduced to scales or none or
rarely alternate; simple, flattened, and
succulent.
Flowers: generally solitary, stalkless,
large, bisexual, generally regular,
slightly fleshy; many sepals and petals
and intermediates and numerous
stamens; the petals generally yellow,
white, or pink; 1 pistil with inferior
1-celled ovary, style, and 2 to many
stigmas.
Fruit: a spiny berry, often juicy and
edible, with many black seeds.

312, 375 Saguaro
"Giant Cactus"
Cereus giganteus Engelm.

Description: Giant, *leafless, columnar tree cactus* with
massive, *spiny trunk* and usually 2–10
stout, nearly erect, spiny branches.
Height: 20–35' (6–10.7 m).
Diameter: 1–2' (0.3–0.6 m),
sometimes larger.
Trunk and branches: *cylindrical, yellow-
green,* smooth, with fleshy, *vertical
ridges,* bearing clusters of spreading,
sharp, *gray spines* ½–2" (1.2–5 cm)
long.
Wood: a framework of vertical, light
brown, lightweight ribs around thick,
whitish, succulent, bitter pith. Ribs
exposed as a skeleton after death.

Flowers: 4–4½" (10–11 cm) long, 2–3" (5–7.5 cm) wide; *funnel-shaped,* with *many waxy, white petals* and stamens at end of greenish, fleshy tube; stalkless; numerous, near tops of branches from cluster of spines; with odor of melon; in late spring, sometimes again in late summer.

Fruit: 2–3½" (5–9 cm) long; *egg-shaped berry, spineless, red,* fleshy, sweet and edible; splitting open along 3 or 4 lines and resembling flowers; maturing in early summer; many rounded, shiny brown, tiny seeds.

Habitat: Rocky or gravelly soils of desert foothills, especially on south-facing slopes; often with paloverdes.

Range: Arizona south to Sonora, Mexico; very local in SE. California; at 700–3500' (213–1067 m).

Indians made use of the entire cactus: they ate the fruit both fresh and dried and made it into preserves and beverages; the framework of ribs provided wood for shelters, fences, and kindling. Saguaro (pronounced "sah-WAH-ro"), the largest native cactus, is the state flower of Arizona and a symbol of desert landscapes. Well-adapted to its hot, dry climate, Saguaro is leafless. Food is manufactured in the green stems, and rainwater is absorbed quickly by the shallow roots and stored in the succulent trunks and branches. The thick, spreading spines offer protection against animals. Gila woodpeckers and gilded flickers make round holes near the tops of branches for nests that are used afterwards by elf owls, cactus wrens, and other birds. Wildlife, especially white-winged doves, consume quantities of the seeds.

311, 391 Jumping Cholla
"Cholla"
Opuntia fulgida Engelm.

Description: A very *spiny cactus,* commonly a shrub, occasionally a small tree with short trunk. The stout, *jointed, cylindrical,* irregular spreading to slightly drooping branches are leafless except when young.
Height: 15' (4.6 m).
Diameter: 6" (15 cm).
Leaves: single at tubercles; ½–1" (1.2–2.5 cm) long; *narrowly cylindrical,* long-pointed, light, green, fleshy, soon falling.
Bark: on both trunk and larger branches blackish, rough, scaly, spineless. Segments or joints 3–8" (7.5–20 cm) long, 1¼–2" (3–5 cm) in diameter; *pale green,* fleshy, bearing many *egg-shaped tubercles,* each with *2–12 large, brown spines* ¾–1¼" (2–3 cm) long and covered with shiny, *straw-colored sheaths.*
Flowers: 1" (2.5 cm) long and broad; with *5–8 pink or white petals* streaked with lavender, stalkless; scattered near ends of joints and on fruit; in late spring and summer.
Fruit: 1–1⅜" (2.5–3.5 cm) long and ¾" (19 mm) in diameter; *pear-shaped*

berries; green, tubercled, spineless, fleshy, with many seeds. Some remain attached several years and bear new flowers and fruit annually, often without seeds. These fruit clusters hang in *long, branched chains.*
Habitat: Dry, sandy soils of valleys, plains, and slopes, forming dense "cactus forests" in deserts.
Range: Central and S. Arizona and NW. Mexico; to 4000' (1219 m).

The common name Cholla (pronounced "CHAW-ya" or "CHO-ya"), meaning "skull" or "head" in Spanish, is applied to various shrubby cacti with jointed branches, of which this species is the largest. Its segments or joints, easily detached by touching, adhere to clothing and skin. They are fancifully said to jump out and attack passersby, especially when one's back is turned, as the common name implies. Their sharp, barbed spines can cause painful wounds and are not easily removed. The dead, weathered, woody skeletons of the stems, forming hollow cylinders with many holes, are used in making novelties. Fruit and seeds of cacti are consumed in great quantities by wildlife of many kinds, especially rodents. Detached branches will root and start new plants, spreading on some rangelands.

ELAEAGNUS FAMILY
(Elaeagnaceae)

Many-branched shrubs, sometimes small trees, often with spiny twigs and with dense covering of silvery, golden-yellow or brownish tiny scales or star-shaped hairs on twigs, lower leaf surfaces, and fruits. About 50 species in north temperate regions, mostly in plains and along coasts; 4 native shrub species including 1 which is sometimes a small tree (omitted from this guide) and 1 naturalized species reaching tree size in North America.

Leaves: deciduous; alternate, opposite or whorled; short-stalked, simple, without teeth, often slightly thickened, without stipules.

Flowers: small, single or in clusters at leaf bases; bisexual (or male and female on same or separate plants) with tubular, 4-lobed calyx and no corolla, 4–8 stamens inserted in tube and 1 pistil with 1-celled ovary, 1 ovule, and 1 style.

Fruit: drupelike, dry, with fleshy cover not opening, 1-seeded.

87, 322 **Russian-olive**
"Oleaster"
Elaeagnus angustifolia L.

Description: Introduced shrub or small tree with trunk often crooked or leaning, dense crown of low branches, *silvery foliage,* and sometimes *spiny twigs.*
Height: 20' (6 m).
Diameter: 4" (10 cm)
Leaves: 1½–3¼" (4–8 cm) long, ⅜–¾" (10–19 mm) wide. *Lance-shaped* or oblong, without teeth, short-stalked. Dull gray-green with obscure veins above, *silvery,* scaly, and *brown-dotted* beneath.
Bark: gray-brown, thin, fissured and *shredding in long strips.*

Twigs: *silvery,* scaly when young,
becoming reddish-brown; long and
slender; often ending in short *spine.*
Flowers: ⅜″ (10 mm) long; *bell-shaped,*
with 4 calyx lobes, yellow inside, *silvery*
outside (petals absent); fragrant; short-
stalked; scattered along twigs at leaf
bases; in late spring or early summer.
Fruit: ⅜–½″ (10–12 mm) long;
berrylike, *elliptical, yellow to brown with
silvery scales,* becoming shiny; thin,
yellow, mealy, *sweet,* edible pulp; large
brown stone; scattered along twig;
maturing in late summer and autumn.

Habitat: Moist soils, from salty to alkaline;
spreading in valleys.

Range: Native of S. Europe and Asia. Planted
and naturalized from British Columbia
east to Ontario and from New England
west to California; to 5000′ (1524 m)
or above.

The fruit is consumed by songbirds,
such as cedar waxwings, robins, and
grosbeaks, and by pheasants and quail.
Tolerant of cold, drought, and city
smoke, Russian-olive is a popular
ornamental. The plants sprout and
spread from roots, sometimes becoming
pests. It is not related to Olive (*Olea
europaea* L.), which also has narrow gray
leaves.

LOOSESTRIFE FAMILY
(Lythraceae)

Widespread, about 500 species of
herbs, shrubs, and a few tropical trees;
no native tree species, 1 common,
introduced, small tree and many native
herb species in North America.
Leaves: simple, generally opposite or
whorled and not toothed, without
stipules or with tiny stipules.
Flowers: usually in branched clusters,
bisexual, generally regular, with
tubular or cuplike base that bears on
the border generally 4, 6 or 8 wrinkled
petals with very narrow base, stamens
double or equal to number of petals and
inserted within tube, and 1 pistil with
superior 2- to 6-celled ovary, generally
with many ovules and 1 style.
Fruit: a capsule with many seeds.

118, 392 Crapemyrtle
Lagerstroemia indica L.

Description: Cultivated ornamental shrub or small
tree often branching near base, with
slightly angled and curved or crooked
trunks and open, spreading, rounded
crown.
Height: 20' (6 m).
Diameter: 4" (10 cm).
Leaves: deciduous or evergreen in
tropical climate; *mostly opposite* or upper
leaves alternate; often appearing in 2
rows; 1–2" (2.5–5 cm) long, ½–⅞"
(12–22 mm) wide. *Elliptical,* without
teeth, nearly stalkless. Dull green
above, paler and sometimes hairy on
midvein beneath.
Bark: *mottled gray and brown; smooth* and
flaking off in patches.
Twigs: light green, turning light
brown; long and slender, slightly 4-
angled, hairless or nearly so.
Flowers: 1¼–1½" (3–4 cm) wide; with
6 spreading, rounded, crapelike, fringed,

stalked petals, commonly pink but
varying from white to red, purple, and
bluish; odorless; *abundant* in showy
masses in upright, branched clusters
2½–6″ (6–15 cm) long; in mid- to late
summer.

Fruit: ⅜–½″ (10–12 mm) in diameter;
rounded, brown capsule; splitting into 6
parts; many small, winged seeds;
maturing in autumn and remaining
attached.

Habitat: Around houses and long persisting at
old home sites, sometimes escaped but
not naturalized; in humid, warm
temperate to tropical regions.

Range: Native of China and nearby SE. Asia.
Planted from Maryland to Florida and
Texas and on the Pacific Coast.

Crapemyrtle is a popular ornamental
for its profuse, showy blossoms. Many
blossoms in late summer. Many
varieties with different flower colors are
grown from cuttings as well as from
seed. It was named by Linnaeus for his
Swedish friend, Magnus von
Lagerstroem (1696–1759). The
common name refers to the wrinkled
petals. It is not related to Myrtle
(*Myrtus communis* L.) of the
Mediterranean region.

MYRTLE FAMILY
(Myrtaceae)

Trees, often large, and shrubs; about
3000 species in tropical regions,
especially America and Australia; 8
native tree species (all in Florida) and 4
naturalized in North America.
Leaves: opposite (alternate in *Eucalyptus*
and other Old World genera), simple,
mostly small, not toothed, leathery,
aromatic, with gland dots, evergreen,
without stipules.
Flowers: generally many clustered
together; large and showy, commonly
white, bisexual, regular; the calyx
generally of 4–5 sepals separate or
united at base and commonly persistent
at tip of fruit; generally 4–5 petals,
very many long, threadlike stamens,
and 1 pistil with 1- to 5-celled ovary,
many ovules, and long slender style.
Fruit: a berry, often edible, or a capsule
with few to many seeds.

86, 352, 449 **Bluegum Eucalyptus**
"Bluegum" "Tasmanian Bluegum"
Eucalyptus globulus Labill.

Description: Very tall, straight-trunked, introduced
tree with narrow, irregular crown of
drooping, evergreen foliage, with *odor
of camphor.*
Height: 120' (37 m).
Diameter: 3' (0.9 m).
Leaves: *evergreen;* 4–12" (10–30 cm)
long, 1–2" (2.5–5 cm) wide. *Narrowly
lance-shaped,* long-pointed, usually
curved; without teeth, *thick and leathery,*
hairless; *dull green* on both surfaces.
Leaves on young plants opposite, ovate,
stalkless, with bluish or whitish bloom
beneath.
Bark: mottled gray, brown, and
greenish; smooth, peeling in strips; at
base becoming gray, thick, rough,
furrowed, and shaggy.

Twigs: yellow-green, slender, angled, hairless, drooping.

Flowers: 2" (5 cm) wide; with *many white stamens* and without petals; with odor of camphor; scattered, *single* and almost stalkless at leaf base; in winter and spring.

Fruit: ¾–1" (2–2.5 cm) wide; broad, *top-shaped, angled capsules,* warty and *bluish-white,* with 3–5 narrow openings at top and many tiny seeds; in spring.

Habitat: Moist soils in subtropical regions.

Range: Native of Australia. Widely planted in California and becoming naturalized.

This is one of the most extensively cultivated species of *Eucalyptus* in subtropical regions of the world and the most common eucalyptus in California among the numerous introduced species. It grows very rapidly and sprouts from stumps. It is used as a street tree, for windbreaks and screens, and in forest plantations for fuel, pulpwood, and construction timber. A medicinal oil used as an expectorant and decongestant is distilled from the aromatic leaves. This species is the floral emblem of the island of Tasmania.

100, 450   Red-ironbark Eucalyptus
"Red-ironbark" "Mugga"
Eucalyptus sideroxylon A. Cunn. ex Benth.

Description: Introduced evergreen tree with straight trunk, irregular crown, and drooping leaves; crushed foliage aromatic, with camphorlike odor.

Height: 50' (15 m).

Diameter: 2' (0.6 m).

Leaves: *evergreen;* 2½–5" (6–13 cm) long, ⅜–¾" (10–19 cm) wide. *Lance-shaped,* long-pointed at both ends, not toothed, hairless, with faint side veins. *Dull gray-green* or blue-green on both surfaces; drooping on long, slender, reddish leafstalks.

Bark: dark brown or blackish; very thick, hard, deeply furrowed into long, narrow ridges, containing shiny crystals of eucalyptus gum or kino.

Twigs: yellow-green, pink-tinged, very slender, hairless.

Flowers: ¾–1" (2–2.5 cm) wide; with *many* spreading *pink stamens* (white or red in varieties) and no corolla, from egg-shaped buds with conical lid; 3–7 clustered on curved stalks at leaf bases; in spring and summer.

Fruit: ⅜" (10 mm) long; *reddish-brown, elliptical* capsules, opening along 5 *lines* on sunken top; many tiny seeds; maturing in late summer.

Habitat: Moist soils in subtropical regions.

Range: Native of SE. Australia. Introduced in California.

This species is planted along highways for shade and for the large masses of pink flowers. It is among the most frost-hardy species of eucalyptus. In Australia the wood is used for general construction, beams, and railroad cross-ties.

DOGWOOD FAMILY
(Cornaceae)

About 120 species of shrubs and trees in north and south temperate zones and tropical mountains; rarely herbs; 15 species of native trees and shrubs, sometimes becoming trees, including dogwood (*Cornus*) and tupelo (*Nyssa*); about 10 native shrubs and 2 native herbs in North America.

Leaves: deciduous or evergreen, opposite or alternate, generally not toothed, without stipules.

Flowers: bisexual, or male and female usually on separate plants; tiny or small; with calyx of 4–5 sepals (none in female flowers of silktassel, *Garrya*) and corolla of usually 4–5 petals (none in *Garrya*), 4–10 stamens, and 1 pistil with inferior ovary usually 1- or 2-celled with 1 style (2 in *Garrya*).

Fruit: usually a drupe (sometimes a berry), sour or bitter, 1- or 2-seeded.

111, 384 **Pacific Dogwood**
"Flowering Dogwood"
"Mountain Dogwood"
Cornus nuttallii Audubon

Description: Tree with dense, conical or rounded crown of often horizontal branches and with beautiful white flower clusters.
Height: 50' (15 m).
Diameter: 1' (0.3 m), rarely larger.
Leaves: opposite; 2½–4½" (6–11 cm) long, 1¼–2¾" (3–7 cm) wide. *Elliptical*, edges slightly wavy, with 5–6 *long, curved veins* on each side of midvein. *Shiny green* and nearly hairless above, paler with woolly hairs beneath; turning orange and red in autumn.
Bark: reddish-brown, thin, smooth or scaly.
Twigs: slender; light green and hairy when young, becoming dark red or blackish.

Flowers: ¼" (6 mm) wide; with 4 *greenish-yellow petals;* many crowded together in a *head* 1" (2.5 cm) wide; bordered by *usually* 6 (sometimes 4–7) large, elliptical, short-pointed, *white* (sometimes pinkish), *petal-like bracts* 1½–2½" (4–6 cm) long, altogether forming a huge "flower" 4–6" (10–15 cm) wide; in spring and early summer, often again in late summer or autumn.

Fruit: ½" (12 mm) long; *elliptical,* shiny *red or orange;* thin, mealy, bitter pulp; stone containing 1–2 seeds; *many crowded together in head* 1½" (4 cm) across; maturing in autumn.

Habitat: Moist soils in mountains in understory of coniferous forests.

Range: SW. British Columbia south to W. Oregon and in mountains to S. California; to 6000' (1829 m).

Pacific Dogwood is one of the most handsome native ornamental trees on the Pacific Coast, with very showy flowers and fruit. The head of flowers with surrounding, petal-like bracts resembles a huge flower and is commonly so called. The "flower" is larger than that of the eastern Flowering Dogwood (*Cornus florida* L.), usually having 6 bracts instead of 4. John James Audubon (1780–1851), the American ornithologist and artist, who painted this tree in his famous work *Birds of America,* named it for its collector, Thomas Nuttall (1786–1859), the British-American botanist and ornithologist.

---

110 Red-osier Dogwood
"Kinnikinnik" "Red Dogwood"
Cornus stolonifera Michx.

Description: Large, spreading, thicket-forming shrub with several stems, clusters of small white flowers, and small whitish fruit; rarely a small tree.

Height: commonly 3–10′ (0.9–3 m),
rarely to 15′ (4.6 m).

Diameter: 3″ (7.5 cm).

Leaves: *opposite;* 1½–3½″ (4–9 cm)
long, ⅝–2″ (1.5–5 cm) wide. *Elliptical*
or ovate, short- or long-pointed,
without teeth; 5–7 *long, curved, sunken
veins* on each side of midvein. *Dull green*
above, whitish green and covered with
fine hairs beneath; turning reddish in
autumn.

Bark: gray or brown, smooth or slightly
furrowed into flat plates.

Twigs: *purplish-red,* slender, hairy when
young, with rings at nodes.

Flowers: ¼″ (6 mm) wide; with 4
spreading, white petals; many, crowded in
upright, *flattish clusters* 1¼–2″ (3–5
cm) wide; in late spring and early
summer.

Fruit: ¼–⅜″ (6–10 mm) in diameter;
whitish, juicy, with 2-seeded stone;
maturing in late summer.

Habitat: Moist soils, especially along streams;
forming thickets and in understory of
forests.

Range: Central Alaska east to Labrador and
Newfoundland, south to N. Virginia,
and west to California; also N. Mexico;
to 5000′ (1524 m); to 9000′ (2743 m)
in the Southwest.

One of the most common and
widespread shrubs across Canada and
the northern states, it is planted as an
ornamental, especially for the showy
twigs in winter. Red-osier Dogwood is
useful for erosion control on stream
banks. The common name recalls the
resemblance of the reddish twigs to
those of some willows, called osiers,
used in basketry. The Latin species
name, meaning "bearing stolons,"
refers to the rooting of branch tips
touching the ground and forming new
shoots.

120 Wavyleaf Silktassel
"Tasseltree" "Quininebush"
Garrya elliptica Dougl. ex Lindl.

Description: Evergreen shrub or small tree with tassel-like clusters of flowers and fruit and paired, leathery, *wavy-edged leaves.* Foliage and other parts have bitter taste.
Height: 20′ (6 m).
Diameter: 4″ (10 cm).
Leaves: *evergreen;* opposite; 2–3¼″ (5–8 cm) long. *Elliptical, thick; shiny green* and nearly hairless above, paler with thick coat of woolly hairs beneath.
Bark: gray, smooth, becoming finely fissured and slightly scaly.
Twigs: 4-angled and densely hairy when young; greenish, becoming brown or blackish; *bitter.*
Flowers: tiny, scaly, greenish, and without petals; many crowded together in *drooping, narrow, catkinlike clusters* 2–5″ (5–13 cm) long; *male and female on separate plants;* in late winter and very early spring.

Fruit: ⅜″ (10 mm) in diameter; *rounded, berrylike, dark purple to black,* densely covered with *white hairs;* becoming dry, with *bitter* pulp; 1-seeded; maturing in summer and splitting open.
Habitat: Dry slopes and ridges; often in thickets, chaparral, and mixed evergreen forests.
Range: W. Oregon south to S. California and Santa Cruz Island; to 2000′ (610 m).

This is the only native species of *Garrya* reaching tree size. This distinct genus of shrubs and small trees is often placed separately in the Silktassel Family (Garryaceae). Although various parts have a bitter taste, as the name "Quininebush" suggests, goats browse the foliage.

576

HEATH FAMILY
(Ericaceae)

Widespread, especially on acid soils,
about 1500 species, mostly shrubs,
sometimes trees; 15 native tree and
many native shrub species in North
America.
Leaves: usually alternate, simple,
elliptical, and not toothed; often thick
and evergreen, without stipules.
Flowers: small or large and showy,
bisexual; regular or slightly irregular;
with 4- to 7-lobed calyx, generally
persisting on fruit; corolla of 4–7 lobes
or petals, often bell- or funnel-shaped,
8–10 stamens from a disk, and 1 pistil
with superior or inferior ovary of
usually 5 cells, many ovules, and 1 style.
Fruit: a capsule, berry or drupe.

91 Arizona Madrone
"Madroño"
Arbutus arizonica (Gray) Sarg.

Description: Medium-sized evergreen tree with
rounded crown of stout, crooked,
smooth red branches, showy white
flowers, and orange-red fruit.
Height: 40′ (12 m).
Diameter: 1½′ (0.5 m).
Leaves: *evergreen;* 1½–3″ (4–7.5 cm)
long, ½–1″ (1.2–2.5 cm) wide. *Lance-
shaped, short-pointed* at both ends;
without teeth or sometimes saw-
toothed; thick and stiff; with short,
slender stalks. *Shiny light green* above,
paler beneath.
Bark: *red-brown,* smooth, thin, and
peeling off in thin, papery scales on
branches; *light gray* or whitish and
divided into square plates on
trunks.
Twigs: whitish when young, turning
red-brown; finely hairy.
Flowers: ¼″ (6 mm) long; with *jug-
shaped* or urn-shaped, *white* or pink

corolla; short-stalked; many together in branched clusters about 2½" (6 cm) long and wide at twig ends; in spring and summer.

Fruit: ⅜" (10 mm) in diameter; *berrylike, orange-red, finely warty;* pulp mealy and sweetish, with large stone and many flattish seeds; maturing in autumn.

Habitat: Oak woodland with other evergreen trees.

Range: SE. Arizona, extreme SW. New Mexico, and NW. Mexico; at 4000–8000' (1219–2438 m).

Arbutus is the classical Latin name of Strawberry-tree (*Arbutus unedo* L.), a related species of southern Europe. *Madroño,* the Spanish name of that tree, was applied to similar trees of Mexico and California by the early Spanish *padres.*

114, 371, 488 Pacific Madrone
"Madrone" "Madroño"
Arbutus menziesii Pursh

Description: Handsome evergreen tree with tall, reddish-brown trunk and open, narrow, rounded or irregular crown of stout, smooth red branches.
Height: 20–80' (6–24 m).
Diameter: 2' (0.6 m).
Leaves: *evergreen;* 2–4½" (5–11 cm) long, 1–3" (2.5–7.5 cm) wide. *Elliptical,* blunt at tip, not toothed or sometimes saw-toothed; *thick* and leathery; hairless except when young. *Shiny dark green* above, paler or whitish beneath; turning red before falling.
Bark: *red, smooth,* thin, and peeling off in thin, papery scales on branches; dark *reddish-brown* and divided into square plates on trunks.
Twigs: light red or green, turning *reddish-brown;* hairless.
Flowers: ¼" (6 mm) long; *jug-shaped* or

urn-shaped, *white* or pink-tinged
corolla; short-stalked; in branched
clusters 2–6″ (5–15 cm) long and wide
at twig ends; in early spring.
Fruit: ⅜–½″ (10–12 mm) in diameter;
berrylike, orange-red, finely warty; with
mealy pulp, large stone, and many
flattened seeds; maturing in autumn.

Habitat: Upland slopes and canyons; in oak and
coniferous forests, often in understory.

Range: Pacific Coast from SW. British
Columbia south to W. Oregon and in
Coast Ranges to S. California; also
Sierra Nevada of central California and
Santa Cruz Island; to 5000′ (1524 m);
sometimes to 6000′ (1829 m).

Pacific Madrone is one of the most
beautiful broadleaf flowering
evergreens, with its glossy foliage, large
clusters of small white flowers, orange-
red fruits, and very showy, reddish,
peeling bark. It is the northernmost
New World tree of its family, ranging
to Canada. The wood can be used for
weaving shuttles. California Indians ate
the fruit raw and cooked; however,
overeating causes cramps. Deer and
birds also consume the fruit, and the
flowers are a source of honey. The
scientific name honors the discoverer,
Archibald Menzies (1754–1842), a
Scottish physician and naturalist.

159, 372 Texas Madrone
"Texas Madroño"
Arbutus texana Buckl.

Description: Small evergreen tree with short, often
crooked trunk and rounded crown of
stout, crooked, spreading branches; or a
shrub.

Height: 20′ (6 m).
Diameter: 8″ (20 cm).
Leaves: *evergreen;* 1–3½″ (2.5–9 cm)
long, ⅜–1½″ (1.5–4 cm) wide.
Elliptical or ovate; without teeth or

sometimes wavy- or saw-toothed; *thick and stiff;* with slender, hairy leafstalks. *Shiny green* above, paler and slightly hairy beneath.

Bark: on *branches, pinkish to reddish-brown, smooth,* thin, peeling off in thin, papery scales; on trunks, dark brown and divided into square plates.

Twigs: *red,* densely covered with hairs when young; becoming dark reddish-brown and scaly.

Flowers: ¼" (6 mm) long; *jug-shaped* or urn-shaped, *white or pink-tinged* corolla; short-stalked; in upright, branched clusters about 2½" (6 cm) long and wide; in early spring.

Fruit: ⅜" (10 mm) in diameter; *berrylike,* dark *red* to yellowish-red, finely warty; mealy, sweetish pulp; large stone containing many flattish seeds; maturing in autumn.

Habitat: Canyons and rocky slopes of mountains, in oak woodlands, and on rocky plains.

Range: Central Texas (Edwards Plateau) to Trans-Pecos Texas, SE. New Mexico (Guadalupe Mountains), and NE. Mexico; at 2000–6000' (610–1829 m).

It is reported that the fruit of this uncommon species is edible and that those of related European species have narcotic properties. The wood has been used locally for tool handles. The local names, "Naked Indian" and "Lady's Leg," refer to the smooth flesh-colored bark.

105, 370, 478 **Common Manzanita**
"Parry Manzanita"
Arctostaphylos manzanita Parry

Description: Large evergreen shrub, sometimes a small tree commonly branching near base, with stout, *crooked, twisted* trunks and branches and dense, rounded crown as broad as high.

Height: 20′ (6 m) or more.

Diameter: 6″ (15 cm).

Leaves: *evergreen;* 1–1¾″ (2.5–4.5 cm) long, ¾–1¼″ (2–3 cm) wide. *Elliptical* or nearly round; without teeth; *thick;* short-stalked; shiny or dull green on both surfaces, sometimes finely hairy.

Bark: *dark reddish-brown, smooth.*

Twigs: crooked, densely covered with gray hairs.

Flowers: ⁵⁄₁₆″ (8 mm) long; *jug-shaped white or pale pink corolla* ending in 5 tiny lobes; numerous, in many-branched clusters, drooping at end of twig; in late winter and early spring.

Fruit: ⁵⁄₁₆–½″ (8–12 mm) in diameter; *berrylike, round* or slightly flattish, white, turning deep *reddish-brown;* with mealy pulp and several nutlets; maturing in late summer.

Habitat: Dry slopes and in mountain canyons, chaparral, foothill and oak woodlands, and in Ponderosa Pine forest.

Range: N. and central California in north Coast Ranges and foothills of Sierra Nevada; at 300–4000′ (91–1219 m).

Whether manzanitas should be considered trees is debatable; a few of the approximately 40 native shrubby species mainly in California (including this one) reach tree size. However, they generally branch or fork near the ground, thus lacking the single trunk of a tree. *Manzanita* is a Spanish word meaning "little apple." The mealy berries are consumed in great quantities by wildlife of many kinds and were eaten by Indians, who also made them into manzanita cider. The dense evergreen foliage provides shelter for birds and small mammals; deer and goats browse the leaves and twigs. The handsome reddish-brown branches become twisted into odd shapes which are trimmed into collectors' items called "mountain driftwood."

SAPODILLA (OR SAPOTE) FAMILY
(Sapotaceae)

Trees and few shrubs with white latex or milky sap. About 700 species mostly in tropical and warm temperate regions; 8 native and 1 naturalized tree and 2 native shrub species in North America.
Leaves: alternate, simple, generally not toothed, thick, usually without stipules.
Flowers: small; generally white, green, or light brown; crowded or solitary at base of leaves or below at nodes; bisexual; regular, with hairy calyx of 4–8 overlapping lobes, corolla with short tube and 4–8 short lobes, stamens very short, generally 4–8 (to many) inserted on corolla opposite the lobes, often with sterile stamens alternate, and 1 pistil with superior ovary containing generally 4–5 (sometimes 1–14) cells with 1 ovule and 1 short style.
Fruit: a berry with 1 (sometimes few) large, elliptical, shiny seed with large scar and milky pulp, sometimes edible.

---

94 Gum Bumelia
"Woolly Buckthorn" "Chittamwood"
Bumelia lanuginosa (Michx.) Pers.

Description: Tree or thicket-forming shrub with straight trunk and narrow crown of short, stiff branches, often with stout *spines.*
Height: 50' (15 m).
Diameter: 1' (0.3 m).
Leaves: alternate or *clustered on short side twigs;* 1–3" (2.5–7.5 cm) long, ⅜–1" (1–2.5 cm) wide. *Elliptical* or obovate, rounded at tip, *widest beyond middle,* tapering toward long-pointed base; without teeth; slightly thickened. Shiny dark green above, *densely covered with gray or rust-colored hairs* beneath; falling irregularly in late autumn.

Bark: dark gray, furrowed into narrow, scaly ridges.

Twigs: slender, often zigzag; covered with gray or rust-colored hairs when young; often ending in straight *spines,* with single spines to ¾" (19 mm) long at base of some leaves; gummy, milky sap.

Flowers: ⅛" (3 mm) wide; with *bell-shaped, 5-lobed, white corolla;* clustered on slender stalks at leaf bases; in summer.

Fruit: ⅜–½" (10–12 mm) long; *elliptical, purplish-black berry;* sweetish pulp; 1 seed; maturing in autumn.

Habitat: Valleys and rocky slopes of uplands in hardwood forests; forming shrubby thickets in the Southwest.

Range: E. and S. Kansas and central Missouri southeast to central Florida and west to S. and W. Texas; local in SW. New Mexico and SE. Arizona; also N. Mexico; to 2500' (762 m); a southwestern variety to 5000' (1524 m).

Gum from cuts in the trunk is sometimes chewed by children. The fruit is edible but if eaten in quantity may cause dizziness and stomach disturbances. *Bumelia* is an ancient Greek name for European Ash (*Fraxinus excelsior* L.); the scientific name, meaning "woolly," describes the young leaves. The wood has been used locally for making tool handles and cabinets.

OLIVE FAMILY
(Oleaceae)

Widespread, especially in Asia; about
500 species of trees and shrubs,
sometimes woody vines; 22 native and
3 naturalized tree and about 10 native
shrub species in North America.
Leaves: opposite, simple (pinnately
compound in ash, *Fraxinus*), generally
without teeth and thick; without
stipules.
Flowers: usually small, sometimes
showy, commonly in clusters; generally
bisexual or male and female on separate
plants (in ash, *Fraxinus*); regular with
4-lobed calyx and tubular corolla
generally 4-lobed, 2 stamens inserted
on corolla; and 1 pistil with superior
ovary of 2 cells, each usually with 2
ovules, 1 style, and 1–2 stigmas.
Fruit: a berry, drupe, capsule or winged
key (samara).

Desert-olive Forestiera
"Desert-olive" "Wild-olive"
Forestiera phillyreoides (Benth.) Torr.

Description: Many-branched, thicket-forming shrub
or small tree, evergreen or nearly so,
with small, paired leaves.
Height: 20′ (6 m).
Diameter: 6″ (15 cm).
Leaves: *more or less evergreen; opposite;* ⅜–
1″ (1.5–2.5 cm) long, ⅛–¼″ (3–6
mm) wide. *Reverse lance-shaped,* rounded
to short-pointed at tip, short-pointed at
base; edges rolled under; *slightly
thickened;* almost stalkless; green and
finely hairy on both surfaces.
Bark: gray or blackish, smooth.
Twigs: gray, slender, generally paired
and short, hairless.
Flowers: ¼″ (6 mm) long; greenish,
without petals; in small clusters along
twigs; in winter and early spring. Male
and female on separate plants.

Fruit: ¼–⅜″ (6–10 mm) long; *egg-shaped, 1-sided,* with narrow point at tip; brown with thin pulp and stone; maturing in late spring.

Habitat: Dry, rocky slopes and canyons in deserts.

Range: Desert mountains of S. Arizona; also Mexico; at 2500–4500′ (762–1372 m).

The common names refer to the slight resemblance of the fruit and foliage to cultivated olives. The genus *Forestiera* is named after Professor Charles Le Forestier, an early 19th century French physician and naturalist. The specific name means "resembling *Phillyrea*," a related genus of evergreen shrubs and small trees of the Mediterranean region.

121 Singleleaf Ash
"Dwarf Ash"
Fraxinus anomala Torr. ex Wats.

Description: Shrub or small tree with short trunk and rounded crown of stout, curved branches.
Height: 25′ (7.6 m).
Diameter: 6″ (15 cm).
Leaves: opposite; *simple* or occasionally pinnately compound, with 2–3 leaflets; 1½–2″ (4–5 cm) long, 1–2″ (2.5–5 cm) wide (leaflets slightly smaller). *Broadly ovate or nearly round,* rounded or short-pointed at tip, blunt or notched at base; inconspicuously wavy-toothed or without teeth; *slightly thick* and leathery; becoming hairless; long-stalked. *Dark green* above, paler beneath.
Bark: dark brown, furrowed into narrow ridges.
Twigs: brown, hairless, *4-angled* or slightly winged.
Flowers: ⅛″ (3 mm) long; greenish, without petals; many together in small,

hairy clusters on previous year's twigs back of leaves in spring; bisexual and female.
Fruit: ¾" (19 mm) long; a *flattened key* with *elliptical, rounded wing* ⅜" (10 mm) wide, extending to base; maturing in summer.

Habitat: Dry canyons and hillsides including rocky slopes in upper desert; woodland and Ponderosa Pine forest.

Range: W. Colorado to E. California, Arizona, and extreme NW. New Mexico; usually at 2000–6500' (610–1981 m); in California to 11,000' (3353 m).

Singleleaf Ash is common in Grand Canyon National Park. As the scientific name suggests, this species is distinct in its genus, which is characterized by pinnately compound leaves.

275, 530 Fragrant Ash
"Flowering Ash"
Fraxinus cuspidata Torr.

Description: Many-branched shrub or small tree with fragrant, showy white flowers and long-pointed, sharply saw-toothed leaflets.
Height: 20' (6 m).
Diameter: 8" (20 cm).
Leaves: *opposite;* pinnately compound; 3–7" (7.5–18 cm) long. *Usually 7 leaflets* (sometimes 3–9), paired except at end, long-stalked; 1½–2½" (4–6 cm) long, ¼–¾" (6–19 mm) wide. Lance-shaped or ovate, long-pointed, sharply *saw-toothed* or not toothed. *Shiny dark green* above, paler and slightly hairy when young beneath.
Bark: gray, smooth, becoming fissured into scaly ridges.
Twigs: gray, slender, hairless.
Flowers: about ⅝" (15 mm) long; with *4 very narrow, white corolla lobes;* fragrant; drooping on slender stalks; many, in branched clusters 3–4"

(7.5–10 cm) long at end of side twigs; in spring.

Fruit: ¾–1" (2–2.5 cm) long; an *oblong key;* with rounded or slightly notched wing extending nearly to base of flattened body; maturing in summer.

Habitat: Rocky slopes of canyons and mountains in oak woodlands.

Range: SW. and Trans-Pecos Texas, New Mexico, Arizona, and N. Mexico; at 4500–7000' (1372–2134 m).

Fragrant Ash can be seen at Grand Canyon National Park. The scientific name refers to the sharp points or cusps of the leaflets. It is planted as an ornamental for the abundant showy flowers and attractive foliage.

---

277 Two-petal Ash
"Flowering Ash" "Foothill Ash"
Fraxinus dipetala Hook. & Arn.

Description: Shrub or small tree with showy white flowers composed of 2 petal-like lobes.
Height: 20' (6 m).
Diameter: 4" (10 cm).
Leaves: *opposite;* pinnately compound; 1½–4½" (4–11 cm) long. *Leaflets 3–7* (sometimes 9), paired except at end; ¾–1½" (2–4 cm) long, ¼–⅝" (6–15 mm) wide; *elliptical* or obovate, *blunt or short-pointed* at tip, short-pointed at base; *coarsely saw-toothed;* becoming hairless. Dark green above, paler beneath.
Bark: light gray, rough, scaly.
Twigs: green, slender, slightly 4-angled when young, usually hairless.
Flowers: ³⁄₁₆" (5 mm) long; with *2 broad, white corolla lobes;* drooping on slender stalks; many, in branched clusters to 4" (10 cm) long on twigs of previous year back of leaves; in spring.

Fruit: ¾–1" (2–2.5 cm) long; an *oblong* key with *broad wing* often slightly notched and extending nearly to base of flattish body; maturing in early summer.

Habitat: Dry slopes in foothills, chaparral, and woodland.

Range: California; also in N. Baja California; to 3500′ (1067 m).

This species is characterized by the white corolla with 2 long, narrow lobes divided almost into 2 separate petals. It is planted as an ornamental shrub for the showy white flowers.

---

278 Gooding Ash
Fraxinus gooddingii Little

Description: Shrub or small tree, usually evergreen, with *small, elliptical leaflets* along narrowly winged axis.
Height: 20′ (6 m).
Diameter: 4″ (10 cm).
Leaves: usually *evergreen;* opposite; pinnately compound; 1–3¼″ (2.5–8 cm) long, axis with *narrow wing.*
Usually 7 leaflets (sometimes 5 or 9), paired except at end; ⅜–1″ (1–2.5 cm) long, ¼–½″ (6–12 mm) wide; *elliptical,* short-pointed at ends; *wavy* or finely saw-toothed beyond middle; thin or slightly thickened; stalkless. Slightly shiny *brownish-green* above, paler with scalelike hairs beneath.
Bark: gray, smooth.
Twigs: gray, slender, finely hairy when young.
Flowers: less than ⅛″ (3 mm) long; without petals, few together in small clusters less than 1½″ (4 cm) long on twigs; in early spring before new leaves.
Fruit: ½–1″ (1.2–2.5 cm) long, a *narrow,* yellow-brown *key* with *broad wing,* mostly *rounded* at tip and extending nearly to base; maturing in late spring.

Habitat: Dry, rocky slopes and ridges of canyons; in desert and oak woodland.

Range: Mountains of SE. Arizona and NE.

Sonora, Mexico; at 3600–5000' (1097–1524 m).

This rare and local shrubby ash of restricted distribution is protected within the Goodding Research Natural Area of the Coronado National Forest, west of Nogales, Arizona. It was named in 1952 for Leslie Newton Goodding (1880–1967), a botanist of the U.S. Department of Agriculture, who discovered it in 1934.

Gregg Ash
"Littleleaf Ash" "Dogleg Ash"
Fraxinus greggii Gray

Description: Nearly evergreen, many-branched shrub or sometimes small tree with very small, narrow, thick leaflets along narrowly winged axis.
Height: 20' (6 m).
Diameter: 8" (20 cm).
Leaves: *nearly evergreen;* opposite; pinnately compound; 1–2½" (2.5–6 cm) long; axis with *narrow wing. Leaflets usually 5 or 7* (sometimes 3), paired except at end; ½–1¼" (1.2–3 cm) long, ⅛–¼" (3–6 mm) wide; *reverse lance-shaped* or obovate, rounded at tip, widest beyond middle; a few small, wavy teeth beyond middle or none; *thick,* with *obscure veins;* nearly stalkless. *Dark green* above, paler with tiny black dots beneath.
Bark: gray, thin, smooth or scaly.
Twigs: slender; dark green and finely hairy when young, becoming gray.
Flowers: less than ⅛" (3 mm) long; without petals; few, in hairy clusters ¾" (19 mm) long, on twigs; in early spring before new leaves; partly male and female and partly bisexual.
Fruit: ½–¾" (12–19 mm) long; a light brown, *narrow key* with *broad, notched wing* twice as long as body; maturing in late summer.

Habitat: Limestone bluffs and hills and in arroyos and canyons.

Range: Along the Rio Grande in Trans-Pecos Texas and NE. Mexico; at 1000–2000′ (305–610 m).

This species can be seen in Big Bend National Park. It is named for its discoverer, Josiah Gregg (1806–50), an explorer and trader in the West and northern Mexico and author of *The Commerce of the Prairies* (1844).

276 Oregon Ash
Fraxinus latifolia Benth.

Description: Tree with long, straight trunk and usually narrow, dense crown.
Height: 80′ (24 m).
Diameter: 2′ (0.6 m).
Leaves: opposite; pinnately compound; generally 5–12″ (13–30 cm) long; axis usually hairy. *Leaflets 5 or 7* (sometimes 9), paired except at end; 2–5″ (5–13 cm) long, 1–1½″ (2.5–4 cm) wide; *elliptical,* short-pointed at ends; *without teeth* or slightly saw-toothed; with prominent network of veins; *stalkless* or nearly so. *Light green* above, paler and *hairy* beneath; turning yellow or brown in autumn.
Bark: dark gray or brown, thick, furrowed into forking, scaly ridges.
Twigs: stout, covered with soft hairs.
Flowers: ⅛″ (3 mm) long; without corolla; male yellowish and female greenish on separate trees; many together in small clusters on twigs; before leaves in early spring.

Fruit: 1¼–2″ (3–5 cm) long; a light brown *key* with *broad, rounded wing extending nearly to base* of slightly flattened body; many hanging in dense clusters; maturing in early autumn.

Habitat: Wet soils along streams and in canyons; with Red Alder, Black Cottonwood, willows, and Oregon White Oak.

Range: W. Washington, W. Oregon, and
south in Coast Ranges and Sierra
Nevada to central California; to 5500'
(1676 m).

Oregon Ash is the only ash native to
the Northwest and the only western ash
with commercially important wood; it
is used for furniture, flooring,
millwork, paneling, boxes, and fuel. It
is also planted as a shade tree along the
Pacific Coast. According to an old
superstition in the Northwest,
poisonous snakes are unknown where
this ash grows, and rattlesnakes will
not crawl over a branch or stick from
the tree.

298 Lowell Ash
Fraxinus lowellii Sarg.

Description: Many-branched shrub or sometimes a
small tree with short trunk and
rounded crown.
Height: 25' (7.6 m).
Diameter: 6" (15 cm).
Leaves: opposite; pinnately compound;
3½–6" (9–15 cm) long. *3, 5 or 7
leaflets* 2¼–3" (6–7.5 cm) long, 1–1½"
(2.5–4 cm) wide; paired except at end,
short-stalked, ovate, long- or short-
pointed, *saw-toothed,* slightly leathery;
yellow-green and hairless or nearly so
on both surfaces.
Bark: brown or blackish, deeply
furrowed.
Twigs: brown, *4-angled,* often slightly
winged, hairless.
Flowers: ⅛" (3 mm) long; greenish,
without petals; in clusters ½–1½"
(1.2–4 cm) long; on twigs; before
leaves in early spring. Male and female
on separate trees.

Fruit: ¾–1¼" (2–3 cm) long; light
brown, *flattened key* with *elliptical,
rounded wing* ⅜" (10 mm) wide,
extending to base; maturing in summer.

Habitat: Local in moist soils along streams and canyons; in oak woodland and upper desert zones.

Range: N. and central Arizona; at 3200–6500' (975–1981 m).

This species was named in 1917 for its discoverer, Percival Lowell (1855–1916), an American astronomer. It is closely related to Singleleaf Ash and is sometimes regarded as a variety of that species. Where the ranges meet, some trees have an intermediate number of leaflets. The reduction to 1 leaflet in Singleleaf Ash is unusual for ashes and is apparently an adaptation to a dry climate.

---

279 Chihuahua Ash
Fraxinus papillosa Lingelsh.

Description: Small tree with leaflets whitish beneath.
Height: 30' (9 m).
Diameter: 1' (0.3 m).
Leaves: opposite; pinnately compound; 3–6" (7.5–15 cm) long; 5–9 *leaflets* 1¼–2¾" (3–7 cm) long, ⅜–1¼" (1–3 cm) wide; paired except at end; *elliptical* or lance-shaped, pointed at ends; finely *saw-toothed* or not toothed; *stalkless* or nearly so. *Dull green* above, *whitish* beneath.
Bark: light gray, furrowed into narrow ridges.

Twigs: brown, stout, hairless.
Flowers: apparently in small clusters on twigs; in early spring.
Fruit: 1–1¼" (2.5–3 cm) long; a narrow *key* with *narrow wing* blunt or notched at tip, *not extending to base* of thick body; maturing in autumn.
Habitat: Moist soils of canyons in mountains; in oak woodland.
Range: Trans-Pecos Texas, SW. New Mexico, SE. Arizona, and N. Mexico; at 5000–7000' (1524–2134 m).

Chihuahua Ash extends northward across the Mexican border into several mountainous areas and may be seen in Coronado National Forest in southeastern Arizona. It is easily recognized by the whitish lower surfaces of the leaflets. Under a microscope, the lower surfaces of the leaves appear as a solid mass of whitish beads, which are the minute projections or papillae of each epidermal cell. The Latin species name refers to this characteristic, present also in Texas Ash (*Fraxinus texensis* (A. Gray) Sarg.) and White Ash (*Fraxinus americana* L.), two related eastern trees. Apparently no botanical description of the flowers of this local species has been published.

---

274   **Velvet Ash**
"Arizona Ash" "Desert Ash"
Fraxinus velutina Torr.

Description: Tree with open, rounded crown of spreading branches and leaflets quite variable in shape and hairiness.
Height: 40' (12 m).
Diameter: 1' (0.3 m).
Leaves: opposite; pinnately compound; 3–6" (7.5–15 cm) long. Commonly 5 *leaflets* (to 9) 1–3" (2.5–7.5 cm) long, ⅜–1¼" (1–3 cm) wide; paired except at end; *lance-shaped to elliptical,* pointed at ends, slightly *wavy-toothed* or sometimes without teeth; often slightly thickened and leathery; short-stalked. *Shiny green* above, paler and densely covered with soft hairs or sometimes hairless beneath; turning yellow in autumn.
Bark: gray, deeply furrowed into broad, scaly ridges.
Twigs: gray or brown, often hairy when young.
Flowers: ⅛" (3 mm) long; without corolla; male yellowish and female greenish on separate trees; many

together in small clusters on twigs; before leaves in early spring.
Fruit: ¾–1¼″ (2–3 cm) long; a light brown, *narrow key* with long *wing not extending to base;* many hanging in dense clusters; maturing in summer and early autumn.

Habitat: Moist soils of stream banks, washes, and canyons, mainly in mountains, desert, desert grassland; in oak woodland and Ponderosa Pine forest.

Range: Trans-Pecos Texas west to extreme SW. Utah, S. Nevada, and S. California and south to N. Mexico; at 2500–7000′ (762–2134 m).

This variable species is the common ash in the Southwest, where it is planted as a shade and street tree. It is hardy in alkaline soils and fast-growing. In the desert, ash trees indicate a permanent underground water supply. The leaflets of different shapes are often covered with velvety hairs beneath, as the scientific and common names imply, but also may be hairless. Modesto Ash is a rapidly growing, cultivated variety, widely planted as a street tree in dry areas (including alkaline soils) in California and the Southwest.

83, 497 **Olive**
"Common Olive" "European Olive"
Olea europaea L.

Description: Small evergreen, planted tree with short, stout trunk, rounded to irregular, dense crown, and the familiar olive as fruit.
Height: 25′ (7.6 m).
Diameter: 1′ (0.3 m), larger with age.
Leaves: *evergreen;* opposite; 1½–3″ (4–7.5 cm) long, ⅜–¾″ (10–19 mm) wide. *Lance-shaped* or narrowly elliptical; edges slightly turned under; *thick.* Gray-green with veins not visible above, silvery scales beneath.

Bark: brownish-gray, furrowed.
Twigs: light gray; slender; covered with whitish, scaly hairs when young.
Flowers: ³⁄₁₆″ (5 mm) long; with *4-lobed, whitish corolla;* fragrant; in short, branched clusters at leaf bases; in spring.
Fruit: 1″ (2.5 cm) long; *elliptical;* olive *green* when immature, turning *shiny black;* bitter pulp; 1 seed; maturing in autumn.

Habitat: Subtropical or Mediterranean climates, especially hot, dry regions under irrigation.

Range: Native probably in the eastern Mediterranean region but cultivated from prehistoric times and widespread. Planted in California, Arizona, and Florida.

Olive is a handsome ornamental and street tree with attractive gray-green foliage, reaching great age. It is grown commercially in the interior valleys of California. The bitter fruit becomes edible after soaking in salt water or a lye solution and thorough washing. Green and ripe fruit are eaten as pickles in brine. Olive oil, obtained by crushing and pressing the mature fruit, is used for salads, in cooking, and in soaps. The wild form has thorny twigs and small fruit with thin pulp.

FIGWORT FAMILY
(Scrophulariaceae)

Herbs and small shrubs, sometimes shrubs and trees; worldwide, about 3000 species; numerous species of native herbs and a few shrubs in southern North America. No native trees; 1 naturalized tree species.
Leaves: generally deciduous, alternate or opposite, simple, toothed, lobed, or not toothed; without stipules.
Flowers: irregular, with deeply 4- or 5-lobed calyx, tubular corolla, often bell-shaped and 2-lipped, with 4–5 lobes, 4–5 stamens on corolla, and 1 pistil with superior 2-celled ovary containing many ovules, 1 style, and 2-lobed stigma.
Fruit: a many-seeded capsule..

124, 233, 395   **Royal Paulownia**
"Princess-tree" "Empress-tree"
Paulownia tomentosa (Thunb.) Steud.

Description: Naturalized tree with short trunk, broad, open crown of stout, spreading branches, *very large leaves,* and *showy purple flowers.*

Height: 50′ (15 m).
Diameter: 2′ (0.6 m).
Leaves: *opposite;* 6–16″ (15–41 cm) long, 4–8″ (10–20 cm) wide. *Broadly ovate,* long-pointed at tip, with *several veins from notched base;* sometimes slightly 3-toothed or 3-lobed. Dull light green and slightly hairy above, paler and densely covered with hairs beneath. Leafstalks 4–8″ (10–20 cm) long.
Bark: gray-brown, with network of irregular, shallow fissures.
Twigs: light brown, stout, densely covered with soft hairs when young.
Flowers: 2″ (5 cm) long; with *bell-shaped,* pale violet corolla ending in 5 rounded, *unequal lobes;* fragrant; in

upright clusters 6–12″ (15–30 cm) long on stout, hairy branches; in early spring from rounded, brown, hairy buds formed previous summer.

Fruit: 1–1½″ (2.5–4 cm) long; *egg-shaped capsule,* pointed, brown, thick-walled, *splitting into 2 parts;* many tiny, winged seeds; maturing in autumn and remaining attached.

Habitat: Waste places, roadsides, and open areas.

Range: Native of China. Cultivated and naturalized from S. New York south to N. Florida, west to S. Texas, and north to Missouri.

This handsome, rapid-growing ornamental and shade tree resembles a catalpa. Vigorous shoots, with enormous leaves 2′ (0.6 m) or more in length and width, can be produced by pruning back almost to the base. The soft, lightweight, whitish wood of this weed tree is exported to Japan for furniture and special uses, such as for sandals. It is named for Anna Paulowna (1795–1865) of Russia, princess of the Netherlands and ancestor of the present queen.

BIGNONIA FAMILY
(Bignoniaceae)

Mostly woody vines, also shrubs and often large trees, rarely herbs; about 700 species in tropical and warm temperate regions; 5 species of native trees and 2 species of native woody vines in North America.

Leaves: mostly opposite, sometimes alternate; palmately or pinnately compound or bipinnate, sometimes simple; without stipules.

Flowers: generally large, showy, in clusters, bisexual; slightly irregular, with tubular, 5-toothed or 5-lobed calyx, large, funnel- or bell-shaped; tubular corolla commonly yellow, pink, or whitish with 5 unequal lobes sometimes in 2 lips; usually 4 large stamens in pairs and 1 inserted in tube and on disk, 1 pistil with superior 2-celled ovary containing many ovules, long, thin style, and 2 stigmas.

Fruit: usually a capsule, often long and podlike, splitting into 2 parts, with many winged seeds; or a berry.

125, 378 Southern Catalpa
"Catawba" "Indian-bean"
Catalpa bignonioides Walt.

Description: Introduced tree with broad, rounded crown of spreading branches, large, heart-shaped leaves, large clusters of white flowers, and long, beanlike fruit.
Height: 50' (15 m).
Diameter: 2' (0.6 m).
Leaves: *3 at a node* (whorled) and opposite; 5–10" (13–25 cm) long, 4–7" (10–18 cm) wide. Ovate, *abruptly long-pointed* at tip, notched at base, without teeth. Dull green above, paler and covered with soft hairs beneath; turning blackish in autumn; with unpleasant odor when crushed. Slender leafstalk 3½–6" (9–15 cm) long.

Bark: brownish-gray, scaly.

Twigs: green, turning brown; stout, hairless or nearly so.

Flowers: 1½″ (4 cm) long and wide; with *bell-shaped corolla* of 5 *unequal, rounded, fringed lobes, white* with 2 orange stripes and many purple spots and stripes inside; slightly fragrant; in upright, branched clusters to 10″ (25 cm) long and wide; in late spring.

Fruit: 6–12″ (15–30 cm) long, ⁵⁄₁₆–³⁄₈″ (8–10 mm) in diameter; narrow, cylindrical, *dark brown capsule, cigarlike,* thin-walled, splitting into 2 parts; many flat, light brown seeds with 2 papery wings; maturing in autumn, remaining attached in winter.

Habitat: Moist soils in open areas such as roadsides and clearings.

Range: Original range uncertain; probably native in SW. Georgia, NW. Florida, Alabama, and Mississippi; widely naturalized from S. New England south to Florida, west to Texas, and north to Michigan; at 100–500′ (30–152 m). Also cultivated in the West.

Catalpa is the American Indian name, while the species name refers to a related vine with flowers of similar shape. It is planted for the abundant, showy flowers, cigarlike pods, and coarse foliage..

126, 377   Northern Catalpa
"Hardy Catalpa" "Indian-bean"
Catalpa speciosa Warder ex Engelm.

Description: Tree with rounded crown of spreading branches and large, heart-shaped leaves, large, showy flowers, and long, beanlike fruit; introduced.

Height: 50–80′ (15–24 m).

Diameter: 2½′ (0.8 m).

Leaves: *3 at a node* (whorled) and opposite; 6–12″ (15–30 cm) long, 4–8″ (10–20 cm) wide. *Ovate, long-pointed,* straight to notched at base; *without teeth*. Dull green above, paler and covered with soft hairs beneath; turning blackish in autumn. Slender leafstalk 4–6″ (10–15 cm) long.

Bark: brownish-gray, smooth, becoming furrowed into scaly plates or ridges.

Twigs: green, turning brown; stout; becoming hairless.

Flowers: 2–2¼″ (5–6 cm) long and wide; with *bell-shaped corolla* of 5 *unequal, rounded, fringed lobes, white* with 2 orange stripes and purple spots and lines inside; in branched, upright clusters 5–8″ (13–20 cm) long and wide; in late spring.

Fruit: 8–18″ (20–46 cm) long, ½–⅝″ (12–15 mm) in diameter; narrow, cylindrical, *dark brown capsule, cigarlike,* thick-walled, splitting into 2 parts; many flat, light brown seeds with 2 papery wings; maturing in autumn, remaining attached in winter.

Habitat: Moist valley soils by streams; naturalized in open areas, such as roadsides and clearings.

Range: Original range uncertain; native apparently from SW. Indiana to NE. Arkansas; widely naturalized in SE. United States; at 200–500′ (61–152 m). Cultivated in the West.

Northern Catalpa is the northernmost New World example of its tropical family and is hardier than Southern Catalpa, which blooms later and has slightly smaller flowers and narrower, thinner-walled capsules. Both are called "Cigartree" and "Indian-bean."

85, 376, 387   **Desert-willow**
"Desert-catalpa"
Chilopsis linearis (Cav.) Sweet

Description: Large shrub or small tree, often with leaning trunk; open, spreading crown;

narrow, *willowlike leaves; large, showy flowers;* and *very long, narrow, beanlike fruit.*

Height: 25′ (7.6 m).

Diameter: 6″ (15 cm).

Leaves: *opposite and alternate;* 3–6″ (7.5–15 cm) long, ¼–⅜″ (6–10 mm) wide. Linear, straight or slightly curved, very long-pointed at ends; not toothed, drooping, short-stalked; light green, sometimes hairy or sticky.

Bark: dark brown, furrowed into scaly ridges.

Twigs: brown, very slender, sometimes hairy or sticky.

Flowers: 1¼″ (3 cm) long and wide; *bell-shaped corolla* with 5 unequal lobes, whitish tinged with *pale purple or pink* and with yellow in throat; fragrant; in usually unbranched clusters to 4″ (10 cm) long at ends of twigs; from late spring to early summer.

Fruit: 4–8″ (10–20 cm) long, ¼″ (6 mm) in diameter; a dark brown, *cigarlike* capsule; maturing in autumn, splitting into 2 parts, and remaining attached in winter; many flat, light brown seeds with 2 papery, hairy wings.

Habitat: Moist soils of stream banks and drainages in plains and foothills, desert and desert grassland zones, often forming thickets.

Range: SW. Texas and New Mexico west to extreme SW. Utah and S. California; also in N. Mexico; at 1000–5000′ (305–1524 m).

Desert-willow is important in erosion control and is planted also as an ornamental. Propagated from cuttings or seeds, it grows rapidly and sprouts after being cut. Indians made bows from the stiff, durable wood, which is also suitable for fenceposts. Despite its name, this species is not related to willows.

MADDER FAMILY
(Rubiaceae)

One of the largest plant families, worldwide, mostly tropical and subtropical, with about 6000 species of trees, shrubs, and few herbs; 7 native tree species and many native herb and shrub species in North America.

Leaves: opposite, sometimes whorled, simple, not toothed, with paired stipules that form buds and leave ring scars at nodes.

Flowers: generally in clusters of many flowers each, small to large, bisexual; regular, with 4- or 5-lobed calyx often persistent, tubular corolla usually 4- or 5-lobed, generally colored and often showy, 4–5 stamens alternate and inserted in tube, and 1 pistil with inferior ovary commonly 2-celled and containing many ovules (sometimes as few as 1), 1 style, and 2 stigmas.

Fruit: a capsule or berry, sometimes a drupe.

109, 353, 539   Buttonbush
"Honey-balls" "Globe-flowers"
Cephalanthus occidentalis L.

Description: Spreading, many-branched shrub or sometimes small tree with numerous branches (often crooked and leaning), irregular crown, balls of white flowers resembling pincushions, and buttonlike balls of fruit.

Height: 20′ (6 m).

Diameter: 4″ (10 cm).

Leaves: *opposite or 3 at a node* (whorled); 2½–6″ (6–15 cm) long, 1–3″ (2.5–7.5 cm) wide. *Ovate or elliptical,* pointed at tip, rounded at base; without teeth. Shiny green above, paler and sometimes hairy beneath; at southern limit nearly evergreen.

Bark: gray or brown, becoming deeply furrowed into rough, scaly ridges.

Twigs: *mostly in 3's,* reddish-brown, stout, sometimes hairy, with rings at nodes.

Flowers: ⅝″ (15 mm) long; with narrow, tubular, white, 4-lobed corolla and long, threadlike style; fragrant; stalkless; crowded in upright, long-stalked *white balls* of many flowers each 1–1½″ (2.5–4 cm) in diameter; from late spring through summer.

Fruit: ¾–1″ (2–2.5 cm) in diameter; compact, *rough brown balls,* composed of many small, narrow, dry nutlets ¼″ (6 mm) long, each 2-seeded; maturing in autumn.

Habitat: Wet soils bordering streams and lakes.

Range: S. Quebec and SW. Nova Scotia south to S. Florida, west to Texas, and north to SE. Minnesota; to 3000′ (914 m); in Arizona and California to 5000′ (1524 m); also in Mexico, Central America, and Cuba.

The poisonous foliage of this abundant and widespread species is unpalatable to livestock. The bitter bark has served in home remedies, but its medicinal value is doubtful. Buttonbush is a handsome ornamental suited to wet soils and is also a honey plant. Ducks and other water birds and shorebirds consume the seeds.

HONEYSUCKLE FAMILY
(Caprifoliaceae)

About 400 species of shrubs, sometimes woody vines and small trees, rarely herbs; widespread, mostly in north temperate regions and in tropical mountains; 11 native tree species and many native shrub species in North America.

Leaves: opposite, usually simple (pinnately compound in *Sambucus*), stipules absent or minute.

Flowers: usually small, sometimes showy, often in clusters, bisexual; regular or irregular, with minute calyx of 5 (sometimes 4) teeth or lobes, tubular corolla of 5 (4) lobes, 5 (4) stamens alternate and inserted in tube, and 1 pistil with inferior ovary of 1–5 cells each with 1 ovule, style, and stigma often lobed.

Fruit: a berry or drupe.

282, 346, 487 **Pacific Red Elder**
"Coast Red Elder" "Red Elderberry"
Sambucus callicarpa Greene

Description: Clump-forming shrub or sometimes small tree with many *small, white flowers* in clusters and bright red berries; flowers and crushed foliage have unpleasant odor.

Height: 20′ (6 m).

Diameter: 6″ (15 cm).

Leaves: opposite; pinnately compound; 5–10″ (13–25 cm) long; with *unpleasant odor; 5 or 7 leaflets* 2–5″ (5–13 cm) long, 1–2″ (2.5–5 cm) wide; paired except at end, *lance-shaped* or elliptical, finely and *sharply saw-toothed.* Green and nearly hairless above, paler and hairy beneath.

Bark: light to dark gray or brown; smooth, becoming fissured into small, scaly or shaggy plates.

Twigs: gray, stout, hairy when young;

with *ringed nodes* and *thick, whitish pith*
becoming yellow-orange or brown.
Flowers: ¼" (6 mm) wide; with *white,
5-lobed corolla;* in upright, many-
branched clusters to 4" (10 cm) long; in
spring and early summer.
Fruit: ⁵⁄₁₆" (8 mm) in diameter; a *round
berry, bright red* or sometimes orange,
juicy, with 1-seeded, *poisonous nutlets;*
maturing in summer.

Habitat: Moist soils, especially in open areas
such as cutover coniferous forests.

Range: SW. Alaska southeast along coast to W.
central California; to 2000' (610 m).

The red fruit, inedible when raw, can
be made into wine and is also eaten by
birds and mammals. The seeds are
considered poisonous, causing diarrhea
and vomiting. The specific name means
"beautiful fruit."

281, 348, 460   **Blue Elder**
"Blue Elderberry" "Blueberry Elder"
Sambucus cerulea Raf.

Description: Large, many-branched, thicket-forming
shrub or small tree often with several
trunks with compact, rounded crown,
numerous small, whitish flowers in
large clusters, and bluish fruit.
Height: 25' (7.6 m).
Diameter: 1' (0.3 m).
Leaves: opposite; pinnately compound;
5–7" (13–18 cm) long; sometimes
nearly evergreen southward; 5–9 *leaflets*
1–5" (2.5–13 cm) long, ⅜–1½"
(1–4 cm) wide; paired except at end,
narrowly ovate or *lance-shaped,* long-
pointed at tip, short-pointed and
unequal at base; sharply saw-toothed;
short-stalked. Yellow-green above,
paler and often hairy beneath.
Bark: gray or brown, furrowed.
Twigs: green, stout, angled, often
hairy; with ringed nodes and *thick,
white pith.*

Flowers: nearly ¼" (6 mm) wide; with *yellowish-white, 5-lobed corolla;* fragrant; in upright, *flat-topped,* many-branched clusters 4–8" (10–20 cm) wide; in summer.

Fruit: nearly ¼" (6 mm) in diameter; dark *blue* berry with *whitish bloom,* thick, juicy, sweet pulp, and 3 1-seeded nutlets; many clusters; maturing in summer and autumn.

Habitat: Moist soils along streams and canyons of mountains, in open areas in coniferous forests; also along roadsides, fencerows, and clearings.

Range: S. British Columbia south along coast to S. California, east in mountains to Trans-Pecos Texas, and north to W. Montana; also in NW. Mexico; to 10,000' (3048) m).

The sweetish, edible berries are used in preserves and pies. Lewis and Clark first reported Blue Elder as an "alder" with "pale, sky blue" berries. Whistles can be made by removing the pith from cut twigs; California Indians made flutes in a similar fashion. A remedy for fever has been concocted from the bark. Blue Elder is planted as an ornamental for the numerous whitish flowers and bluish fruits.

280, 347   Mexican Elder
"Arizona Elder" "Desert Elderberry"
Sambucus mexicana Presl

Description: Small evergreen tree with short trunk and compact, rounded crown and with many *small, yellowish-white flowers* in showy clusters and blue berries.
Height: 25' (7.6 m).
Diameter: 1' (0.3 m).
Leaves: opposite; pinnately compound; 5–7" (13–18 cm) long; mostly evergreen, often deciduous in dry seasons. 3–5 leaflets 1¼–3" (3–7.5 cm) long, ⅜–1½" (1–4 cm) wide;

paired except at end, *elliptical or ovate,*
pointed at tip and base; *finely saw-
toothed;* slightly *thick and leathery;*
usually hairless. Green above, paler
beneath.
Bark: gray or light brown; furrowed
into long, narrow, scaly ridges.
Twigs: light green and usually hairless,
becoming light brown; stout, with
ringed nodes and *thick pith.*
Flowers: ¼" (6 mm) wide; with
yellowish-white, 5-lobed corolla; in
upright, flat-topped, many-branched
clusters 2–8" (5–20 cm) wide;

blooming nearly throughout the year.
Fruit: nearly ¼" (6 mm) in diameter;
blackish or dark blue berry; with whitish
bloom; juicy, sweet, edible pulp; 3 to 5
1-seeded nutlets; many in cluster;
maturing nearly throughout the year.

Habitat: Along streams and drainages in
woodlands, deserts, and desert
grasslands.

Range: Mountains of SW. New Mexico west to
W. Nevada and N. California and
south to coast and islands of S.
California; also in Mexico; to 5000'
(1524 m).

One of the largest native elders,
Mexican Elder is characterized by the
stout trunks, evergreen foliage, and its
partially lowland habitat. It is planted
in dry areas as an ornamental for the
numerous flower clusters and evergreen
foliage. The berries can be made into
pies and jellies and are eaten by birds;
Indians dried and stored them like
raisins.

Part III
Appendices

GLOSSARY

Achene A small, dry, seedlike fruit with a thin wall that does not open.

Acorn The hard-shelled, 1-seeded nut of an oak, with a pointed tip and a scaly cup at the base.

Aggregate fruit A fused cluster of several fruits, each one formed from an individual ovary, but all derived from a single flower, as in a magnolia.

Alternate Arranged singly along a twig or shoot, and not in whorls or opposite pairs.

Anther The terminal part of a stamen, containing pollen in 1 or more pollen-sacs.

Axis The central stalk of a compound leaf or flower cluster.

Bark The outer covering of the trunk and branches of a tree, usually corky, papery, or leathery.

Berry A fleshy fruit with more than 1 seed.

Bipinnate With leaflets arranged on side branches off a main axis; twice-pinnate; bipinnately compound.

Bisexual With male and female organs in the same flower.

Blade The broad, flat part of a leaf.

Bract A modified and often scalelike leaf, usually located at the base of a flower, a fruit, or a cluster of flowers or fruits.

Bud A young and undeveloped leaf, flower, or shoot, usually covered tightly with scales.

Calyx Collective term for the sepals of a flower.

Capsule A dry, thin-walled fruit containing 2 or more seeds and splitting along natural grooved lines at maturity.

Catkin A compact and often drooping cluster of reduced, stalkless, and usually unisexual flowers.

Cell A cavity in an ovary or fruit, containing ovules or seeds.

Compound leaf A leaf whose blade is divided into 3 or more smaller leaflets.

Cone A conical fruit consisting of seed-bearing, overlapping scales around a central axis.

Cone-scale One of the scales of a cone.

Conifer A cone-bearing tree of the Pine family, usually evergreen.

Corolla Collective term for the petals of a flower.

Crown The mass of branches, twigs, and leaves at the top of a tree, with particular reference to its shape.

Cultivated Planted and maintained by man.

Deciduous Shedding leaves seasonally, and leafless for part of the year.

Drupe A fleshy fruit with a central stonelike core containing 1 or more seeds.

Elliptical Elongately oval, about twice as long as wide, and broadest at the middle; like an ellipse.

Entire Smooth-edged, not lobed or toothed.

Escaped Spread from cultivation and now growing and reproducing without aid from man.

Family A group of related genera.

Filament The threadlike stalk of a stamen.

Fleshy fruit A fruit with juicy or mealy pulp.

Flower The reproductive structure of a tree or other plant, consisting of at least 1 pistil or stamen, and often including petals and sepals.

Follicle A dry, 1-celled fruit, splitting at maturity along a single grooved line.

Fruit The mature, fully developed ovary of a flower, containing 1 or more seeds.

Genus A group of closely related species. Plural, genera.

Gland-dot A tiny, dotlike gland or pore, usually secreting a fluid.

Habit The characteristic growth form or general shape of a plant.

Herb A plant with soft, not woody, stems, dying to the ground in winter.

Hybrid A plant or animal of mixed parentage, resulting from the interbreeding of 2 different species.

Intergradation   The gradual merging of 2 or more distinct forms or kinds, through a series of intermediate forms.

Introduced   Intentionally or accidentally established in an area by man, and not native; exotic or foreign.

Irregular flower   A flower with petals of unequal size.

Keel   A sharp ridge or rib, resembling the keel of a boat, found on some fruits and seeds.

Key   A dry, 1-seeded fruit with a wing; a samara.

Lanceolate   Shaped like a lance, several times longer than wide, pointed at the tip and broadest near the base.

Leader   The highest, terminal shoot of a plant.

Leaflet   One of the leaflike subdivisions of a compound leaf.

Linear   Long, narrow, and parallel-sided.

Lobed   With the edge of the leaf deeply but not completely divided.

Male cone   The conical, pollen-bearing male element of a conifer.

Midvein   The prominent, central vein or rib in the blade of a leaf; midrib.

Multiple fruit   A fused cluster of several fruits, each one derived from a separate flower, as in a mulberry.

Native   Occurring naturally in an area and not introduced by man; indigenous.

Naturalized   Successfully established and reproducing naturally in an area where not native.

Needle The very long and narrow leaf of pines and related trees.

Node The point on a shoot where a leaf, flower, or bud is attached.

Nut A dry, 1-seeded fruit with a thick, hard shell that does not split along a natural grooved line.

Nutlet One of several small, nutlike parts of a compound fruit, as in a sycamore; the hard inner core of some fruits, containing a seed and surrounded by softer flesh, as in a hackberry.

Oblanceolate Reverse lanceolate; shaped like a lance, several times longer than wide, broadest near the tip and pointed at the base.

Oblong With nearly parallel edges.

Obovate Reverse ovate; oval, with the broader end at the tip.

Opposite Arranged along a twig or shoot in pairs, with 1 on each side, and not alternate or in whorls.

Ovary The enlarged base of a pistil, containing 1 or more ovules.

Ovate Oval, with the broader end at the base.

Ovule A small structure in the cell of an ovary, containing the egg and ripening into a seed.

Palmate With leaflets attached directly to the end of the leafstalk and not arranged in rows along an axis; palmately compound; digitate.

Palmate-veined With the principal veins arising from the end of the leafstalk and radiating toward the edge of the leaf, and not branching from a single midvein.

Parallel-veined With the veins running more or less parallel toward the tip of the leaf.

Persistent Remaining attached, and not falling off.

Petal One of a series of flower parts lying within the sepals and next to the stamens and pistil, often large and brightly colored.

Pinnate With leaflets arranged in 2 rows along an axis; pinnately compound.

Pinnate-veined With the principal veins branching from a single midvein, and not arising from the end of the leafstalk and radiating toward the edge of the leaf.

Pistil The female structure of a flower, consisting of stigma, style, and ovary.

Pith The soft, spongy, innermost tissue in a stem.

Pod A dry, 1-celled fruit, splitting along natural grooved lines, with thicker walls than a capsule.

Pollen Minute grains containing the male germ cells and released by the stamens.

Pollen-sac The part of an anther containing the pollen.

Pome A fruit with fleshy outer tissue and a papery-walled, inner chamber containing the seeds.

Regular flower A flower with petals all of equal size.

Resin A plant secretion, often aromatic, that is insoluble in water but soluble in ether or alcohol.

Ring scar A ringlike scar left on a twig after a leaf falls away.

Samara A dry, 1-seeded fruit with a wing; a key.

Scale One of the very short, pointed, and overlapping leaves of some conifers; the leaflike covering of a bud or of the cup of an acorn.

Scurfy Covered with tiny, broad scales.

Seed A fertilized and mature ovule, containing an embryonic plant.

Sepal One of the outermost series of flower parts, arranged in a ring outside the petals, and usually green and leaflike.

Sheath A tubular, surrounding structure; in some conifers, the papery tube enclosing the base of a bundle of needles.

Shoot A young, actively growing twig or stem.

Shrub A woody plant, smaller than a tree, with several stems or trunks arising from a single base; a bush.

Simple fruit A fruit developed from a single ovary.

Simple leaf A leaf with a single blade, not compound or composed of leaflets.

Sinus The space between 2 lobes of a leaf.

Solitary Borne singly, not in pairs or clusters.

Species A kind or group of plants or animals, composed of populations of individuals that interbreed and produce similar offspring.

Spur A short side twig, often bearing a cluster of leaves.

Stamen One of the male structures of a flower, consisting of a threadlike filament and a pollen-bearing anther.

Stigma The tip of a pistil, usually enlarged, that receives the pollen.

Stipule A leaflike scale, often paired, at the base of a leafstalk in some trees.

Style The stalklike column of a pistil, arising from the ovary and ending in a stigma.

Toothed With an edge finely divided into short, toothlike projections.

Tree line The upper limit of tree growth at high latitudes or on mountains; timberline.

Trunk The major woody stem of a tree.

Tubular With the petals partly united to form a tube.

Unisexual With male or female organs, but not both, in a single flower.

Vein One of the riblike vessels in the blade of a leaf.

Whorled Arranged along a twig or shoot in groups of three or more at each node.

Wing A thin, flat, dry, shelflike projection on a fruit or seed, or along the side of a twig.

Wood The hard, fibrous inner tissue of the trunk and branches of a tree or shrub.

PICTURE CREDITS

The numbers in parentheses are plate numbers. Some photographers have pictures under agency names as well as their own. Agency names appear in boldface.

William Aplin (331, 354, 379, 381)
G. W. Argus (101 right, 143 right, 144 right)
Craig Blacklock (206 left)
Gary Braasch (52 right)
Sonja Bullaty and Angelo Lomeo (9 right, 10 right, 11 left, 12 left and right, 17 left and right, 18 left, 28 left and right, 29 left, 30 left and right, 31 left and right, 32 left and right, 40 right, 41 right, 43 left and right, 48 left, 54 right, 56 left and right, 59 left and right, 60 right, 61 left and right, 94 left and right, 95 left, 96 right, 109 left and right, 110 left and right, 112 left and right, 113 left and right, 123 left and right, 124 left and right, 125 left and right, 126 left, 130 left and right, 137 right, 138 right, 145 left and right, 146 right, 147 left and right, 149 left and right, 150 left and right, 151 right, 154 left and right, 155 left and right, 161 left and right, 162 left and right, 184 left and right, 189 left and right, 193 right, 194 left and right, 195 left and right, 196 left and right, 197 left and right, 204 right, 205 left and right, 206 right, 208 left and right, 210 left and right, 213 left, 214 left and right, 215 left and right, 220 left and right, 221 left and right, 227 left and right, 228 left and right, 232 right, 233 left and right, 234 right, 235 left and right,

right, 230 left and right, 231 left and
right, 232 left, 234 left, 237 left, 238
left and right, 241 left and right, 242
left and right, 244 left and right, 245
left and right, 246 right, 248 left and
right, 249 left and right, 250 left and
right, 254 left and right, 255 left, 257
left and right, 259 left, 260 left, 261
left and right, 265 left and right, 266
left, 267 left and right, 271 left and
right, 272 right, 274 left, 276 left and
right, 277 left, 279 right, 280 left and
right, 281 left and right, 282 left and
right, 283 left, 284 left and right, 290
left and right, 291 left and right, 294
left and right, 296 left, 297 left and
right, 299 right, 302 left and right,
308 left and right, 310 left and right,
313 left and right, 324, 326, 327,
328, 330, 346, 347, 348, 349, 350,
357, 363, 364, 376, 382, 386, 390,
392, 396, 398, 401, 402, 403, 404,
408, 410, 411, 412, 413, 414, 415,
416, 417, 418, 419, 420, 423, 424,
425, 426, 429, 430, 431, 433, 434,
436, 437, 438, 439, 440, 441, 442,
443, 444, 445, 446, 447, 448, 449,
450, 451, 452, 453, 454, 455, 456,
459, 467, 468, 471, 473, 478, 479,
481, 483, 487, 489, 496, 497, 498,
516, 535, 536)
Norden H. Cheatham (21 right, 36
right, 166 right, 321, 351, 504, 505)
Scooter Cheatham (13 right, 71 left
and right, 76 left and right, 152 left
and right, 159 left, 170 left and right,
174 left and right, 181 left and right,
199 left and right, 202 left and right,
239 right, 240 left and right, 247 left
and right, 256 left and right, 259
right, 260 right, 262 left and right,
264 left and right, 270 left and right,
272 left, 275 right, 279 left, 283
right, 285 left and right, 286 left and
right, 307 right, 332, 397, 399, 432,
470, 480, 482, 484, 502, 514, 515,
522, 524, 530)

Bruce Coleman, Inc.
Jane Burton (337)

Stephen Collins (527, 528)
Steve Crouch (427, 462, 476, 501, 512, 513, 517)
Thase Daniel (539)
Kent and Donna Dannen (203 right)
E. R. Degginger (126 right, 309 left, 325, 500)
Wilbur H. Duncan (490)
Diane Ensign-Caughey (81 right, 101 left)
Jeff Foott (312 right)
Lois Theodora Grady (138 left, 322, 339, 464)
Farrell Grehan (318, 319)
Pamela J. Harper (15 left, 95 right, 160 right, 203 left, 213 right, 236 right, 252 right, 320, 384, 388, 395, 428, 466, 485, 526, 529, 538)
Grant Heilman Photography (421, 507)
W. Richard Hildreth (42 right, 201 left)
Walter H. Hodge (243 left, 269 left, 344, 353, 356, 358, 365, 389, 400, 405, 435, 465, 491, 492)
Charles C. Johnson (374)
Bill Jordan (102 left and right, 218 left and right)
Gill C. Kenny (394)
J. James Kilbaso (477, 532)
Donald J. Leopold (316, 409)
C. D. Luckhart (26 right, 35 right, 50 right, 68 right, 69 right, 72 right, 88 right, 100 left and right, 127 right, 188 right, 198 left, 296 right)
John A. Lynch (531)
C. Allan Morgan (370)
David Muench (64 right, 65 right, 66 right, 165 right, 169 left and right)
Helen M. Mulligan (488)
Dr. Theodore Niehaus (25 right, 338, 355, 359, 361, 503)

Photo Researchers, Inc.
A-Z Collection Limited (341), A. W. Amber (345), Tim Branch (472), H. F. Flanders (304 left), Michael P. Gadomski (362), Bob Grant (509), Robert C. Hermes (305 left), Anne E.

Hubbard (475), Charlie Ott (312 left), Miriam Reinhart (45 left), Alvin E. Staffan (342, 486, 510), Vermont Institute of Natural Science (537), V. P. Weinland (461, 511)

Betty Randall and Robert Potts (1 left and right, 7 left and right, 8 right, 18 right, 19 right, 23 right, 24 left and right, 27 right, 29 right, 34 left and right, 37 left and right, 38 left and right, 39 left and right, 46 right, 47 left and right, 49 left and right, 55 left and right, 57 right, 58 right, 67 right, 70 right, 74 left and right, 82 left and right, 87 left and right, 90 left, 91 left and right, 99 right, 104 right, 108 left and right, 121 left and right, 122 right, 128 left and right, 132 right, 133 right, 134 left and right, 135 left and right, 136 right, 137 left, 142 left and right, 143 left, 144 left, 157 left and right, 158 left and right, 163 left and right, 164 left and right, 168 right, 172 left and right, 173 left, 177 right, 182 right, 186 right, 190 right, 201 right, 207 right, 209 left and right, 211 left, 222 right, 223 left and right, 226 left, 237 right, 263 right, 269 right, 274 right, 277 right, 292 right, 298 left and right, 315 left and right, 329, 391, 457, 458, 460, 474, 493, 494, 506, 518, 523)
Bill Ratcliffe (323)
Edward S. Ross (333)
Clark Schaack (75 left, 79 left and right, 84 left and right, 131 right, 185 left and right, 200 left and right, 246 left, 251 left and right, 258 left and right, 275 left, 306 right, 309 right, 311 right, 508, 520, 521, 533)
John Shaw (8 left, 75 right, 292 left)
Benny Simpson (307 left)
Arlo I. Smith (311 left)
Norman F. Smith (367)
Richard Spellenberg (9 left, 23 left, 51 right, 62 left and right, 73 right, 77 right, 78 right, 81 left, 114 right, 127 left, 151 left, 159 right, 191 right,

INDEX

All species are listed by both the scientific and the accepted common names. Many species also have local names which are given in quotation marks and refer the reader to the description under the accepted name. Numerals in bold-face type refer to plate numbers. Numerals in italics refer to page numbers. Circles preceding the accepted common names make it easy for you to keep a record of the trees you have seen.

THE AUDUBON SOCIETY FIELD GUIDE SERIES

Also available in this unique all-color, all-photographic format:

Birds (two volumes: *Eastern Region* and *Western Region*)

Butterflies

Fishes, Whales, and Dolphins

Fossils

Insects and Spiders

Mammals

Mushrooms

Reptiles and Amphibians

Rocks and Minerals

Seashells

Seashore Creatures

Trees (*Eastern Region*)

Wildflowers (two volumes: *Eastern Region* and *Western Region*)

STAFF

Prepared and produced by Chanticleer Press, Inc.

Publisher: Paul Steiner
Editor-in-Chief: Gudrun Buettner
Executive Editor: Susan Costello
Managing Editor: Jane Opper
Guides Editor: Susan Rayfield
Project Editor: Olivia Buehl
Natural Science Editor: John Farrand, Jr.
Production: Helga Lose
Art Director: Carol Nehring
Picture Library: Edward Douglas
Symbols and Range Maps: Paul Singer
Drawings: Bobbi Angell
Silhouettes: Dolores R. Santoliquido
Design: Massimo Vignelli